Studies in Fuzziness and Soft Computing

Volume 343

Series editor

Janusz Kacprzyk, Polish Academy of Sciences, Warsaw, Poland
e-mail: kacprzyk@ibspan.waw.pl

About this Series

The series "Studies in Fuzziness and Soft Computing" contains publications on various topics in the area of soft computing, which include fuzzy sets, rough sets, neural networks, evolutionary computation, probabilistic and evidential reasoning, multi-valued logic, and related fields. The publications within "Studies in Fuzziness and Soft Computing" are primarily monographs and edited volumes. They cover significant recent developments in the field, both of a foundational and applicable character. An important feature of the series is its short publication time and world-wide distribution. This permits a rapid and broad dissemination of research results.

More information about this series at http://www.springer.com/series/2941

Cengiz Kahraman · Özgür Kabak
Editors

Fuzzy Statistical Decision-Making

Theory and Applications

 Springer

Editors
Cengiz Kahraman
Management Faculty, Industrial Engineering
 Department
Istanbul Technical University
Istanbul
Turkey

Özgür Kabak
Management Faculty, Industrial Engineering
 Department
Istanbul Technical University
Istanbul
Turkey

ISSN 1434-9922   ISSN 1860-0808   (electronic)
Studies in Fuzziness and Soft Computing
ISBN 978-3-319-81793-4   ISBN 978-3-319-39014-7   (eBook)
DOI 10.1007/978-3-319-39014-7

To the honourable people of past, today and future.

Prof. Cengiz Kahraman

To my father, Nayil Kabak, who has passed away suddenly on May 4, 2015.

With his words "Let love never fall away."

Sevgiler hiç eksilmesin...

Assoc. Prof. Özgür Kabak

Preface

Statistical decision-making helps us learn to analyze data and use methods of statistical inference in making business decisions. Statistical inference is the process of drawing conclusions about the population based on information from a sample of that population. Statistical inference methods include point and interval estimations, hypothesis testing, clustering, etc. However, descriptive statistics is used in the inferential statistics as an input to conclude about the population.

Fuzziness is a kind of uncertainty that everything is a matter of degree. The sources of this uncertainty may be the incomplete information or insufficient data. Fuzzy statistics is a complementary statistics in these cases where classical statistics has almost nothing to do. In this book, fuzzy decision-making techniques are presented by their theory and applications. Fuzzy interval estimation, fuzzy hypothesis testing, fuzzy regression and correlation, and fuzzy process control are some of these techniques involved in this book.

The first chapter presents an introduction to fuzzy decision-making. The authors survey the literature of fuzzy statistics and fuzzy statistical decision-making and present the results by graphical illustrations.

The second chapter presents discrete fuzzy probability distributions. The fuzzy expectation theory is introduced. Fuzzy Bayes theorem, fuzzy binomial distribution, and fuzzy Poisson distribution are derived and numerical examples are given.

The third chapter presents continuous fuzzy probability distributions. Fuzzy continuous expectation, fuzzy continuous uniform distribution, fuzzy exponential distribution, fuzzy Laplace distribution, fuzzy normal distribution, and fuzzy log-normal distribution are developed and numerical examples are given.

The fourth chapter explains the generalized Bayes theorem in handling fuzzy a priori information and fuzzy data. Individual measurement results also contain another kind of uncertainty, which is called fuzziness. The combination of fuzziness and stochastic uncertainty calls for a generalization of Bayesian inference, i.e., fuzzy Bayesian inference.

The fifth chapter converts the classical central tendency measures to their fuzzy cases. Fuzzy mean, fuzzy mode, and fuzzy median are explained by numerical examples. Fuzzy frequency distribution is another subtitle of this chapter. Classical graphical illustrations are examined under fuzziness. A numerical example for each central tendency measure is given.

The sixth chapter develops the fuzzy versions of classical dispersion measures namely, standard deviation and variance, mean absolute deviation, coefficient of variation, range, and quartiles. Initially it summarizes the classical dispersion measures and then develops their fuzzy versions for triangular fuzzy data. A numerical example for each fuzzy dispersion measure is given.

The seventh chapter introduces a new approach for the estimation of a parameter in the statistical models, based on fuzzy sample space. Two basic concepts of the point estimation, i.e., sufficiency and completeness, are extended to the fuzzy data case. Then, the unbiased estimator and the uniformly minimum variance unbiased (UMVU) estimator are defined for such situations. The properties of these estimators are investigated, and some procedures are provided to obtain the UMVU estimators, based on fuzzy data.

The eighth chapter is on fuzzy confidence regions. The construction is explained for classical statistics as well as for Bayesian analysis. A numerical example is also given.

The ninth chapter focuses on analyzing the works on fuzzy point and interval estimations (PIE) for the years between 1980 and 2015. In this chapter, the literature is reviewed through Scopus database and the review results are given by graphical illustrations. The chapter also uses the extensions of fuzzy sets such as interval-valued intuitionistic fuzzy sets (IVIFS) and hesitant fuzzy sets (HFS) to develop the confidence intervals based on these sets. The chapter also includes numerical examples to increase the understandability of the proposed approaches.

The tenth chapter generalizes the p-value concept for testing fuzzy hypotheses on the basis of Zadeh's probability measure of fuzzy events. The authors prove that the introduced p-value has uniform distribution over (0,1) when the null fuzzy hypothesis is true. Then based on such a p-value, a procedure is illustrated to test various types of fuzzy hypotheses. Several applied examples are given to show the performance of the method.

The eleventh chapter has two objectives. It critically exposes the most relevant fuzzy linear regression methods and remarks the most relevant actuarial applications of fuzzy regression and also develops one recurrent application: the estimation of the public debt yield curve as a basis for fuzzy financial pricing of insurance contracts.

The twelfth chapter deals with fuzzy correlation and fuzzy nonlinear regression analyses. Both correlation and regression analyses that are useful and widely employed statistical tools are redefined in the framework of fuzzy set theory in order to comprehend relation and to model observations of variables collected as either qualitative or approximately known quantities which are no longer being utilized directly in classical sense. While extension principle based methods are utilized in the computational procedures for fuzzy correlation coefficient, the distance based methods preferred rather than mathematical programming ones are employed in parameter estimation of fuzzy regression models. Illustrative examples in detail for fuzzy correlation analysis are given.

The thirteenth chapter discusses Interactive Dichotomizer 3 (ID3) algorithm and supervised learning in quest (SLIQ) decision tree algorithm. These algorithms generate fuzzy decision trees. Their performances are tested using simple training sets from the literature.

The fourteenth chapter presents the Shewhart process control techniques under fuzziness. Variable and attribute control charts are extended to their fuzzy versions. Numerical examples are also given.

The fifteenth chapter develops exponentially weighted moving averages (EWMA) and cumulative sum (CUSUM) control charts having the ability of detecting small shifts in the process mean. Numerical illustrations of fuzzy EWMA and CUSUM control charts are also given.

The sixteenth chapter presents a new method to test linear hypothesis using a fuzzy set statistic produced by a set of confidence intervals with non-equal tails. A fuzzy significance level is used to evaluate the linear hypothesis. One-way ANOVA based on fuzzy test statistic and fuzzy significance level is developed. Numerical examples are given for illustration.

The seventeenth chapter summarizes and reviews the fuzzy ANOVA where the collected data considered fuzzy rather than crisp numbers. A real case study is also presented.

The eighteenth chapter presents different types of fuzzy data mining approaches including Apriori-based fuzzy data mining, tree-based fuzzy data mining, and genetic-fuzzy data mining approaches.

We hope that this book will provide a useful resource of ideas, techniques, and methods for the development of fuzzy statistics. We are grateful to the referees whose valuable and highly appreciated works contributed to select the high-quality chapters published in this book. We would also like to thank Prof. Janusz Kacprzyk, the editor of Studies in Fuzziness and Soft Computing at Springer for his supportive role in this process.

Istanbul, Turkey

Cengiz Kahraman
Özgür Kabak

Contents

Fuzzy Statistical Decision-Making

Cengiz Kahraman and Özgür Kabak

Abstract The classification of decision-making methods can be based on the types of the data in hand. If the data are given as a decision matrix with discrete values, you can use multiple attribute decision-making. If the data are given as unit cost or profit values together with budget or capacity constraints and if you have more than one objective, then you can use multiple objective decision-making in a continuous space. If the data are given as the parameters of certain probability distributions, then you can use statistical decision-making, generally through hypothesis tests. If the data are not exactly known, the fuzzy sets based approaches are incorporated into these decision-making methods. Fuzzy statistical decision-making is one of the most often used methods when insufficient statistical data exist in hand. Fuzzy hypothesis tests, fuzzy variance analysis, and fuzzy design of experiments are the examples of fuzzy statistical decision-making techniques. In this chapter, we survey the literature of fuzzy statistics and fuzzy statistical decision-making and present the results by graphical illustrations.

Keywords Fuzzy statistics · Statistical decision-making · Fuzzy event · Classification

1 Introduction

The probability theory is based on the data obtained from sufficient observations for a certain event. From these data, you can determine the probability distribution of the event and calculate its occurrence probability and make probabilistic estimations. Without having sufficient observed data, these probabilistic calculations cannot be made. Then, an expert would try to produce a subjective possibility distribution based on few data or few observations.

C. Kahraman (✉) · Ö. Kabak
Department of Industrial Engineering, Istanbul Technical University,
34367 Macka, Istanbul, Turkey
e-mail: kahramanc@itu.edu.tr

© Springer International Publishing Switzerland 2016
C. Kahraman and Ö. Kabak (eds.), *Fuzzy Statistical Decision-Making*,
Studies in Fuzziness and Soft Computing 343,
DOI 10.1007/978-3-319-39014-7_1

1

In classical statistical decision-making, we have crisp parameter values and their probability distributions are clearly known to make decision and estimations. In fuzzy statistical decision-making, the values of parameters are not certain. Hence, possibility distributions instead of probability distributions are preferred.

Zadeh's [27] following famous example clearly explains the difference between possibility and probability distributions.

Consider the statement "Hans ate X eggs for breakfast", with X taking values in $U = \{1, 2, 3, 4, \ldots\}$. We may associate a possibility distribution with X by interpreting $\pi_X(u)$ as the degree of ease with which Hans can eat u eggs. We may also associate a probability distribution with X by interpreting $p_X(u)$ as the probability of Hans eating u eggs for breakfast. Assuming that we employ some explicit or implicit criterion for assessing the degree with which Hans can eat u eggs for breakfast, the values of $\pi_X(u)$ and $p_X(u)$ might be as shown in Table 1.

It can be easily seen that, whereas the possibility that Hans may eat three eggs for breakfast is 1, the probability that he may do so might be quite small, e.g., 0.1. Thus, a high degree of possibility does not imply a high degree of probability, nor does a low degree of probability imply a low degree of possibility. However, if an event is impossible, it is bound to be improbable.

The probability measure for crisp events can be extended to a probability measure for fuzzy events. Probabilistic fuzzy systems are based on the concept of the probability of a fuzzy event [26]. The probability of a continuous fuzzy event $\tilde{A}$ is obtained by taking the expectation of the membership function as in Eq. (1).

$$Pr(\tilde{A}) = \int\limits_{-\infty}^{+\infty} u_{\tilde{A}}(x)f(x)dx = E\left(u_{\tilde{A}}(x)\right) \tag{1}$$

where $f(x)$ is the probability density function and $u_{\tilde{A}}(x)$ is the continuous membership function of the fuzzy event $\tilde{A}$.

In the literature, there are about 4,000 publications on fuzzy statistics including journal manuscripts, conference papers, book chapters, etc. One of the most-referenced papers on fuzzy statistics is by Ahmed et al. [1]. They present a novel algorithm for fuzzy segmentation of magnetic resonance imaging (MRI) data and estimation of intensity inhomogeneities using fuzzy logic. MRI intensity inhomogeneities are attributed to imperfections in the radio-frequency coils or to problems associated with the acquisition sequences. Another most-referenced paper is by Krishnapuram and Keller [16]. They proposed the possibilistic C-means algorithm (PCM) to address the drawbacks associated with the constrained memberships used in algorithms such as the fuzzy C-means (FCM).

Table 1 The possibility and probability distributions associated with X

u	1	2	3	4	5	6	7	8
$\pi_X(u)$	1	1	1	1	0.8	0.6	0.4	0.2
$p_X(u)$	0.1	0.8	0.1	0	0	0	0	0

The history of fuzzy statistics goes back to 1970s. Stallings [22] developed formalism for describing the syntactic pattern recognition of capital letters. Next, the fuzzy set approach based on the concepts of Bayesian statistics was described. Finally, the two approaches were compared. Later, Jain [13] indicated that the comparison was improper as it compared Bayesian statistics with a particular case of the fuzzy set theory. He showed that other interpretation of fuzzy connectives may result in entirely different results. Kandel and Byatt [15] introduce a bibliography of 570 items, all classified with the fuzzy set theory and its applications, which will help the readers to come to grips with the literature explosion on the subjects of fuzzy sets, fuzzy algebra, fuzzy statistics, and closely related applications. Kandel [14] made some fuzzy statistics applications for computer system security. Hong et al. [12] studied fuzzy conceptions, fuzzy measures, fuzzy statistics, and multi-dimension attribute spaces.

Buckley [6] investigated the use of fuzzy data in a fuzzy decision problem. An optional decision is a decision function from the data into the set of actions with maximum membership in some fuzzy sets. Fuzzy estimation is when the set of actions and the states of nature are identical. Fuzzy hypothesis testing is when an action states that the state of nature belongs to one of two disjoint sets. He defined a class of optional decisions identified under general assumptions.

Based upon Peizhuang's theory of set-valued statistics and Yager's theory of bags, Baowen et al. [3] gave a new form to the definition of fuzzy bag and its properties as well as operations in set-valued statistics. Zeng et al. [28] presented a review of noise levels of turbine generator sets in power stations, together with the occurrences encountered in 125 and 300 mw units which are summarized from the measured data of 17 units in power stations. Xihe [25] demonstrated 'stability of membership frequencies' in a fuzzy statistical model. Bialasiewicz [5] presented an approach to the selection of essential features of objects, which is based on sufficient and ε-sufficient statistics. It is shown how sufficient and ε-sufficient statistics can be used to construct partitions of the space of outcomes of an experiment in order to simplify the pattern recognition process. After 1990s, the number of publications on fuzzy statistics gained a significant acceleration. Taheri [23] reviewed essential works on fuzzy estimation, fuzzy hypotheses testing, fuzzy regression, fuzzy Bayesian statistics, and some relevant fields. The rest of the literature will be classified with respect to their publication years, authors, countries, etc. in Sect. 2.

The organization of this chapter is as follows. Section 2 gives the literature review with graphical illustrations. Section 3 includes the books on fuzzy statistics in the literature. Section 4 gives a numerical application of fuzzy statistical decision-making. Section 5 concludes the chapter.

2   Literature Review: Graphical Illustrations

In this section, we have searched the term "fuzzy statistics" in Scopus database. When the term *fuzzy statistics* is searched in the title, abstracts, and keywords of articles, the number of the documents that the Scopus found is 3,940. When the term *fuzzy statistics* is searched in the title of articles, the number of the documents that the Scopus found became 151. In the following, 3,940 documents are classified with respect to their publication years, their subject areas, their authors, source universities and countries.

2.1   Classification with Respect to Publication Years

Figure 1 gives the graph *publication year versus frequency*. The publications on fuzzy statistics appear at the end of 1970s. We observe a slight increasing trend between the years 1980 and 1990 whereas a stronger increasing trend exists between the years 1990 and 2000. After the year 2000, we observe an exponential increase in the number of publications on fuzzy statistics.

2.2   Classification with Respect to Affiliations

Figure 2 presents the graph *affiliation versus frequency*. University of Oviedo is by far the first university among all, which produce studies on fuzzy statistics. Those institutes compose of the second class producing publications on fuzzy statistics: Tsingua University, IEEE, University of Tehran, Institute of the Polish Academy of Sciences, Harbin Institute of Technology, Beihang University, and Shanghai Jiaotong University. The third class involves Zhejiang University, National Chengchi University, Islamic

Fig. 1 Frequencies of fuzzy statistics publications with respect to the years

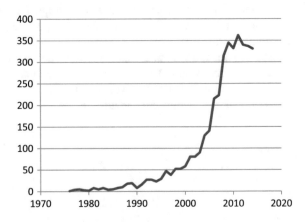

Fig. 2 Universities
producing the papers of fuzzy
statistics

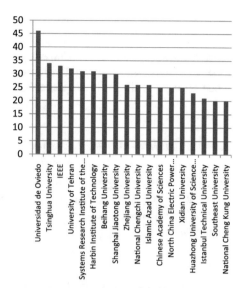

Azad University, Chinese Academy of Sciences, North China Electric Power University, and Xidian University whereas the fourth class involves Huazhong University of Science and Technology, Istanbul Technical University, Southeast University (China), and National Cheng Kung University. There are 12 Chinese universities among the first 18 universities producing publications on fuzzy statistics.

2.3 Classification with Respect to Authors

Figure 3 shows the graph authors versus frequency. The first 13 researchers are Ozgur Kisi (Turkey), Berlin Wu (Taiwan), Maria Angeles Gil (Spain), Olgierd Hryniewicz (Poland), Witold Pedrycz (Poland), Abraham Kandel (US), Ihsan Kaya (Turkey), Cengiz Kahraman (Turkey), Junzo Watada (Japan), Didier Dubois

Fig. 3 Fuzzy statistics:
number of publications with
respect to their authors

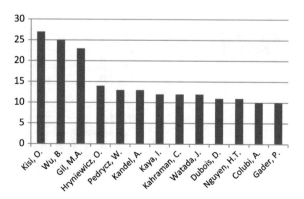

(France), Hung T. Nguyen (US), Ana Colubi (Spain), and Paul D. Gader (US), respectively. It is interesting that there is no Chinese researcher among the first 13 researchers even there are 12 Chinese universities among the first 18 universities producing fuzzy statistics publications. The reason is that Chinese researchers' mode on fuzzy statistics is low even their number is high.

2.4   Classification with Respect to Countries

Figure 4 presents the graph *countries versus frequency*. China has the first rank, followed by US, Taiwan, Iran, India, Spain, Turkey, and the others. The sum of fuzzy statistics publications from China and US is also equal to the sum of all other countries.

2.5   Classification with Respect to Document Type

Figure 5 gives the graph *document type versus frequency*. Almost all of the fuzzy statistics publications have been published as journal articles or conference papers. As it is clearly seen from the graph, the number of book publications on fuzzy statistics is very few when compared with articles and conference papers. This proves the need for new books on fuzzy statistics.

2.6   Classification with Respect to Subject Areas

Figure 6 presents the graph *subject areas versus frequency*. The first three subject areas, which fuzzy statistics are used, are computer science, engineering, and

Fig. 4 Countries producing the studies of fuzzy statistics

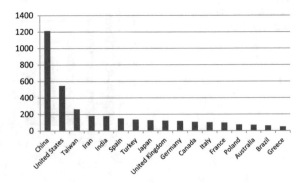

Fig. 5 Fuzzy statistics
publications with respect to
the document types

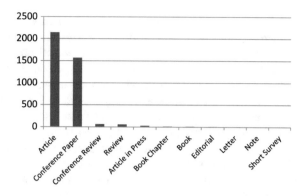

Fig. 6 Subject areas of fuzzy
statistics publications

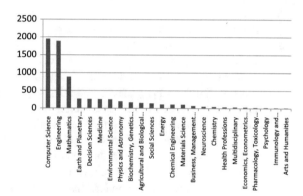

mathematics, respectively. The sum of the publications on computer science and
engineering are much larger than the sum of all other subject areas.

3 Books on Fuzzy Statistics

There are few books on fuzzy statistics, which have been published by various
publishers. Negoita and Ralescu [17] offer the first detailed descriptions of fuzzy
controllers, fuzzy statistics, and fuzzy-set theory. Ayyub and Gupta [2] handle
uncertainty analysis in engineering and sciences. They develop some models using
fuzzy logic, statistics, and neural network approach. Slowinski [21] cover
decision-making, mathematical programming, statistics and data analysis, and
reliability, maintenance and replacement in his book. He accounts for advances in
fuzzy data analysis, fuzzy statistics, and applications to reliability analysis. Ross
et al. [20] provide clear descriptions of fuzzy logic and probability, as well as the
theoretical background, examples. Grzegorzewski et al. [11] present some "soft-
ening" approaches, which utilize concepts and techniques developed in theories
such as fuzzy sets theory, rough sets, possibility theory, theory of belief functions
and imprecise probabilities.

Bertoluzza et al. [4] connect probability theory/statistics and fuzzy set theory in different ways. Several probabilistic studies are developed, as well as techniques and criteria to get descriptive and inferential statistical conclusions from fuzzy data. Buckley [8] introduces elementary fuzzy statistics based on crisp data. In the introductory chapters the book presents a very readable survey of fuzzy sets including fuzzy arithmetic and fuzzy functions. The book develops fuzzy estimation and demonstrates the construction of fuzzy estimators for various important and special cases of variance, mean and distribution functions. It is shown how to use fuzzy estimators in hypothesis testing and regression, which leads to a comprehensive presentation of fuzzy hypothesis testing and fuzzy regression as well as fuzzy prediction. Nguyen and Wu [18] present basic foundational aspects for a theory of statistics with fuzzy data, together with a set of practical applications. Fuzzy data are modeled as observations from random fuzzy sets. Theories of fuzzy logic and of random closed sets are used as basic ingredients in building statistical concepts and procedures in the context of imprecise data, including coarse data analysis. The monograph also aims at motivating statisticians to look at fuzzy statistics to enlarge the domain of applicability of statistics in general. Buckley [9] combines material from his previous books FP [7] and FS [8], plus has about one third new results. From FP he has material on basic fuzzy probability, discrete (fuzzy Poisson, binomial) and continuous (uniform, normal, exponential) fuzzy random variables. From FS he includes chapters on fuzzy estimation and fuzzy hypothesis testing related to means, variances, proportions, correlation and regression. New material includes fuzzy estimators for arrival and service rates, and the uniform distribution, with applications in fuzzy queuing theory. Also, new to this book, is three chapters on fuzzy maximum entropy (imprecise side conditions) estimators producing fuzzy distributions and crisp discrete/continuous distributions. Other new results are: (1) two chapters on fuzzy ANOVA (one-way and two-way); (2) random fuzzy numbers with applications to fuzzy Monte Carlo studies; and (3) a fuzzy nonparametric estimator for the median. Viertl [24] presents the foundations of the description of fuzzy data, including methods on how to obtain the characterizing function of fuzzy measurement results. Furthermore, statistical methods are then generalized to the analysis of fuzzy data and fuzzy a-priori information. Nguyen et al. [19] show how to compute statistics under interval and fuzzy uncertainty. The resulting methods are applied to computer science (optimal scheduling of different processors), to information technology (maintaining privacy), to computer engineering (design of computer chips), and to data processing in geosciences, radar imaging, and structural mechanics.

4   Fuzzy Statistical Decision-Making: An Example

In classical inferential statistics all the parameters of a mathematical model and the data are exactly known. However, if all these parameters and data are not obtained, the classical models are valid only under some additional assumptions that might be not fulfilled. We face such a situation when our experimental data are of a linguistic type [10]. We present an example of fuzzy hypothesis tests in the following [9].

A sample of 144 units gave a sample mean of 214.452 and a sample standard deviation of 25.6. The significance level is 0.02. Make a fuzzy hypothesis test for H_0: $\mu \leq 210$ versus H_a: $\mu > 210$ using α-cut levels 0.4 and 0.9.

This is a right-tailed hypothesis test. First of all, from the standard normal distribution (SND) table, we find $Z_0 = Z_{\beta=0.02} = 2.05375$. From the given data, we calculate the Z_c value from the sample statistics as follows:

$$Z_c = \frac{\bar{x} - \mu}{(\sigma/\sqrt{n})} = \frac{214.452 - 210}{(25.6/\sqrt{144})} = 2.086875 \tag{2}$$

Table 2 gives the triangular fuzzy Z values obtained from both SND table and sample statistics. The classical Z_0 value from SND table is 2.05375. This Z_0 value may be changed depending on the risk perception of the decision maker. Hence, we represent the Z_0 value as a triangular fuzzy number. Since we consider the possible slight changes in the sample mean and sample standard deviation, the Z_c value calculated from the sample may change accordingly.

Table 2 Comparison of fuzzy Z values

Fuzzy Z values from SND table				Calculated fuzzy Z values			
α	$z_{\alpha/2}$	$z_0 - z_{\alpha/2}$	$z_0 + z_{\alpha/2}$	α	$z_{\alpha/2}$	$z_c - z_{\alpha/2}$	$z_c + z_{\alpha/2}$
0.01	−2.576	−0.522	4.630	0.01	−2.576	−0.489	4.663
0.02	−2.326	−0.273	4.380	0.02	−2.326	−0.239	4.413
0.03	−2.170	−0.116	4.224	0.03	−2.170	−0.083	4.257
0.04	−2.054	0.000	4.107	0.04	−2.054	0.033	4.141
0.05	−1.960	0.094	4.014	0.05	−1.960	0.127	4.047
0.06	−1.881	0.173	3.935	0.06	−1.881	0.206	3.968
0.07	−1.812	0.242	3.866	0.07	−1.812	0.275	3.899
0.08	−1.751	0.303	3.804	0.08	−1.751	0.336	3.838
0.09	−1.695	0.358	3.749	0.09	−1.695	0.391	3.782
0.1	−1.645	0.409	3.699	0.1	−1.645	0.442	3.732
⋮	⋮	⋮	⋮	⋮	⋮	⋮	⋮
0.35	−0.935	1.119	2.988	0.35	−0.935	1.152	3.021
0.36	−0.915	1.138	2.969	0.36	−0.915	1.172	3.002
0.37	−0.896	1.157	2.950	0.37	−0.896	1.190	2.983
0.38	−0.878	1.176	2.932	0.38	−0.878	1.209	2.965
0.39	−0.860	1.194	2.913	0.39	−0.860	1.227	2.946
0.4	−0.842	1.212	2.895	0.4	−0.842	1.245	2.928
0.41	−0.824	1.230	2.878	0.41	−0.824	1.263	2.911
0.42	−0.806	1.247	2.860	0.42	−0.806	1.280	2.893

(continued)

Table 2 (continued)

Fuzzy Z values from SND table				Calculated fuzzy Z values			
α	$z_{\alpha/2}$	$z_0 - z_{\alpha/2}$	$z_0 + z_{\alpha/2}$	α	$z_{\alpha/2}$	$z_c - z_{\alpha/2}$	$z_c + z_{\alpha/2}$
0.43	−0.789	1.265	2.843	0.43	−0.789	1.298	2.876
0.44	−0.772	1.282	2.826	0.44	−0.772	1.315	2.859
0.45	−0.755	1.298	2.809	0.45	−0.755	1.331	2.842
⋮	⋮	⋮	⋮	⋮	⋮	⋮	⋮
0.85	−0.189	1.865	2.243	0.85	−0.189	2.276	1.820
0.86	−0.176	1.877	2.230	0.86	−0.176	2.263	1.832
0.87	−0.164	1.890	2.217	0.87	−0.164	2.251	1.845
0.88	−0.151	1.903	2.205	0.88	−0.151	2.238	1.858
0.89	−0.138	1.915	2.192	0.89	−0.138	2.225	1.870
0.9	−0.126	1.928	2.179	0.9	−0.126	2.213	1.883
0.91	−0.113	1.941	2.167	0.91	−0.113	2.200	1.896
0.92	−0.100	1.953	2.154	0.92	−0.100	2.187	1.908
0.93	−0.088	1.966	2.142	0.93	−0.088	2.175	1.921
0.94	−0.075	1.978	2.129	0.94	−0.075	2.162	1.933
0.95	−0.063	1.991	2.116	0.95	−0.063	2.150	1.946
0.96	−0.050	2.004	2.104	−0.050	2.037	2.137	−0.050
0.97	−0.038	2.016	2.091	−0.038	2.049	2.124	−0.038
0.98	−0.025	2.029	2.079	−0.025	2.062	2.112	−0.025
0.99	−0.013	2.041	2.066	−0.013	2.074	2.099	−0.013
1	0.000	2.054	2.054	1	0.000	2.087	2.087

The intersection value of the left side representation of Z_c and the right side representation of Z_0 is between α = 0.98 and α = 0.99. For both α-cut level = 0.4 and α-cut level = 0.9, no decision can be given. The classical decision would be reject the null hypothesis.

5   Conclusion

In this chapter, we presented the literature review results for fuzzy statistics and fuzzy statistical decision-making. The literature review reveals that after the year 2000, fuzzy statistics attracts the researchers. The researchers in Chinese universities produce most of the publications on fuzzy statistics. However, the largest frequency per researcher (mode) of these publications does not belong to Chinese researchers. The researchers having the largest frequencies are from Turkey, Taiwan, Spain, Poland, US, Japan, and France. The total publications on fuzzy statistics from China and USA are larger than the sum of all other countries. Almost all of the publications on fuzzy statistics have been published in Journals or at conferences. The subject areas of most of these publications are on computer science, engineering, and mathematics.

The extensions of fuzzy sets such as type-2 fuzzy sets, intuitionistic fuzzy sets, and hesitant fuzzy sets present new opportunities to extend fuzzy statistics. For further research, these extensions are suggested to expand the fuzzy statistical decision-making techniques such as intuitionistic fuzzy hypothesis tests and hesitant fuzzy hypothesis tests.

References

1. Ahmed, M.N., Yamany, S.M., Mohamed, N., Farag, A.A., Moriarty, T.: A modified fuzzy C-means algorithm for bias field estimation and segmentation of MRI data. IEEE T. Med. Imaging. **21** (3), 193–199 (2002)
2. Ayyub, B., Gupta, M.M. (eds.): Uncertainty Analysis in Engineering and Sciences: Fuzzy Logic, Statistics, and Neural Network Approach, International Series in Intelligent Technologies, Springer (1998)
3. Baowen, L., Peizhuang, W., Xihui, L., Yong, S.: Fuzzy bags and relations with set-valued statistics. Comput. Math. Appl. **15** (10), 811–818 (1988)
4. Bertoluzza, C., Gil, M.A., Ralescu, D.A. (eds.): Statistical Modeling, Analysis and Management of Fuzzy Data, vol. 87. Physica-Verlag, Heidelberg (2002)
5. Bialasiewicz, J.: Sufficient and c-Sufficient statistics in pattern recognition and their Relation to fuzzy techniques. IEEE T. Syst. Man. Cyb. **19** (5), 1261–1263 (1989)
6. Buckley, J.J.: Fuzzy decision-making with data: applications to statistics. Fuzzy Sets Syst. **16** (2), 139–147 (1985)
7. Buckley, J.J.: Fuzzy Probabilities: New Approach and Applications. Physica-Verlag, Heidelberg (2003)
8. Buckley, J.J.: Fuzzy Statistics, Studies in Fuzziness and Soft Computing, vol. 149. Springer, New York (2004)
9. Buckley, J.J.: Fuzzy Probability and Statistics, Studies in Fuzziness and Soft Computing, vol. 196. Springer, Heidelberg (2006)
10. Grzegorzewski, P., Hryniewicz, O.: Testing statistical hypotheses in fuzzy environment. Mathware Soft Comput. **4**, 203–217 (1997)
11. Grzegorzewski, P., Hryniewicz, O.G., Maria, A. (eds.): Soft Methods in Probability, Statistics and Data Analysis. Advances in Intelligent and Soft Computing, volume 16 (2002)
12. Hong, H.-L., Sang, L.-G., Mei, Z.-W.: Method of fuzzy statistics and applications. IFAC Proc. Series. 349–354 (1984)
13. Jain, R.: Fuzzy set theory versus bayesian statistics. IEEE T. syst. Man. Cyb. **8** (4), 332–333 (1978)
14. Kandel, A.: Fuzzy statistics and forecast evaluation. IEEE T. Syst. Man. Cyb. SMC. **8** (5), 396–401 (1978)
15. Kandel, A., Byatt, W.J., Fuzzy sets, fuzzy algebra, and fuzzy statistics. P. IEEE. **66** (12), 1619–1639 (1978)
16. Krishnapuram, R., Keller, J.M.: The possibilistic C-means algorithm: Insights and recommendations. IEEE T. Fuzzy Syst. **4** (3), 385–393 (1996)
17. Negoita, C.V., Ralescu, D.: Simulation, knowledge-based computing, and fuzzy statistics, Van Nostrand Reinhold electrical/computer science and engineering series (1987)
18. Nguyen, H.T., Wu, B.: Fundamentals of Statistics with Fuzzy Data, Studies in Fuzziness and Soft Computing, vol. 198. Springer, Berlin (2006)

19. Nguyen, H.T., Kreinovich, V., Wu, B., Xiang, G.: Computing Statistics under Interval and Fuzzy Uncertainty: Applications to Computer Science and Engineering. Springer, Berlin (2012)
20. Ross, T.J., Booker, J.M., Parkinson W.J. (eds.): Fuzzy Logic and Probability Applications: Bridging the Gap, ASA-SIAM Series on Statistics and Applied Probability (Book 11). Society for Industrial and Applied Mathematics (2002)
21. Slowinski, R.: Fuzzy Sets in Decision Analysis, Operations Research and Statistics. Springer, New York 1998
22. Stallings, W.: Fuzzy set theory versus bayesian statistics. IEEE T. Syst. Man. Cyb. SMC. **7** (3), 216–219 (1977)
23. Taheri, S.M.: Trends in fuzzy statistics. Austrian J. Stat. **32**(3), 239–257 (2003)
24. Viertl, R.: Statistical Methods for Fuzzy Data. Wiley, New York (2011)
25. Xihe, L.: Stability of random membership frequency and fuzzy statistics. Fuzzy Sets Syst. **29** (1), 89–102 (1989)
26. Zadeh, L.A.: Probability measures of fuzzy events. J. Math. Anal. Appl. **23**, 421–427 (1968)
27. Zadeh, L.A.: Fuzzy sets as a basis for a theory of possibility. Fuzzy Sets Syst. **1**, 3–28 (1978)
28. Zeng, H., Du, J.-G., Wu, M.-R.: Noise level statistics and fuzzy classification of 125MW and 300MW turbine-generator sets. Dongli Gongcheng/Pow. Eng. **8** (5), 37–43 (1988)

Fuzzy Probability Theory I: Discrete Case

I. Burak Parlak and A. Cağrı Tolga

Abstract This chapter introduces the underlying theory of Fuzzy Probability and Statistics related to the differences and similarities between discrete probability and possibility spaces. Fuzzy Probability Theory for Discrete Case starts with the fundamental tools to implement an immigration of crisp probability theory into fuzzy probability theory. Fuzzy random variables are the initial steps to develop this theory. Different models for fuzzy random variables are designated regarding the fuzzy expectation and fuzzy variance. In order to derive the observation related to fuzzy discrete random variables, a brief summary of alpha-cuts is introduced. Furthermore, essential properties of fuzzy probability are derived to present the measurement of fuzzy conditional probability, fuzzy independency and fuzzy Bayes theorem. The fuzzy expectation theory is studied in order to characterize fuzzy probability distributions. Fuzzy discrete distributions; Fuzzy Binomial and Fuzzy Poisson are introduced with different examples. The chapter is concluded with further steps in the discrete case.

Keywords Fuzzy random variable · Fuzzy conditional probability · Fuzzy independency · Fuzzy Bayes theorem · Fuzzy Hypergeometric distribution · Fuzzy Binomial distribution · Fuzzy Poisson distribution

1 Introduction

When certainties occur people typically look back to gathered data and try to estimate future events. Traditionally, one of the methodologies; well known as probability theory has met these requirements in dealing with uncertainty and

I. Burak Parlak (✉)
Department of Computer Engineering, Galatasaray University, Istanbul, Turkey
e-mail: bparlak@gsu.edu.tr

A. Cağrı Tolga
Department of Industrial Engineering, Galatasaray University, Istanbul, Turkey
e-mail: ctolga@gsu.edu.tr

© Springer International Publishing Switzerland 2016
C. Kahraman and Ö. Kabak (eds.), *Fuzzy Statistical Decision-Making*,
Studies in Fuzziness and Soft Computing 343,
DOI 10.1007/978-3-319-39014-7_2

13

imprecision. However in full-uncertainty cases, probability theory may not be considered sufficient and it should be integrated with fuzzy logic to enhance its robustness. Full-uncertainty can be described as no one has data on the occurrence of possible cases moreover in some cases nobody know anything about these becoming true possible events. For example, think about space missions: likewise landing of the Rosetta spacecraft to the comet 67P. Additionally think about Mars relocation mission. Of course scientists can compute all the probabilities however despite all the observations made by *Phoenix* and *Curiosity* at Mars the events what will happen in the near future in the Mars mission contain deep vaguenesses. By these contingent events, people try to estimate the various events and additionally their probabilities of course.

Probability measures in fuzzy sets were first revealed by Zadeh [1]. In his work, Zadeh stated that an extension by fuzzy sets might eventually broaden the domain of practicability of probability theory, notably in those fields in which fuzziness is an expansive phenomenon. Then, in his another paper published after 10 years from the previous one, he claimed that the imprecision which is intrinsic in natural languages is, in the main, possibilistic rather than probabilistic in nature [2]. Fuzzy random variables (FRVs) were defined by Kwakernaak and he put forward several theories about independent fuzzy variables for the first time in the literature [3]. Then, he added algorithms about fuzzy random variable after 1 year, and also he gave examples for the discrete case [4].

Liu and Liu offered a new concept of fuzzy expected values related with Choquet integral occurring by chance with random variables (RVs) [5]. They contemplated a fuzzy simulation technique in order to calculate the expected value of general fuzzy variable. Also, a new description of scalar expected value operator for fuzzy random variables was given initially in their paper [6]. Buckley developed fuzzy probabilistic definitions and theorems about fuzzy probability in his pioneer books [7–9]. Nguyen and Wu presented some mathematical background of probability theory for linguistic fuzzy data and introduced several practical examples in their book [10]. Recently, Shapiro reviewed the fuzzy probability theory and summarized the application fields in order to represent fuzzy random variables and the variations between probability and possibility spaces [11–13].

In a nutshell, a fuzzy random variable is a random variable (RV) which is defined using a membership function related to a fuzzy set. However, this definition could be interpreted within the *Probability Space* and the *Possibility Space: Ps*. Let Ω be the sample space, $\mathcal{F}$ be the σ-algebra of subsets of Ω and P be the probability on Ω. $(\Omega, \mathcal{F}, P)$ is the 3-tuple which is called the *Probability Space*. On the other hand, let Θ be the sample space, $P(\Theta)$ be power set of Θ and Ps be the possibility on Θ. Then, $(\Theta, P(\Theta), Ps)$ is the 3-tuple which is called the *Possibility Space*.

In order to compare these triples, let us consider the following example; in a football match of a national league, total number of goals is determined between 1 and 8. The question is to illustrate the meaning between the probable number of goals and the possible number of goals. This difference is represented in Fig. 1. The left chart describes the probability values for the number of goals. On the other hand, the right chart shows the possibility of the number of goals. It is remarkable

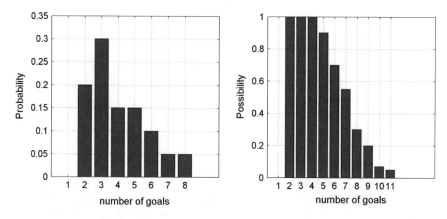

Fig. 1 The probability and possibility of having a certain number of goals in a football match

Table 1 Properties of probability and possibility spaces

Probability space	Possibility space
$(\Omega, \mathcal{F}, P)$	$(\Theta, P(\Theta), Ps)$
Ω: Sample space	Θ: Sample space
$\mathcal{F}$: σ-algebra of subsets	$P(\Theta)$: power set of Θ
P: Probability of Ω	Ps: Possibility on Θ
$P(\Omega) = 1$	$Ps(\Theta) = 1$
$P(A) \geq 0$	$Ps(\emptyset) = 0$
$P\{\bigcup_{i=1}^{\infty} A_i\}$ A_i : disjoint events	$Ps\{\bigcup_i A_i\} = sup_i\, Ps\{A_i\}$

that having one or two goals have different probabilities, but they are equally possible in a football match.

The differences between probability space $(\Omega, \mathcal{F}, P)$ and possibility space $(\Theta, P(\Theta), Ps)$ are detailed within the studies of Shapiro [11–13]. As a summary, Table 1 condenses the properties for both spaces as follows;

Discrete possibility space is plotted in Fig. 2 as a graph of membership functions for each discrete event.

In this chapter, fuzzy random variables will be defined and related results will be presented to link them with discrete fuzzy random variables. The discrete fuzzy probability function and its related expectation will be given also. While deepening in fuzzy random variables α-cuts need to be investigated. The consequent section will contain this alpha-cuts topic. Fuzzy probability will be discussed in Sect. 4. In Sect. 5, we will penetrate to discrete fuzzy expectation topic. Section 6 will make mention of fuzzy conditional probability. Fuzzy independency and fuzzy Bayes formula will be investigated in Sects. 7 and 8 successively. Then Fuzzy Hypergeometric distribution will be investigated in Sect. 9. After that, in Sect. 10 Fuzzy Binomial distribution will be expressed. And Fuzzy Poisson distribution,

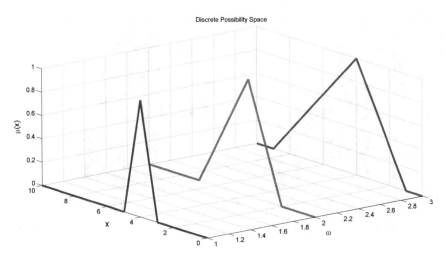

Fig. 2 The illustration of discrete possibility space

ending of the discrete probability distributions, will be explained in Sect. 11. All these Fuzzy discrete distributions will be intensified by illustrative examples. Finally, we will complete this chapter by an inclusive section.

2 Fuzzy Random Variables

The statement of fuzzy random variable (FRV) was firstly developed by Kwakernaak [3], as '*random variables whose values are not real, but fuzzy numbers*'. He defined a FRV as an ambiguous cognition of a crisp but unobservable random variable. For example, let's think about assigning an age to people in a conference. Consider X as their existing age that is an unexceptional random variable on the real line, of course at the positive side. But, someone can simply conceive a random variable x through a set of values as follows: "young", "middle age" and "old". Which means; someone conceive fuzzy sets as noteable results since the genuine X is not remarkable.

As Shapiro [12] remarked; Kwakernaak [4] introduced the fundamentals for a fuzzy random variable model. However before Kwakernaak's study there were essential studies made by various scientist on random sets [1, 2, 14–17]. Puri and Ralescu [18] developed new idea to generate the fuzziness and they stated that the expected value could be fuzzy but the variance should be scalar. Liu and Liu [6] asserted the law of truth conservation and they insisted on that possibility measure was inconsistent with the law of excluded middle and the law of contradiction. They maintained that the expected value and variance of any FRV should be scalar both. However, in contradiction to all these ideas in his books Buckley [7–9] developed another aspect that is generating the fuzziness and the model where the

Table 2 Summary of milestones on fuzzy probabilistic studies

Researchers	State of art
Zadeh 1968 [1]	Fuzzy probability measures
Kendall 1974 [14]	Random sets
Matheron 1975 [15]	Random sets
Fron 1976 [16]	Random fuzzy sets
Zadeh 1978 [2]	Possibility theory
Nguyen 1978 [17]	Random sets and belief functions
Kwakernaak 1978 [3], 1979 [4]	Fuzzy random variable
Puri and Ralescu 1986 [18]	Fuzzy random variables
Buckley 2004 [7], 2005 [8], 2006 [7]	Fuzzy discrete and continuous probability distributions
Couso et al. 2014 [24]	Ill perceived random sets

expected value and the variance are both fuzzy values. Table 2 summarizes fundamental studies on fuzzy probability theory and applications.

A fuzzy membership function would have different values in an interval. Therefore, a fuzzy random variable will be a random variable whose value would be a set using the fuzzy membership function.

Let x_i, $i \in \mathbb{N}$ be a discrete fuzzy random variable. If the values of x_i are denumarable, x_i is called a discrete fuzzy random variable. The probability function of a discrete fuzzy random variable is the representation of the discrete values and their respective probabilities. Furthermore, the fuzzy values of x_i are denoted as $\mu(x_i)$.

In discrete case, this membership function will be represented using fuzzy probability mass function whereas it will be the fuzzy probability density function in continuous case.

In order to generalize the FRV in discrete case, let A be a fuzzy subset of Ω. If $A(x) \neq 0$ for n times of x values in Ω, this subset A could be identified as a discrete fuzzy set. Let us suppose $A(x) \neq 0$ for $1 \leq i \leq n$ in Ω. Therefore, we may write the fuzzy set as follows

$$\tilde{A} = \mu_1, \ldots, \mu_i, \ldots, \mu_n \tag{1}$$

Here μ_i are called the membership values of x_i. In this chapter we will adapt the following expression $\tilde{A}(x_i) = \tilde{\mu}_i$, $1 \leq i \leq n$. In the generalized form, discrete fuzzy subsets could be any space in Ω. Therefore, we may note that α-cuts of discrete fuzzy sets of $\mathbb{R}$, the set of real numbers, do not produce closed, bounded intervals.

Let $X = x_1, \ldots, x_n$ be a finite set and suppose a probability function denominated by P should depicted on all subsets of X with $P(x_i) = a_i$, $1 \leq i \leq n$, $0 \leq a_i \leq 1$ for all i. As we recall from probability theory the summation of a_i ($1 \leq i \leq n$) values should be equal to 1. The relation between X and P is identified as discrete probability distribution.

Even the probabilistic a_i values should be already known, they are generally estimated, or are observed by experiments. In order to immigrate the fuzzy case, let us start with the assumption of the uncertainty for some a_i and let us model them using fuzzy numbers as described recently. In practice, we may write some a_i as fuzzy numbers and the others as crisp number. However, we should apply fuzzy notation for both a_i numbers in order to facilitate the nomenclature and the calculations.

Moreover, we may write the uncertain a_i values as $\tilde{a}_i$; fuzzy values, and we may apply the probability theory in a similar way; $\forall\ a_i$ and nominate that $0 < \tilde{a}_i < 1$, $1 \leq i \leq n$. Throughout the rest of this chapter, this fuzzy nomenclature is used for given or estimated probabilities.

The probability value a_i is expressed as $\tilde{a}_i = a_i$. However, $\tilde{a}_i$ might be omitted and $\tilde{A}$ might be preferred in order to express the whole distribution in some cases. Therefore, FRV values $X = x_i$ coupled with fuzzy probabilistic values; $\tilde{A} = a_i$ are called a discrete fuzzy probability distribution. Finally, fuzzy P is $\tilde{P}$ and intrinsically $\tilde{P}(x_i) = \tilde{a}_i$, $1 \leq i \leq n$, $0 \leq \tilde{a}_i \leq 1$ could be written.

Consequently, in order to satisfy the summation of probability values, the following restriction should be taken into account when a missed observation is estimated regarding the fuzzy case. We can estimate a_i in $\tilde{a}_i[\alpha]$, all α, and $a_i \in \tilde{a}_i[1]$ should satisfy $\sum_{i=1}^{n} a_i = 1$.

2.1   The FRV Model by Puri and Ralescu

Let $(\Omega, \mathcal{F}, P)$ be a probability space, $A(\mathbb{R}^n)$ emphasize the set of fuzzy subsets, $x : \mathbb{R}^n \to [0, 1], X : \Omega \to A(\mathbb{R}^n)$ be defined by $X_\alpha(\omega) = (x \in \mathbb{R}^n : X(\omega)(x) \geq \alpha)$, and $\mathcal{B}$ denote the Borel subsets of $\mathbb{R}^n$. An FRV by Puri and Ralescu is a function $X : \Omega \to A(\mathbb{R}^n)$ such that, for every $\alpha \in [0, 1]$:

$$\{(\omega, x) : x \in X_\alpha(\omega)\} \in \mathcal{F} \times \mathcal{B}$$

The most wonted measure is the Aumann-type mean for digitizing the central tendency of the distribution of an FRV model by Puri and Ralescu. In his work, Aumann [19] widened the real-valued variable's mean and maintained its principal essential characters and attitude. Before giving the statement of expected value and variance a definition of integrably boundedness has to be made:

A ϑ is an FRV as can be mentioned an integrably bounded FRV related with the probability space $(\Omega, \mathcal{F}, P)$ iff $\|\vartheta_0\| \in L^1(\Omega, \mathcal{F}, P)$, where, for the function f, $L^1(\Omega, \mathcal{F}, P) = \{f \mid f : \Omega \to \mathbb{R}, \mathcal{F}\text{-measurable}, \int |f|^1 dP < \infty\}$.

Let ϑ be an integrably bounded FRV related with $(\Omega, \mathcal{F}, P)$, and $S(A)$ be a nonempty bounded set as regard to the $L^1(P)$-norm, the expected value of ϑ is the unpaired fuzzy set $\tilde{E}(\vartheta \mid P)$ of $\mathbb{R}^n$ such that

$$\left(\tilde{E}(\vartheta \mid P)\right)_\alpha = \int_\Omega \vartheta_\alpha dP \quad \text{for all } \alpha \in [0, 1], \tag{2}$$

where $\int_\Omega \vartheta_\alpha dP = \left\{ \int_\Omega f dP \mid f \in S(\vartheta_\alpha) \right\}$ is the Aumann integral of ϑ_α as regard to P.

Variance of a Puri and Ralescu type FRV is argued by Feng et al. [20]. They claimed that, the variance should be used to observe the spread or deployment of the FRV around its expected value (EV) just like under the circumstances of real random variables. Ultimately, they illustrated the Puri and Ralescu type variance as a scalar shown below:

$$Var(\tilde{X}) = \frac{1}{2} \int_0^1 [V(\underline{X}_\alpha) + V(\bar{X}_\alpha)] d\alpha \tag{3}$$

Also in the literature, there are some other offers for scalar variance. Premier one is considering a numerical element of every fuzzy realization of the FRV as the midpoint of the support and then computing the deployment of these representative values.

2.2 The FRV Model by Buckley Based on Kwakernaak

As given in the previous subsection, let $(\Omega, \mathcal{F}, P)$ be a probability space and let $A(\mathbb{R})$ emphasize the all fuzzy numbers' set in the real numbers set, $\mathbb{R}$. Particularly, $A(\mathbb{R})$ depicts the class of normal convex fuzzy subsets of $\mathbb{R}$ which has the severe α-levels for $\alpha \in [0, 1]$. Assignment class could be defined as U, and $U : \mathbb{R} \to [0, 1]$, i.e., $U(u) \in [0, 1]$, for all $u \in \mathbb{R}$, such that U_α is a non-empty severe interval, where

$$U_\alpha = \begin{cases} \{x \in \mathbb{R} \mid U(x) \le \alpha\} & \text{if } \alpha \in (0, 1] \\ cl(supp U) & \text{if } \alpha = 0 \end{cases}$$

An FRV is an assignment $\vartheta : \Omega \to A(\mathbb{R})$ thus for each $\alpha \in [0, 1]$ and all $\omega \in \Omega$ the real-valued assignment:

inf $\vartheta_\alpha : \Omega \to \mathbb{R}$, ensuring inf $\vartheta_\alpha(\omega) = $ inf $(\vartheta(\omega))_\alpha$, and

sup $\vartheta_\alpha : \Omega \to \mathbb{R}$, ensuring inf $\vartheta_\alpha(\omega) = $ inf $(\vartheta(\omega))_\alpha$, are real valued RVs.

The central tendency of the distribution of an FRV model by Buckley based on Kwakernaak can be widened to the real-valued variable's mean and can be calculated as below:

$$\mu_{E(U)}(\vartheta) = \sup \{\mu_\vartheta(U) \mid U \in \mathcal{U}_A, E(U) = \vartheta\}, \vartheta \in \mathbb{R} \tag{4}$$

where E indicates the usual expectation. Similarly, the fuzzy variance of ϑ is a fuzzy set on $[0, \infty)$ with

$$\mu_{V(U)}(\vartheta) = \sup \ \{\mu_\vartheta(U) \mid U \in \mathcal{U}_{\mathcal{A}}, V^2 U = \sigma^2\}, \sigma^2 \in [0, \infty). \qquad (5)$$

where, ϑ is an FRV, $\mathcal{U}_{\mathcal{A}}$ is the collection of all $\mathcal{A}$-measurable RVs of Ω and V indicates the usual variance.

As stated in the previous sections, mean and variance of a fuzzy random variable in Liu and Liu's model are both scalar. One can calculate these values more easily than the cited FRV models above.

3 Alpha-Cuts

In a more universal perception, the random set could be produced by the α-cuts of A as stated in Zadeh's article [21]. To be more en detail, an α-cut, A_α, of A is a non-fuzzy set described by $A_\alpha = \{x \mid \mu_A(x) \geq \alpha\}, 0 \leq \alpha \leq 1$. The α-cuts are employed to be the main components of a random set; with α it is assumed to be uniformly distributed over the interval $[0, 1]$.

A fuzzy set A can be produced from a random set. Essentially, the identical result can be obtained without appearing of randomness. It is clear that a fuzzy set may be reproduced from its α-cuts both discriminating and additively. In order to explain this case, suppose that $\mu_{A_\alpha}(x)$ express the membership function of A_α. As A_α is non-fuzzy, it might be confused with the specific function of A_α. Then, the membership function of A can be displayed in terms of the membership functions of the A_α (a) discriminating as $\mu_A(x) = sup_\alpha(\alpha \wedge \mu_{A_\alpha}(x)), 0 \leq \alpha \leq 1$, where $\wedge$ expresses min, and (b) additively as

$$\mu_A(x) = \int\limits_0^1 \mu_{A_\alpha}(x)d\alpha. \qquad (6)$$

The illustrative representation is depicted in Fig. 3.

4 Fuzzy Probability

Let us start to define two (crisp) subsets A and B be of X. Moreover, we want to compute fuzzy probabilities which are denoted $\tilde{P}(A)$ and $\tilde{P}(B)$, respectively. We have to implement fuzzy algebra to calculate these values. In some a_i values there would be an uncertainty, however in discrete probability distribution there is no uncertainty. Therefore, the probability summation rule $a_1 + \cdots + a_n = 1$ for each a_i values in $\tilde{a}_i[\alpha]$ should be satisfied. This constraint will be served as the basis of our fuzzy algebra.

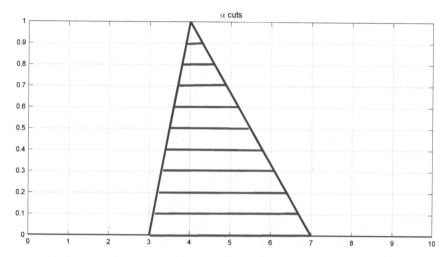

Fig. 3 The representation of alpha cuts for a triangular fuzzy number

We suppose that $A = x_1, \ldots, x_k$, $1 \leq k < n$, then define

$$\tilde{P}(A)[\alpha] = \left\{ \sum_{i=1}^{k} a_i \right\} \tag{7}$$

for $0 \leq \alpha \leq 1$, where $a_i \in \tilde{a}_i[\alpha]$, $1 \leq i \leq n$, $\sum_{i=1}^{n} a_i = 1$ which is related to the fuzzy algebra. First of all, a complete discrete probability distribution using the α-cuts should be determined. Secondly, a probability in Eq. (7) should be calculated.

Let us denote that $\tilde{P}(A)[\alpha]$ is not equal to the sum on fuzzy the intervals $\tilde{a}_i[\alpha]$ with the fuzzy algebra in $1 \leq i \leq k$. Let us try to complete the definition of $\tilde{P}(A)[\alpha]$ by introducing the α-cuts of a fuzzy number $P(A)$.

Initially, we note that $x_l, \ldots, x_m$, $x_i \geq 0$, $\sum_{i=1}^{n} x_i = 1$ and $f(a_1, a_2, \ldots, a_n) = \sum_{i=1}^{k} a_i$. Using these definitions, we might write that

1. If $A \cap B = \emptyset$, then $P(\tilde{A}) + P(\tilde{B}) \geq \tilde{P}(A \cup B)$.
2. If $A \subseteq B$, $\tilde{P}(A)[\alpha] = [p_{a1}(\alpha), p_{a2}(\alpha)]$ and $\tilde{P}(B)[\alpha] = [p_{b1}(\alpha), p_{b2}(\alpha)]$ then $p_{a_i}(\alpha) \leq p_{b_i}(\alpha)$ for $i = 1, 2$ and $0 \leq \alpha \leq 1$
3. $0 \leq \tilde{P}[A] \leq 1$ all A with $\tilde{P}(\emptyset) = 0$, $\tilde{P}(X) = 1$
4. $\tilde{P}[A] + \tilde{P}[A'] \geq 1$, A' is the complement of A.
5. For $A \cap B \neq \emptyset$, $\tilde{P}(A \cup B) \leq \tilde{P}(A) + \tilde{P}(B) - \tilde{P}(A \cup B)$

Example 1 Let n be 5 and the sets are; $X = \{a_1, a_2, a_3\}$ $Y = \{a_4, a_5\}$ and the fuzzy probabilities for each random variables are $\tilde{x}_i = (0.1, 0.2, 0.3)$ $1 \leq i \leq 5$.

Therefore, we may compute that; $\tilde{P}(X)[0] = [0.3, 0.9]$, $\tilde{P}(X)[1] = [0.6, 0.6]$ and $\tilde{P}(Y)[0] = [0.2, 0.6]$, $\tilde{P}(Y)[1] = [0.4, 0.4]$.

Example 2 Let n be 6 and the sets are; $X = \{a_1, a_2, a_3\}$ $Y = \{a_3, a_4\}$ and $Z = \{a_4, a_5, a_6\}$ the fuzzy probabilities for random variables are as follows; $\tilde{x}_i = (0.1, 0.15, 0.2)$ for $1 \leq i \leq 4$ $\tilde{x}_j = (0.15, 0.2, 0.22)$ for $5 \leq j \leq 6$.

Therefore, we may compute that; $\tilde{P}(X)[0] = [0.3, 0.6]$, $\tilde{P}(X)[1] = [0.45, 0.45]$, $\tilde{P}(Y)[0] = [0.2, 0.4]$, $\tilde{P}(Y)[1] = [0.3, 0.3]$ and $\tilde{P}(Z)[0] = [0.4, 0.64]$, $\tilde{P}(Z)[1] = [0.55, 0.55]$.

Furthermore, we may deduce the fuzzy probabilities of the intersection; $\tilde{P}(X \cap Y)[0] = [0.1, 0.2]$, $\tilde{P}(X \cap Y)[1] = [0.15, 0.15]$, $\tilde{P}(Y \cap Z)[0] = [0.1, 0.2]$, $\tilde{P}(Y \cap Z)[1] = [0.15, 0.15]$.

Finally the fuzzy union probabilities are as follows; $\tilde{P}(X \cup Y)[0] = [0.4, 0.8]$, $\tilde{P}(X \cup Y)[1] = [0.6, 0.6]$, $\tilde{P}(Y \cup Z)[0] = [0.5, 0.84]$, $\tilde{P}(Y \cup Z)[1] = [0.7, 0.7]$.

We may note that; $[0.4, 0.8] \neq [0.3, 0.6] + [0.2, 0.4] - [0.1, 0.2]$ and $[0.5, 0.84] \neq [0.2, 0.4] + [0.4, 0.64] - [0.1, 0.2]$.

Therefore, we should remark that $\tilde{P}(X \cup Y)$ and $\tilde{P}(Y \cup Z)$ can be expressed as a subset of $\tilde{P}(X) + \tilde{P}(Y) - \tilde{P}(X \cap Y)$, $\tilde{P}(Y) + \tilde{P}(Z) - \tilde{P}(Y \cap Z)$, respectively.

5 Fuzzy Discrete Expectation

Suppose X and Y are two random variables with joint probability density $f(x, y; \theta)$. where $x \in \mathbb{R}$ and the joint density function's vector of parameters should be $\theta = (\theta_1, \ldots \theta_n)$. Mostly cited parameters are anticipated employing a random sample from the population. These anticipations can be a point estimate or a confidence interval. In lieu of anticipation of a point, a confidence interval for each θ_i could be replaced by the probability density function to obtain an interval joint probability density function. Whereas more general form should be constructed and this necessitates formulation of the indefiniteness in the θ_i by replacing a fuzzy number for θ_i and acquire a joint fuzzy probability density function. For α-cuts of the fuzzy number utilized for θ_i, vide supra Sect. 3. If one wishes to implement the interval probability density functions utilizing the fuzzy numbers with α-cuts of course could be better. The joint fuzzy density functions are fulfilled by substituting fuzzy numbers for the vague parameters. For to clarify the nubiluous definitions fuzzy marginals should be discussed now.

Since the θ_i in θ are uncertain fuzzy numbers $\tilde{\theta}_i$ are replaced for the θ_i, $1 \leq i \leq n$, that provides joint fuzzy density $f(x, y; \tilde{\theta})$. The fuzzy marginal for X is

$$f(x; \tilde{\theta}) = \sum_{k=1}^{\infty} f(x_k, y_k; \tilde{\theta}). \tag{8}$$

The fuzzy marginal for Y could be written by the same way. The α-cuts of the fuzzy marginals $f(x; \tilde{\theta})$ could be calculated by the following equation

$$f(x; \tilde{\theta})[\alpha] = \left\{ \sum_{k=1}^{\infty} f(x_k, y_k; \theta) \mid \theta_i \in \tilde{\theta}_i[\alpha], 1 \le i \le n \right\}, \qquad (9)$$

for $0 \le \alpha \le 1$, and also we can derive an analogous equation for $f(y; \tilde{\theta})[\alpha]$. Equation (9) gives the α-cuts of a fuzzy set for each value of x.

At this juncture let $f(x; \theta)$ be the crisp marginal of x which means non-fuzzy marginal of x. Let's utilize $f(x; \theta)$ to procure the mean $\mu_x(\theta)$ and variance $Var_x(\theta)$ of X. The mean and variance of X are written as functions of θ as they are related with the values of the parameters. Suppose that $\mu_x(\theta)$ and $Var_x(\theta)$ are discrete functions of θ. Deriving the fuzzy mean and variance of the fuzzy marginal could be made by fuzzification of the crisp mean and variance. Subjacent theorem can be deduced from the previous explanations for X and also can be utilized for Y.

Theorem *The fuzzy mean and variance of the fuzzy marginal $f(x; \tilde{\theta})$ are $\mu_x(\tilde{\theta})$ and $Var_x(\tilde{\theta})$* [8].

Proof An α-cut of the fuzzy mean of the fuzzy marginal for X is

$$\mathcal{M}_x(\tilde{\theta})[\alpha] = \left\{ \sum_{k=1}^{\infty} x_k f(x_k; \theta) \mid \theta_i \in \tilde{\theta}_i[\alpha], 1 \le i \le n \right\}, \qquad (10)$$

for $0 \le \alpha \le 1$. Now the sum in Eq. (10) equals $\mathcal{M}_x(\theta)$ for each $\theta_i \in \tilde{\theta}_i$, $1 \le i \le n$. So

$$\mathcal{M}_x(\tilde{\theta})[\alpha] = \{ \mu_x(\theta) \mid \theta_i \in \tilde{\theta}_i[\alpha], 1 \le i \le n \}. \qquad (11)$$

Because of this, the fuzzy mean is $\mathcal{M}_x(\tilde{\theta})$. See the studies explained at the cited references in the various fuzzy distributions parts of this chapter.

The fuzzy variance with α-cuts could be written as follows:

$$Var_x(\tilde{\theta})[\alpha] = \{ \sum_{k=1}^{\infty} (x_k - \mu_{x_k}(\theta))^2 f(x_k; \theta) \mid \theta_i \in \tilde{\theta}[\alpha], 1 \le i \le n \}, \qquad (12)$$

for $0 \le \alpha \le 1$. But the sum in the above equation equals $Var_x(\theta)$ for each $\theta_i \in \tilde{\theta}_i$, $1 \le i \le n$. Due to this

$$Var_x(\tilde{\theta})[\alpha] = \{ Var_x(\theta) \mid \theta_i \in \tilde{\theta}_i[\alpha], 1 \le i \le n \}. \qquad (13)$$

So, the fuzzy variance is just $Var_x(\tilde{\theta})$. $\qquad\square$

6 Fuzzy Conditional Probability

In probability theory, conditional probability serves us to introduce and calculate joint probabilities. In fuzzy probability theory, we will apply the same approach by using fuzzy random variables.

Let $M = x_1, \ldots, x_r$, $N = x_s, \ldots, x_t$ for $1 \leq r \leq s \leq t \leq n$ so that M and N are not disjoint. The fuzzy conditional probability of M given N should be defined naturally. The fuzzy conditional probability might be written as $P(M \mid N)$. Furthermore, the following definitions for fuzzy conditional probability could be presented. At first,

$$\tilde{P}(M \mid N) = \frac{\sum_{i=r}^{s} a_i}{\sum_{i=r}^{t} a_i} \quad a_i \in \tilde{a}_i[\alpha] \quad 1 \leq i \leq n \tag{14}$$

The numerator of the division is the sum of the a_i; in the intersection of M and N, while the denominator is the sum of the a_i in N. Then we may write;

$$\tilde{P}(M \mid N) = \frac{\tilde{P}(M \cap N)}{\tilde{P}(N)} \tag{15}$$

These definitions for fuzzy conditional probability would be considered as the fuzzy version of conditional probability theory. Therefore, we might interpret the fundamental characteristics of fuzzy conditional probability which are:

1. $0 \leq \tilde{P}(M \mid N) \leq 1$
2. $\tilde{P}(N \mid N) = 1$ crisp one
3. $\tilde{P}(M \mid N) = 1$ crisp if $N \subseteq M$
4. $\tilde{P}(M \mid N) = 0$ crisp if $N \cap M = \emptyset$
5. $\tilde{P}(M_1 \cup M_2 \mid N) \leq \tilde{P}(M_1 \mid N) + \tilde{P}(M_2 \mid N)$ if $M_1 \cap M_2 = \emptyset$

Firstly, we may note that the first three properties will be directly related to the initial definition of fuzzy conditional probability. Let us assume an empty sum which will be equal to zero. As the numerator in Eq. 14 is an set empty, the fourth property will be true while these events M and N would be disjoint. For the last property in fuzzy conditional case, let us define that;

$$\tilde{P}(M_1 \cup M_2 \mid N)[\alpha] \subseteq \tilde{P}(M_1 \mid N)[\alpha] + \tilde{P}(M_2 \mid N)[\alpha] \quad 0 < \alpha < 1 \tag{16}$$

We evaluate the expression through α. Let us set $x = \frac{\beta + \gamma}{\theta}$ which belongs to $(M_1 \cup M_2 \mid N)[\alpha]$. For x we may write;

- β is equal to the sum of the a_i for $x_i \in M_1 \cap N$.
- γ is equal to the sum of the a_i for; $x_i \in M_2 \cap N$.
- θ is equal to the sum of the a_i for $x_i \in N$.

As a simplification, the sum of a_i will be equal to one as $a_i \in \tilde{a}_i[\alpha]$. Therefore, we may write that β/δ belongs to $\tilde{P}(M_1 \mid N)[\alpha]$ and γ/δ belongs to $\tilde{P}(M_2 \mid N)[\alpha]$.

7 Fuzzy Independency

In probability theory, the dependency or the independency of the events are crucial to observe the joints probabilities. However, the properties related to the moments are generally related to the independency to simplify the calculations. In the case of fuzzy probability, a similar reasoning would be applied to define the case for two events M and N.

As it is represented in fuzzy conditional probability, we may adapt two definitions; strong and weak independency for the events M and N.

The first expression to define the independency is based on the fuzzy conditional probability. M and N are characterized as strongly independent if

$$\tilde{P}(M \mid N) = \tilde{P}(M) \tag{17}$$

and

$$\tilde{P}(N \mid M) = \tilde{P}(N) \tag{18}$$

These expressions are not always obvious to define the independency. Therefore, we need to introduce a new term which is the weak independency in fuzzy probability theory. We may write that the events M and N are weakly independent if

$$\tilde{P}(M \mid N)[1] = \tilde{P}(M)[1] \tag{19}$$

and

$$\tilde{P}(N \mid M)[1] = \tilde{P}(N)[1] \tag{20}$$

In the second formulation where the events are characterized as weakly independent, we use the $\alpha = 1$ cuts to satisfy the equality. Therefore, events which are strongly independent are obviously weakly independent.

In order to conclude the fuzzy independency, we may start to use the conventional expression of independency based on the crisp way. Initially, the events M and N are said independent when

$$\tilde{P}(M \cap N) = \tilde{P}(M)\tilde{P}(N) \tag{21}$$

8 Fuzzy Bayes Formula

In the probability theory, Bayes formula formulates the relationship between the current and the prior information. It is supported by the theory of conditional probability. In this section, we will show the fuzzy interpretation of Bayes formula.

Let $X = x_1, \ldots, x_n$ be a random variable and let β_i, $1 \le i \le k$, be a partition of X. We assume that β_i are not empty sets, and they are mutually disjoint. The union of β_i is X. In a case where the probability of β_i is not known, we can develop a conditional probability to calculate β_i.

If θ_i are some priors that we may know and

$$p_{ij} = P(\beta_i \mid \theta_j) \tag{22}$$

where p_{ij} will generate the probability of β_i.

In order to calculate p_{ij}, we need the estimates $p_j = P(\theta_j)$. The probability p_j is defined as the prior probability distribution. The probability that the partition β_i is given when the priors θ_j has accomplished as follows;

$$P(\theta_j \mid \beta_i) = \frac{P(\beta_i \mid \theta_j)P(\theta_j)}{\sum_{j=1}^{J} P(\beta_i \mid \theta_j)P(\theta_j)} \qquad 1 \le j \le J \tag{23}$$

Moreover, $p_{kj} = P(\theta_j \mid \beta_i)$ is called the posterior probability distribution. The probability $P(\beta_i)$ might be calculated by integrating p_{ij} and p_j as follows;

$$P(\beta_i) = \sum_{j=1}^{J} P(\beta_i \mid \theta_j)P(\theta_j) \qquad 1 \le i \le k \tag{24}$$

For a specific event β_k which has occured, we might develop the prior for the posterior and calculate the probabilities of β_i as

$$P(\beta_i) = \sum_{j=1}^{J} P(\beta_i \mid \theta_j)P(\theta_j \mid \beta_k) \qquad 1 \le i \le k \tag{25}$$

Finally, Fuzzy Bayes formulation could be written using the α-cuts of fuzzy posterior distribution as follows;

$$\tilde{P}(\theta_j \mid \beta_k)[\alpha] = \frac{p_{ij}p_j}{\sum_{j=1}^{J} p_{ij}p_j} \tag{26}$$

9 Fuzzy Hypergeometric Distribution

In the discrete probability theory, the hypergeometric distribution is considered among the fundamental probability distributions where lot acceptance area is modeled using the probabilistic information. This formulation is developed for fuzzy case in the inspection of geospatial data by Tong and Wang [22].

It is assumed that there is a finite population concerning N units in hypergeometric distribution. Let's say some number D of these units contribute a class of interest which can be a success or a failure ($D \leq N$). This type of probability distribution describes the probability of x interests in n pulls without replacement. Then, x can be depicted as a hypergeometric random variable with the probability distribution as below:

$$P(x) = \frac{\binom{D}{x}\binom{N-D}{n-x}}{\binom{N}{n}} \qquad x = 0,1,2,\ldots,min(n,D) \qquad (27)$$

In order to calculate the fuzzy probability, we need to use the fuzzy algebra for $\tilde{D} = N\tilde{r}$ and $\tilde{n} = N\tilde{l}$ and in addition we should have derive the minimum and the maximum of $\tilde{P}(x)$ using the α-cuts. The fuzzy hypergeometric distribution is characterized as the probability model for a fuzzy random sample selection of $\tilde{n}$ items without replacement from a lot of N items of which $\tilde{D}$ are non-conforming or defective. Therefore, fuzzy hypergeometric probability mass function is derived by using fuzzy numbers for the conforming items based on the approach of Tong and Wang [22] as follows;

$$\tilde{P}(x)[\alpha] = min\left\{ \frac{\binom{\tilde{D}}{x}\binom{N-\tilde{D}}{\tilde{n}-x}}{\binom{N}{\tilde{n}}} \right\} \qquad 0 \leq \alpha \leq 1 \qquad (28)$$

$$\tilde{P}(x)[\alpha] = max\left\{ \frac{\binom{\tilde{D}}{x}\binom{N-\tilde{D}}{\tilde{n}-x}}{\binom{N}{\tilde{n}}} \right\} \qquad 0 \leq \alpha \leq 1 \qquad (29)$$

Moreover, the fuzzy probability $\tilde{P}[\alpha]$ is obtained within $[P(r_1), P(r_2)]$. Therefore,

$$\tilde{P}(x)[\alpha] = \frac{\binom{Nr}{x}\binom{N-Nr}{Nl-x}}{\binom{N}{Nl}} \qquad r \in \tilde{r}[\alpha]\ \ l \in \tilde{l}[\alpha]\ \ 0 \leq \alpha \leq 1 \qquad (30)$$

Example 3 Suppose that a lot contains 2000 items, a fraction of $\tilde{r} = (0.25, 0.3, 0.4)$ which do not conform requirements. If a fraction of $\tilde{l} = (0.04, 0.05, 0.06)$ items is selected at random without replacement, then the fuzzy probability of finding one or fewer nonconforming items in the sample is as follows;

$$\tilde{P}\{x \leq 1\} = \tilde{P}\{x = 0\} + \tilde{P}\{x = 1\}$$

$$min\{\tilde{P}\{x \leq 1\}\} = \cfrac{\left(\binom{2000(0.25 + (0.3 - 0.25)\alpha)}{0}\right)\left(\binom{2000 - 2000(0.25 + (0.3 - 0.25)\alpha)}{2000(0.04 + (0.05 - 0.04)\alpha)}\right)}{\left(\binom{2000}{2000(0.04 + (0.05 - 0.04)\alpha)}\right)}$$
$$+ \cfrac{\left(\binom{2000(0.25 + (0.3 - 0.25)\alpha)}{0}\right)\left(\binom{2000 - 2000(0.25 + (0.3 - 0.25)\alpha)}{(2000(0.04 + (0.05 - 0.04)\alpha) - 1)}\right)}{\left(\binom{2000}{2000(0.04 + (0.05 - 0.04)\alpha)}\right)}$$

$$(31)$$

$$max\{\tilde{P}\{x \leq 1\}\} = \cfrac{\left(\binom{2000(0.4 + (0.3 - 0.4)\alpha)}{0}\right)\left(\binom{2000 - 2000(0.4 + (0.3 - 0.4)\alpha)}{2000(0.06 + (0.05 - 0.06)\alpha)}\right)}{\left(\binom{2000}{2000(0.06 + (0.05 - 0.06)\alpha)}\right)}$$
$$+ \cfrac{\left(\binom{2000(0.4 + (0.3 - 0.4)\alpha)}{0}\right)\left(\binom{2000 - 2000(0.4 + (0.3 - 0.4)\alpha)}{(2000(0.06 + (0.05 - 0.06)\alpha) - 1)}\right)}{\left(\binom{2000}{2000(0.06 + (0.05 - 0.06)\alpha)}\right)}$$

$$(32)$$

$$\tilde{P}\{x \leq 1\} = \left[\cfrac{\left(\binom{600\alpha}{0}\right)\left(\binom{1400\alpha}{100\alpha}\right)}{\left(\binom{2000}{100\alpha}\right)} + \cfrac{\left(\binom{600\alpha}{1}\right)\left(\binom{1400\alpha}{(100\alpha) - 1}\right)}{\left(\binom{2000}{100\alpha}\right)} , \cfrac{\left(\binom{600\alpha}{0}\right)\left(\binom{1400\alpha}{100\alpha}\right)}{\left(\binom{2000}{100\alpha}\right)} + \cfrac{\left(\binom{600\alpha}{1}\right)\left(\binom{1400\alpha}{(100\alpha) - 1}\right)}{\left(\binom{2000}{140\alpha}\right)} \right]$$

$$(33)$$

Fuzzy Mean and Variance of Hypergeometric Distribution

The mean and variance of the hypergeometric distribution could be calculated

$$\tilde{\mu}[\alpha] = \frac{N\tilde{l} * N\tilde{r}}{N} \tag{34}$$

and

$$\tilde{\sigma}^2[\alpha] = \frac{N\tilde{l} * N\tilde{r}}{N}\left(1 - \frac{N\tilde{r}}{N}\right)\left(\frac{N - N\tilde{l}}{N - 1}\right) \tag{35}$$

10   Fuzzy Binomial Distribution

Binomial distribution could be considered as the generalized form of Bernouilli distribution. Fuzzy Binomial distribution is defined with a fuzzy random variable.

In this section, the studies about binomial distribution in fuzzy form were firstly developed by Buckley using the α-cuts [8]. Kahraman and Kaya applied this model into fuzzy sampling by numerous examples [23].

Let $X = \{x_1, x_2, \ldots, x_n\}$ be a discrete random variable. We start to apply the initial definition of Binomial distribution. A number of experiments n is considered independent, the probability of success is p and the probability of failure is $1 - p$ for a single experiment. Therefore X could be defined as a binomial random variable.

In order to generalize this expression, independent experiments should be repeated n times to gather the probability of x_i successes for $i \in [1, n]$. Thus;

$$P(x) = \binom{n}{x}p^x q^{nx} \qquad x = 0, 1, 2, \ldots, n \tag{36}$$

Fuzzy Binomial probability mass function is derived by using fuzzy numbers for the success: $\tilde{p}$ and the failure: $\tilde{q}$. In order to calculate the fuzzy probability, we need to use the fuzzy algebra for $\tilde{p}$ and $\tilde{q}$ and to derive the minimum and the maximum of $\tilde{P}(x)$ using the α-cuts as follows;

$$\tilde{P}(k_1)[\alpha] = min\left\{\binom{n}{x}\tilde{p}^x\tilde{q}^{nx}\right\} \qquad 0 \le \alpha \le 1 \tag{37}$$

$$\tilde{P}(k_2)[\alpha] = max\left\{\binom{n}{x}\tilde{p}^x\tilde{q}^{nx}\right\} \qquad 0 \le \alpha \le 1 \tag{38}$$

where $p \in \tilde{p}[\alpha]$, $q \in \tilde{q}[\alpha]$, $p + q = 1$.

Furthermore, the fuzzy probability $\tilde{P}[\alpha]$ is obtained within $[P(k_1), P(k_2)]$. Thus,

$$\tilde{P}(x) = \binom{n}{x}p^x q^{nx} \qquad 0 \le \alpha \le 1 \tag{39}$$

Example 4 For a determined time period, a shipyard company calculates that their yachts develop squeaks of indoor equipments in a measured percentage interval $(8, 12, 20)$ within the guarantee period. In a randomly delivery, 5 yachts reach the end of the guarantee period without any squeaks. Find the fuzzy probability in this case?

In order to solve this problem, we start to define p and q. Then, $\tilde{q} = 1 - \tilde{p} = 1 - (0.08, 0.12, 0.2) = (0.8, 0.88, 0.92)$. We need to calculate the fuzzy number $\tilde{P}[5]$. Therefore, we can rewrite the Eqs. 37, 38 as follows:

$$P(k_1)[\alpha] = min\left\{ \binom{5}{0} \tilde{p}^0 (1 - \tilde{p})^{(5-0)} \right\} \qquad 0 \leq \alpha \leq 1 \qquad (40)$$

$$P(k_2)[\alpha] = max\left\{ \binom{5}{0} \tilde{p}^0 (1 - \tilde{p})^{(5-0)} \right\} \qquad 0 \leq \alpha \leq 1 \qquad (41)$$

We obtain $P(k_1)[\alpha] = min\left\{ (1-p)^5 \right\}$, $P(k_2)[\alpha] = max\left\{ (1-p)^5 \right\}$. As the derivation $\frac{d(1-p)^5}{dp} \geq 0$, for $\tilde{p}[\alpha]$, where $\alpha = 0$, we may write the probability in the case of 5 yachts as follows;

$$\tilde{P}(5)[\alpha] = \left[(1 - p_1(\alpha))^5, (1 - p_2(\alpha))^5 \right] \qquad (42)$$

Therefore, $\tilde{p}[\alpha] = [p_1(\alpha), p_2(\alpha)] = [0.08 + 0.04\alpha, 0.2 - 0.08\alpha]$ for $0 \leq \alpha \leq 1$.

Fuzzy Mean and Variance of Binomial Distribution

By using the α-cuts the fuzzy mean of Fuzzy Binomial distribution could be calculated as follows;

$$\tilde{\mu}[\alpha] = \sum_{i=1}^{n} x_i k_i \qquad (43)$$

for $k_i \in \tilde{k}_i[\alpha]$, $1 \leq i \leq n$ and $\sum_{i=1}^{n} k_i = 1$.

Therefore, we may write;

$$\tilde{\mu}[\alpha] = \sum_{i=1}^{n} i \binom{n}{i} p^i q^{n-i} \qquad (44)$$

which is equal to $\tilde{\mu}[\alpha] = n\tilde{p}$.

The variance is also calculated using the same principle;

$$\tilde{\sigma}^2[\alpha] = \sum_{i=1}^{n} (x_i - \mu_i)^2 k_i \qquad (45)$$

Finally, we may write the fuzzy variance of the fuzzy binomial distribution as follows; $\tilde{\sigma}^2[\alpha] = n\tilde{p}\tilde{q}$.

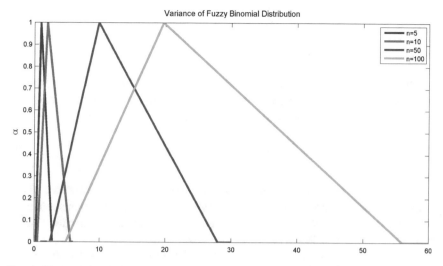

Fig. 4 The representation of variance analysis for fuzzy binomial distribution

Example 5 Let $\tilde{p}$ have a linear triangular fuzzy membership function $(0.14, 0.27, 0.65)$. Calculate the variance of fuzzy binomial distribution.

We know that;

$$\tilde{\sigma}^2[\alpha] = n\tilde{p}\tilde{q} \tag{46}$$

Furthermore, we may write $\tilde{\sigma}^2[\alpha] = n\tilde{p}(1 - \tilde{p})$ and we obtain; $(1 - \tilde{p}) = (0.35, 0.73, 0.86)$. In order to interpret the variance, MATLAB 2014a is used to sketch the variance for the different n values. We obtained the following Fig. 4.

11 Fuzzy Poisson Distribution

Poisson distribution is characterized by the experiments whose outputs are discrete in continuous space. A regular time observation could be considered as discrete values whereas time is continuous. On the other hand, Fuzzy Poisson distribution is represented by a fuzzy random variable.

The studies about Fuzzy Poisson distribution were firstly developed by Buckley using the α-cuts [8]. Kahraman and Kaya applied this model into fuzzy sampling by several examples [23].

In order to generate the fuzzy Poisson distribution we define $X = \{x_1, x_2, \ldots, x_n\}$ which is a discrete random variable. X has also the Poisson probability mass function. When the probability $P(x)$ is defined for the probability that $X = x$, we may write

$$P(x) = \frac{\lambda^x exp(-\lambda)}{x!} \tag{47}$$

where, $x = 0, 1, 2, \ldots, n$, and $\lambda > 0$.

Fuzzy Poisson probability mass function is derived by using fuzzy number $\tilde{\lambda} > 0$. Let us denote $\tilde{P}(x)$ to be the fuzzy probability that $X = x$. Therefore, we can calculate the fuzzy probability using α-cuts algebra,

$$\tilde{P}(x)[\alpha] = \frac{\lambda^x exp(-\lambda)}{x!}, \qquad \lambda \in \tilde{\lambda}[\alpha], \qquad 0 \leq \alpha \leq 1 \tag{48}$$

Furthermore, the fuzzy expression will depend on x where $\tilde{\lambda}[0]$ is observed. For a fixed x, let us calculate $m(\lambda) = \lambda e^{-\lambda}/x!$. When we observe the monotonicity of $m(\lambda)$, we remark that it is an increasing function for $\lambda < x$, the maximum value of $m(\lambda)$ is for $\lambda = x$ and finally, $m(\lambda)$ is a decreasing function for $\lambda > x$.

In order to summarize the way to analyze the Fuzzy Poisson distribution, let $\tilde{\lambda}[\alpha] = [\lambda_1(\alpha), \lambda_2(\alpha)]$ and $0 \leq \alpha \leq 1$. We note that;

1. For $x > \lambda_2(0)$, $\tilde{P}(x)[\alpha] = [m(\lambda_1), m(\lambda_2)]$
2. For $\lambda_1(0) > x$, $\tilde{P}(x)[\alpha] = [m(\lambda_2), m(\lambda_1)]$
3. For $x \in \tilde{\lambda}[0]$, $\beta, \alpha \in [0, 1]$

- $\tilde{P}(x)[\alpha] = [m(\lambda_1), m(x)]\ 0 \leq \alpha \leq \beta$
- $\tilde{P}(x)[\alpha] = [m(x), m(\lambda_2)]\ \beta \leq \alpha \leq 1$

Example 6 In order to illustrate an example of Fuzzy Poisson distributionlet x be the measurement of the defective percentage in a lot: $x = 0.1$ and $\tilde{\lambda} = (0.08, 0.12, 0.18)$. Determine the fuzzy probability $\tilde{P}(0.1)$.

Since $x = 0.1 \in [0.08, 0.18]$, let us start to evaluate the interval for $\tilde{\lambda}[0]$. $\tilde{\lambda}[\alpha] = [0.08 + 0.04\alpha, 0.18 - 0.06\alpha]$. In order to calculate $\tilde{P}(0.1)$, we must interpret the fuzzy intervals;

$$p_1(\alpha) = min\{m(\lambda)\}, \quad p_2(\alpha) = max\{m(\lambda)\} \quad \lambda \in \tilde{\lambda}[\alpha] \tag{49}$$

When the set of Eqs. (49) are examined, we may write;

$$\tilde{P}(0.1)[\alpha] = [m(0.08 + 0.04\alpha), m(0.1)] \quad 0 \leq \alpha \leq 0.5 \tag{50}$$

$$\tilde{P}(0.1)[\alpha] = [m(0.1), m(0.18 - 0.06\alpha)] \quad 0.5 \leq \alpha \leq 1 \tag{51}$$

Fuzzy Mean and Variance of Poisson Distribution

Furthermore, we need to calculate the fuzzy mean and the fuzzy variance of fuzzy Poisson probability distribution. Using the same principle in the fuzzy binomial distribution, we may write;

$$\tilde{\mu}[\alpha] = \left\{ \sum_{k=0}^{\infty} kh(\lambda) \right\} \tag{52}$$

This expression could be simplied into $\tilde{\mu} = \tilde{\lambda}$ which is similar to the crisp case. Let us calculate the variance with the similar way.

$$\tilde{\sigma}^2[\alpha] = \left\{ \sum_{k=0}^{\infty} (k - \mu)^2 h(\lambda) \right\} \tag{53}$$

which gives us the similar representation of crisp case; $\tilde{\sigma}^2 = \tilde{\lambda}$.

12 Conclusion

In an uncertain environment, like placing on the market of a new product, acceptance of defective lot sizes or interplanetary missions the occurrence of some events can not be anticipated through imprecise linguistic data. However, those cases necessitate formulation of probability distributions with fuzzy random variables. Fuzzy expectation in discrete case was provided in this chapter to find means for fuzzy distributions those are Fuzzy Hypergeometric, Fuzzy Binomial and Fuzzy Poisson distributions. And also fuzzy variances of those distributions were provided. We also dealt with fuzzy conditional probability. Independency with fuzzy random variables was offered before Fuzzy Bayes formula which are the basic of fuzzy probability theory. Additionally explanatory examples are employed for more paraphrasing. We tried to state the fuzzy probability theory by discrete form more clearly in a well-organized frame in this chapter. Fuzzy Hypergeometric distribution reinforced with an example was the additional contribution of this chapter to the literature among the books related with this topic.

For further research, general or interval type-2 fuzzy numbers can be integrated in the fuzzy discrete probability theory. Moreover, Fuzzy Hypergeometric, Fuzzy Binomial and Fuzzy Poisson distributions would be expressed within this framework. Finally, hidden Markov models might be extended by considering intuitionistic fuzzy probabilities.

References

1. Zadeh, L.A.: Probability measures of fuzzy events. J. Math. Analysis and Appl. **10**, 421–427 (1968)
2. Zadeh, L.A.: Fuzzy set as the basis for the theory of possibility. Fuzzy Sets Syst. **1**, 3–28 (1978)

3. Kwakernaak, H.: Fuzzy random variables-I. Definitions and theorems. Inf. Sci. **15**, 1–29 (1978)
4. Kwakernaak, H.: Fuzzy random variables-II. Algorithms and examples for the discrete case. Inf. Sci. **17**, 253–278 (1979)
5. Liu, B., Liu, Y.K.: Expected value of fuzzy variable and fuzzy expected value models. IEEE Trans. Fuzzy Syst. **10**, 445–450 (2002)
6. Liu, B., Liu, Y.K.: Fuzzy random variables: A scalar expected value operator. Fuzzy Optim. Decis. Making **2**, 143–160 (2003)
7. Buckley, J.J.: Fuzzy statistics. Springer, Berlin (2004)
8. Buckley, J.J.: Fuzzy probabilities. Springer, Berlin (2005)
9. Buckley, J.J.: Fuzzy probability and statistics. Springer, The Netherlands (2006)
10. Nguyen, H.T., Wu, B.: Fundamentals of statistics with fuzzy data. Springer, The Netherlands (2010)
11. Shapiro, A.F.: Fuzzy random variables. Insur. Math. Econ. **44**, 307–314 (2009)
12. Shapiro, A.F. (2013). Implementing fuzzy random variables: some preliminary observations. In: ARCH 2013.1 Proceedings, pp. 1–15
13. Shapiro, A.F.: Modeling future lifetime as a fuzzy random variable. Insur. Math. Econ. **53**, 864–870 (2013)
14. Kendall, D.G.: Foundations of a theory of random sets. In: Harding, E.F., Kendall, D.G. (eds.) Stochastic Geometry. Wiley, New York (1974)
15. Matheron, G.: Random Sets and Integral Geometry. Wiley, New York (1975)
16. Feron, R.: Ensemble aléatoires flous. C.R. Acad. Sci. Paris Ser. A **282**, 903–906 (1976)
17. Nguyen, H.T.: On random sets and belief functions. J. Math. Anal. Appl. **63**, 532–542 (1978)
18. Puri, M.L., Ralescu, D.A.: Fuzzy random variables. J. Math. Anal. Appl. **114**, 409–422 (1986)
19. Aumann, R.J.: Integrals of set-valued functions. J. Math. Anal. Appl. **12**, 1–12 (1965)
20. Feng, Y., Hu, L., Shu, H.: The variance and covariance of fuzzy random variables and their applications. Fuzzy Sets Syst. **120**, 487–497 (2001)
21. Zadeh, L.A.: Discussion: Probability theory and fuzzy logic are complementary rather than competitive. Technometrics **37**(3), 271–276 (1995)
22. Tong, X., Wang, Z.: Fuzzy acceptance sampling plans for inspection of geospatial data with ambiguity in quality characteristics **48**, 256–266 (2012)
23. Kahraman, C., Kaya, I.: Fuzzy acceptance sampling plans. In: Kahraman, C., Yavuz, M. (eds.) Production Engineering and Management under Fuzziness. Springer, Berlin (2010)
24. Couso, I., Dubois, D., Sanchez, L.: Random sets and random fuzzy sets as ill-perceived random variables. Springer, Heidelberg (2014)

Fuzzy Probability Theory II: Continuous Case

A. Cağrı Tolga and I. Burak Parlak

Abstract Continuous probability density functions are widely used in various domains. The characterization of the fuzzy continuous probability theory is similar to the discrete case. However, the possibility space is continuous and the integration between the minimum and the maximum values would set the fuzzy probability through the alpha-cuts. In this chapter, the foundations of fuzzy probability and possibility theory are described for the continuous case. A brief introduction summarized the key concepts in this area with recent applications. The expectation theory is interpreted using the relationship with fuzzy continuous random variables. Fuzzy continuous applications are enriched with different probability density functions. Therefore, fundamental distributions are detailed within their uses and their properties. In this chapter, fuzzy uniform, fuzzy exponential, fuzzy laplace, fuzzy normal and fuzzy lognormal distributions are examined. Several examples are given for the use of these fuzzy distributions regarding the fuzzy interval algebra. Finally, the future suggestions and applications are discussed in the conclusion.

Keywords Fuzzy uniform distribution · Fuzzy exponential distribution · Fuzzy laplace distribution · Fuzzy normal distribution · Fuzzy lognormal distribution

1 Introduction

Continuous probabilistic approaches are crucial to develop models in order to describe real probabilistic applications. In this chapter, the background of fuzzy probability theory will be inherited from the discrete case. Furthermore, continuous probability theory will be developed to represent fuzzy continuous distributions.

A. Cağrı Tolga (✉)
Department of Industrial Engineering, Galatasaray University, Istanbul, Turkey
e-mail: ctolga@gsu.edu.tr

I. Burak Parlak
Department of Computer Engineering, Galatasaray University, Istanbul, Turkey
e-mail: bparlak@gsu.edu.tr

© Springer International Publishing Switzerland 2016
C. Kahraman and Ö. Kabak (eds.), *Fuzzy Statistical Decision-Making*,
Studies in Fuzziness and Soft Computing 343,
DOI 10.1007/978-3-319-39014-7_3

35

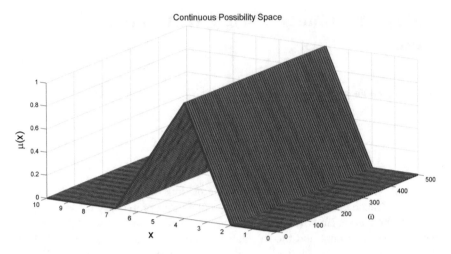

Fig. 1 The illustration of continuous possibility space

Initially, basic fuzzy random variables are considered for continuous domain in order to define the fuzzy continuity and to explain the fuzzy continuous distributions. Figure 1 represents the continuous possibility space for events having similar triangular fuzzy membership functions. Therefore, a triangular surface is depicted for this case. Fuzzy continuous probability functions and its related expectations are given, respectively. Fuzzy events are associated with the occurrences as continuous subsets in the probability space. Furthermore, each fuzzy continuous probability distribution is studied with numerical examples. In fuzzy continuous distributions, crisp numbers are substituted by fuzzy numbers by taking α-cuts to produce continuous fuzzy probability density functions.

As the uniform probability distribution is one of the most common distributions, we started to develop its properties in fuzzy case. The uniform probability density function expressed as $U(m, n), m < n$; it is illustrated by fuzzy random variables; $\tilde{m}$ and $\tilde{n}$. Fuzzy exponential distribution followed the fuzzy uniform distribution. In the fuzzy exponential distribution k will be replaced with $\tilde{k}$ which is a fuzzy number. Then fuzzy normal distribution is interpreted and its theory is applied to derive fuzzy lognormal distribution. The normal probability function $N(\mu, \sigma^2)$ is depicted in fuzzy form by $N(\tilde{\mu}, \tilde{\sigma}^2)$ notation. Finally, the fuzzy lognormal probability function is represented as $L(\tilde{\mu}, \tilde{\sigma}^2)$.

After the definition of probability measures of fuzzy events by Zadeh [1, 2], continuous fuzzy random variables were first defined by Buckley in his books [3–5]. Liu and Liu operated on fuzzy expectations in the context of their studies [6, 7]. Nguyen and Wu investigated convergence of random fuzzy sets with Choquet integral in their study [8]. An important study on process capability indices by using fuzzy normal distribution was annexed to the literature by Kaya and Kahraman [9]. Process capability analysis is a significant area in quality topics used at measurements of manufacturing. By fuzzy normal distribution the upper spec limits and lower spec limits gain more flexible evaluation. Shapiro explained fuzzy continuous distribution in a clear form and applied the continuity into real life applications [10, 11]. Rakus-Andersson expanded the continuity of fuzzy probabilities by the means of continuous events according to the procedure of the approximation, and characterized this approach by an irrelevant cumulative error [12]. Dey and Chakraborty derived a fuzzy random continuous system regarding an annual customer demand which is based on a uniformly distributed continuous fuzzy random variable [13]. They reduced the setup cost capital and ameliorated the process quality which incorporated into the total cost with minimizing object. Montes et al. [14] worked on the area of P_Z-compatibility based on t-norms for Zadeh's probability which fulfills Kolmogorov's axioms in fuzzy occurrences. They characterized a complete description of P_Z-compatible continuous t-norms.

In this chapter, we assume that fuzzy random variables, α-cuts are already known. Fuzzy probability were investigated in the previous section we will not give definitions of these topics for continuous case. In Sect. 2, we will penetrate to continuous fuzzy expectation issue. Section 3 will make mention of Fuzzy Uniform distribution. Then Fuzzy Exponential distribution will be investigated in Sect. 4. Section 5 will contain and depict Fuzzy Laplace distribution. After that, in Sect. 6 Fuzzy Normal distribution will be expressed. Various approximations between distributions in fuzzy manner will be depicted in that section also. And the ending of the continuous probability distributions, Fuzzy Lognormal distribution will be explained in Sect. 7. All these Fuzzy continuous distributions will be reinforced by illustrative examples. Finally, we will complete this chapter by a comprehensive review.

2 Fuzzy Continuous Expectation

Before the definition and the analysis of fuzzy continuous distributions, we need to generalize the idea of expectation in continuous case. As Kwakernaak [15, 16] introduced the mathematical foundations of the fuzzy expectation and the fuzzy variance, Buckley generated a clear point of view for their use through the examples and the applications [3–5]. In this section we will review a summary of these concepts.

Let us start with an initial definition; suppose K and L are two random variables with joint probability density $f(k, l; \theta)$ where $k, l \in \mathbb{R}$ and the joint density function's vector of parameters should be $\theta = (\theta_1, \ldots \theta_n)$.

In continuous probability, a random sample is essential in order to estimate stochastic dynamics of the population. The anticipation related to the population could be a point or an interval. Instead of point estimation, a confidence interval for each θ_i could be replaced by the probability density function to obtain an interval joint probability density function. On the other hand, a general formulation should be expanded and this step requires an expression of the uncertainty based on θ_i by replacing a fuzzy number for θ_i and acquiring a joint fuzzy probability density function.

As a recall, the α-cuts of fuzzy number utilized for θ_i are represented in a similar way in *Fuzzy probability theory I: discrete case: section 3*. We remark that a better formulation could be implemented utilizing α-cuts of the fuzzy numbers for the interval probability density functions. The joint fuzzy density functions are fulfilled by substituting fuzzy numbers for the vague parameters. Therefore, the nubiluous definitions of fuzzy marginals should be discussed now.

Since the θ_i in θ are uncertain, fuzzy numbers $\tilde{\theta}_i$ are replaced for the θ_i, $1 \leq i \leq n$, then it provides joint fuzzy density $f(k, l; \tilde{\theta})$. The fuzzy marginal for K is

$$f(k; \tilde{\theta}) = \int_{-\infty}^{\infty} f(k, l; \tilde{\theta}) dl. \tag{1}$$

A similar formulation could be adapted to derive the fuzzy marginal of L. The α-cuts of the fuzzy marginals $f(k; \tilde{\theta})$ could be calculated by the following equation

$$f(k; \tilde{\theta})[\alpha] = \{ \int_{-\infty}^{\infty} f(k, l; \theta) dl \mid \theta_i \in \tilde{\theta}_i[\alpha], 1 \leq i \leq n \}, \tag{2}$$

for $0 \leq \alpha \leq 1$, and also we can derive an analogous equation for $f(l; \tilde{\theta})[\alpha]$. Equation (2) gives the α-cuts of a fuzzy set for each value of k.

At this juncture let $f(k; \theta)$ be the crisp (not fuzzy) marginal of k. Let's utilize $f(k; \theta)$ to find the mean $\mu_k(\theta)$ and variance $Var_k(\theta)$ of K. The mean and variance of K are written as the functions of θ as they are related with the values of the parameters. Suppose that $\mu_k(\theta)$ and $Var_k(\theta)$ are continuous functions of θ. Deriving the fuzzy mean and variance of the fuzzy marginal could be made by fuzzification of the crisp mean and variance. The following theorem can be deduced from the previous explanations for K and also can be utilized for L.

Theorem *For the fuzzy marginal $f(k; \tilde{\theta})$, the fuzzy mean and the fuzzy variance are $\mu_k(\tilde{\theta})$ and $Var_k(\tilde{\theta})$, respectively* [4].

Proof An α-cut of the fuzzy mean of the fuzzy marginal for K is

$$\mathcal{M}_k(\tilde{\theta})[\alpha] = \{ \int_{-\infty}^{\infty} kf(k; \theta) dk \mid \theta_i \in \tilde{\theta}_i[\alpha], 1 \leq i \leq n \}, \tag{3}$$

for $0 \leq \alpha \leq 1$. Now the integral in Eq. (3) equals $\mathcal{M}_k(\theta)$ for each $\theta_i \in \tilde{\theta}_i$, $1 \leq i \leq n$. So

$$\mathcal{M}_k(\tilde{\theta})[\alpha] = \{\mu_k(\theta) \mid \theta_i \in \tilde{\theta}_i[\alpha], 1 \leq i \leq n\}. \tag{4}$$

Because of this, the fuzzy mean is $\mathcal{M}_k(\tilde{\theta})$. See the "examples" parts in various fuzzy distributions sections of this chapter.

The α-cuts of the fuzzy variance are

$$Var_k(\tilde{\theta})[\alpha] = \{ \int\limits_{-\infty}^{\infty} (k - \mu_k(\theta))^2 f(k; \theta)dk \mid \theta_i \in \tilde{\theta}_i[\alpha], 1 \leq i \leq n\}, \tag{5}$$

for $0 \leq \alpha \leq 1$. But the integral in the above equation equals $Var_k(\theta)$ for each $\theta_i \in \tilde{\theta}_i$, $1 \leq i \leq n$. Due to this

$$Var_k(\tilde{\theta})[\alpha] = \{Var_k(\theta) \mid \theta_i \in \tilde{\theta}_i[\alpha], 1 \leq i \leq n\}. \tag{6}$$

So, the fuzzy variance is just $Var_k(\tilde{\theta})$.

3 Fuzzy Uniform Distribution

The uniform distribution is a probabilistic approach to model an interval where the events are equally probable. In order to implement this distribution, upper and lower parameters are required to generate the interval. For the fuzzy uniform distribution, this interval will be expressed as a fuzzy interval.

Let $U(m, n)$ be the uniform probability density function for $m < n$. Thus,

$$P(x) = \begin{cases} \frac{1}{m-n} & \text{for } m \leq x \leq n \\ 0 & \text{otherwise} \end{cases} \tag{7}$$

Furthermore, the uniform distribution will be transformed into a fuzzy distribution by substituting m, n using fuzzy numbers $\tilde{m}, \tilde{n}$;

$$\tilde{P}[a, b][\alpha] = \frac{1}{m-n} \quad \text{for } m \in \tilde{m}[\alpha] \quad, n \in \tilde{n}[\alpha] \tag{8}$$

However, $\tilde{m}$ and $\tilde{n}$ must be determined in order to calculate the probability value $\tilde{P}[k, l]$. Thus,

$$p_1[k, l][\alpha] = min\left\{\frac{\Lambda}{m-n}\right\} \tag{9}$$

$$p_2[k, l][\alpha] = max\left\{\frac{\Lambda}{m-n}\right\} \tag{10}$$

where $\varLambda$ denotes the length of the interval $[k, l] \cap [m, n]$.

Example 1 Let $\tilde{X}$ be a continuous fuzzy uniform random variable with parameters $[-4, -2]$ and $[2, 4]$. Calculate the probability $P[-3 < X < 3]$.

We start with the calculation of $\tilde{P}[a, b][\alpha] = [p_1(\alpha), p_2(\alpha)]$ where $p_1(\alpha), p_2(\alpha)$ are the minimum and the maximum values of the interval. Thus,

$$p_1(\alpha) = min\left\{\frac{\varLambda(-3, 3; m, n)}{m - n}\right\} \tag{11}$$

$$p_2(\alpha) = max\left\{\frac{\varLambda(-3, 3; m, n)}{m - n}\right\} \tag{12}$$

For all α, the maximum value will be equal to 1. Therefore, we need to calculate the minimum value of the interval. For this purpose, we must study the intervals for β and γ where $\tilde{\beta}[\alpha] = [-4 + \alpha, -2 - \alpha]$ and $\tilde{\gamma}[\alpha] = [2 + \alpha, 4 - \alpha]$. In this case, we need to consider the interval algebra as follows

- $-4 + \alpha \leq n \leq -3,$ $2 + \alpha \leq m \leq 3$
- $-4 + \alpha \leq n \leq -3,$ $3 \leq m \leq 4 - \alpha$
- $-3 \leq n \leq -\alpha - 2,$ $2 + \alpha \leq m \leq 3$
- $-3 \leq n \leq -\alpha - 2,$ $3 \leq m \leq 4 - \alpha$
  Thus,
- $5 + \alpha \leq m - n \leq 7 - \alpha$
- $6 \leq m - n \leq 8 - 2\alpha$
- $4 + 2\alpha \leq m - n \leq 6$
- $5 + \alpha \leq m - n \leq 7 - \alpha$

Therefore, we find that the minimum of the interval is equal to $6/(8 - 2\alpha)$. Finally, the α-cuts of $\tilde{P}[-3, 3]$ are expressed with $6/(8 - 2\alpha)$ and the solution is illustrated in Fig. 2.

Fuzzy Mean and Variance of Uniform Distribution

Finally, we may note that the mean $\tilde{\mu}$ and variance $\tilde{\sigma}^2$ of the uniform distribution could be calculated;

$$\tilde{\mu}[\alpha] = \frac{\tilde{m} + \tilde{n}}{2} \tag{13}$$

$$\tilde{\sigma}^2[\alpha] = \frac{(\tilde{m} - \tilde{n})^2}{12} \tag{14}$$

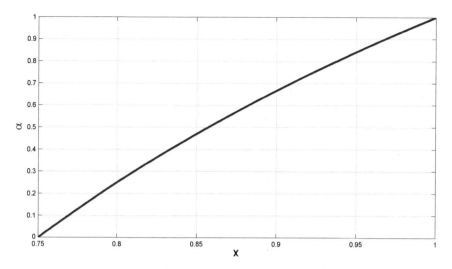

Fig. 2 The probability chart of fuzzy uniform distribution for $\tilde{P}[-3,3]$

4 Fuzzy Exponential Distribution

Exponential distribution is considered as the probabilistic approach which describes the time factor between the events occurring in a Poisson process.

The exponential density function is expressed as;

$$f(t) = ke^{-kt} \quad 0 < t \quad \text{and} \quad 0 < k \tag{15}$$

In order to calculate the fuzzy probability values, we need to define an interval $[a, b]$ in which we calculate the probability $\tilde{P}[a, b]$. Therefore, we may calculate this value as follows;

$$\tilde{P}[a, b][\alpha] = \int_{a}^{b} ke^{-kt} dt \quad \text{for } k \in \tilde{k}[\alpha] \tag{16}$$

Let us denote $\tilde{P}[a, b][\alpha] = [p_1(\alpha), p_2(\alpha)]$. Thus,

$$p_1[a, b][\alpha] = min\left\{ \int_{a}^{b} ke^{-kt} dt \right\} \quad \text{for } k \in \tilde{k}[\alpha] \tag{17}$$

$$p_2[a, b][\alpha] = max\left\{ \int_{a}^{b} ke^{-kt} dt \right\} \quad \text{for } k \in \tilde{k}[\alpha] \tag{18}$$

Table 1 Fuzzy probability values of Example 2

α	$min\{\tilde{P}[2,5][\alpha]\}$	$max\{\tilde{P}[2,5][\alpha]\}$
0	0.1836	0.3257
0.1	0.1963	0.3257
0.2	0.2096	0.3254
0.3	0.2233	0.3248
0.4	0.2373	0.3239
0.5	0.2514	0.3228
0.6	0.2655	0.3215
0.7	0.2792	0.3199
0.8	0.2922	0.3181
0.9	0.3040	0.3161
1	0.314	0.314

Example 2 Let $k = (0.3, 0.4, 0.8)$. Find the fuzzy probability $\tilde{P}[2,5]$.

We need to calculate the minimum and the maximum probabilities using the α-cuts based on the Eqs. (17), and (18). Afterwards, we obtain the following Table 1 where these values are represented.

Finally, we can sketch $\tilde{P}[2,5]$ as it is shown in Fig. 3.

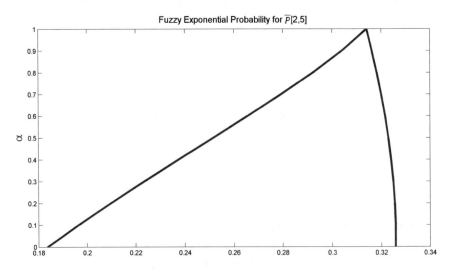

Fig. 3 Fuzzy probability values of Example 2

Fuzzy Mean and Variance of Exponential Distribution

In order to complete the properties of fuzzy exponential distribution we have to review the corresponding fuzzy mean and the fuzzy variance. We may note the mean $\tilde{\mu}$ and variance $\tilde{\sigma}^2$ of the exponential distribution could be calculated as follows;

$$\tilde{\mu}[\alpha] = \frac{1}{\tilde{k}} \tag{19}$$

$$\tilde{\sigma}^2[\alpha] = \frac{1}{\tilde{k}^2} \tag{20}$$

5 Fuzzy Laplace Distribution

The Laplace distribution which also called double exponential distribution due to its supplementary parameter. It is widely used in signal and control theory applications where the coefficient of Discrete Fourier Transformation is considered. Laplace distribution consists of two different parameters μ and $b > 0$. The Laplace density function is characterized as;

$$f(x) = \frac{1}{2b} e^{\frac{-|x-\mu|}{b}} \tag{21}$$

Using the same principle, we may derive the fuzzy probabilistic expression by considering μ and b as fuzzy values. Therefore, we may note an interval $[k, l]$ in which we calculate the fuzzy probability as follows;

$$p_1[k, l][\alpha] = min\left\{\frac{1}{2b} e^{\frac{-|x-\mu|}{b}}\right\} \tag{22}$$

$$p_2[k, l][\alpha] = max\left\{\frac{1}{2b} e^{\frac{-|x-\mu|}{b}}\right\} \tag{23}$$

where $\mu \in \tilde{\mu}[\alpha]$, $b \in \tilde{b}[\alpha]$ and $0 \le \alpha \le 1$.

Therefore, we note the calculation of $\tilde{P}[k, l][\alpha] = [p_1(\alpha)p_2(\alpha)]$ where $p_1(\alpha)$ and $p_2(\alpha)$ are the minimum and the maximum of the interval. Thus,

$$\tilde{P}[k, l][\alpha] = \frac{1}{2b} e^{\frac{-|x-\mu|}{b}} \quad \text{for } \mu \in \tilde{\mu}[\alpha] \quad, b \in \tilde{b}[\alpha] \tag{24}$$

Example 3 In a manufacturing process, daily yields are found distributed symmetrically around 0 and $\tilde{b}$ is found $(35, 45, 55)$. The rate of return in this process is modeled with fuzzy Laplace distribution. Calculate the fuzzy probability of receiving a rate of return bigger than 0.05.

In order to find the solution, we need to resolve the following integration;

$$\tilde{P}[0.05, \infty)[\alpha] = \int\limits_{0.05}^{\infty} \frac{1}{2b} e^{\frac{-|x-\mu|}{b}} dx \quad \mu \in \tilde{\mu}[\alpha], \quad b \in \tilde{b}[\alpha], \quad 0 \leq \alpha \leq 1 \qquad (25)$$

Thus,

$$\tilde{P}[0.05, \infty)[\alpha] = \frac{1}{2b} e^{\frac{\mu}{b}} \int\limits_{0.05}^{\infty} e^{\frac{-x}{b}} dx \quad \mu \in \tilde{\mu}[\alpha], \quad b \in \tilde{b}[\alpha], \quad 0 \leq \alpha \leq 1 \qquad (26)$$

Therefore,

$$\tilde{P}[0.05, \infty)[\alpha] = \frac{1}{2b} e^{\frac{\mu}{b}} \left[-be^{-x/b} \right]_{0.05}^{\infty} \quad \mu \in \tilde{\mu}[\alpha], \quad b \in \tilde{b}[\alpha], \quad 0 \leq \alpha \leq 1 \qquad (27)$$

At this point, we have to calculate the α-cuts for Eq. 27.

Finally, we obtain the fuzzy probability values of $\tilde{P}[0.05, \infty)$ as indicated in Table 2.

Fuzzy Mean and Variance of Laplace Distribution

When we review the corresponding fuzzy mean and the fuzzy variance of Laplace distribution, we should remark that the mean is $\tilde{\mu}$ and the variance is $2\tilde{b}^2$.

Table 2 Fuzzy probability values of Example 3	α	$min\{\tilde{P}[0.05, \infty)[\alpha]\}$	$max\{\tilde{P}[0.05, \infty)[\alpha]\}$
	0	0.4993	0.4995
	0.1	0.4993	0.4995
	0.2	0.4993	0.4995
	0.3	0.4993	0.4995
	0.4	0.4994	0.4995
	0.5	0.4994	0.4995
	0.6	0.4994	0.4995
	0.7	0.4994	0.4995
	0.8	0.4994	0.4995
	0.9	0.4994	0.4995
	1	0.4994	0.4994

6   Fuzzy Normal Distribution

The normal distribution could be considered as one of the most popular distribution in this field. The normal or Gaussian distribution has a bell shape which characterizes that any real observation will be limited by an upper and lower bound. A special case of the normal distribution is called standard normal or unit normal distribution where the mean μ is 0 and the variance σ^2 is 1.

The normal density function is defined as;

$$f(x) = \frac{1}{\sigma\sqrt{2\pi}}e^{-\frac{1}{2}(\frac{x-\mu}{\sigma})^2} \tag{28}$$

where $-\infty < \mu < \infty$ and $\sigma^2 > 0$

Let us denote $\tilde{P}[a,b][\alpha] = [p_1(\alpha), p_2(\alpha)]$. Thus, in order to derive the fuzzy probability, we have to resolve the interval values as follows;

$$p_1[a,b][\alpha] = min\left\{\frac{1}{\sigma\sqrt{2\pi}}e^{-\frac{1}{2}(\frac{x-\mu}{\sigma})^2}\right\} \tag{29}$$

$$p_2[a,b][\alpha] = max\left\{\frac{1}{\sigma\sqrt{2\pi}}e^{-\frac{1}{2}(\frac{x-\mu}{\sigma})^2}\right\} \tag{30}$$

where $\mu \in \tilde{\mu}[\alpha]$ and $\sigma^2 \in \tilde{\sigma}^2[\alpha]$

Fuzzy Mean and Variance of Normal Distribution

As the normal distribution is introduced with its mean and its variance, it is trivial to note that the fuzzy mean is $\tilde{\mu}$ and the fuzzy variance is $\tilde{\sigma}^2$, respectively.

Example 4 Let the mean $\tilde{\mu} = (15, 18, 19)$ and the variance $\tilde{\sigma}^2 = (1, 4, 5)$. Compute the fuzzy probability value for $\tilde{P}[18, 21]$.

In order to calculate this value we should use the following integration;

$$\tilde{P}[18, 21][\alpha] = \left\{\frac{1}{\sqrt{2\pi}}\int_{z_1}^{z_2} e^{-z^2/2}dz\right\} \tag{31}$$

where, $\mu \in \tilde{\mu}[\alpha], \sigma^2 \in \tilde{\sigma}^2$. We set that $z_1 = \frac{18-\mu}{\sigma}$ and $z_2 = \frac{21-\mu}{\sigma}$. Thus, we generate Table 3 using the calculation of the integration in Eq. (31) and with these values one can easily trace the Fig. 4.

The calculation of the mean and the variance are crucial to characterize the parameters of a random process. These parameters are also necessary for the fuzzy normal distribution. Recently, Kaya and Kahraman [9] and Buckley [4, 5] studied fuzzy estimation methods to handle the membership functions of the mean and the variance. We will follow a similar approach in order to estimate them.

Table 3 Fuzzy probability values of Example 4

α	$min\{\tilde{P}[18,21][\alpha]\}$	$max\{\tilde{P}[18,21][\alpha]\}$
0	0.0013	0.4871
0.1	0.0066	0.5037
0.2	0.0223	0.5210
0.3	0.0572	0.5390
0.4	0.1183	0.5578
0.5	0.2066	0.5773
0.6	0.3154	0.5975
0.7	0.4320	0.6183
0.8	0.5408	0.6395
0.9	0.6275	0.6607
1	0.6816	0.6816

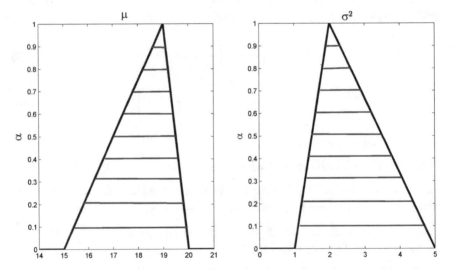

Fig. 4 Fuzzy triangular MFs of the mean and the variance

Let x be a random variable with a normal density function $N(\mu, \sigma^2)$ where the mean; μ is unknown and the variance; σ^2 is known. In order to estimate the mean μ, a random sample $x_1, x_2, \ldots, x_n$ could be considered. The mean of this random sample would be a crisp number k. We recall that k would have a normal density function $N(\mu, \sigma^2/n)$ and standard normal density function $N(0,1)$ would be written using $\frac{k-\mu}{\sigma/\sqrt{n}}$ (Fig. 5). Then;

$$P\left(z_{\Psi/2} \geq \frac{k-\mu}{\frac{\sigma}{\sqrt{n}}} \geq -z_{\Psi/2}\right) = 1 - \beta \qquad (32)$$

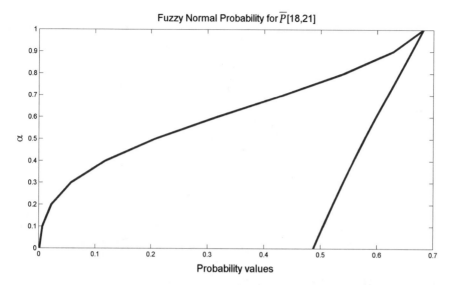

Fig. 5 Fuzzy probability values of Example 4

and we may write;

$$P\left(k + z_{\Psi/2}\frac{\sigma}{\sqrt{n}} \geq \mu \geq k - z_{\Psi/2}\frac{\sigma}{\sqrt{n}}\right) = 1 - \beta \qquad (33)$$

where $z_{\Psi/2}$ is the z value of the probability from $N(0, 1)$ distribution which is above $\Psi/2$. Therefore, $(1 - \beta)$ 100 % confidence interval for μ is as follows;

$$[\tau_1(\beta), \tau_2(\beta)] = \left[k - z_{\Psi/2}\frac{\sigma}{\sqrt{n}}, k + z_{\Psi/2}\frac{\sigma}{\sqrt{n}}\right] \qquad (34)$$

Here, we note that $\int_{-\infty}^{z_{\Psi/2}} N(0, 1)dx = 1 - \beta/2$. In order to interpret the fuzzy case, we may substitute the βs with the α-cuts. Finally we may obtain the fuzzy estimator $\tilde{\mu}$ as follows;

$$[\mu_1(\alpha), \mu_2(\alpha)] = \left[k - z_{\alpha/2}\frac{\sigma}{\sqrt{n}}, k + z_{\alpha/2}\frac{\sigma}{\sqrt{n}}\right] \qquad (35)$$

Example 5 Let x be a random variable with probability density function $N(\mu, 2)$. There is a random sample $x_1, x_2, \ldots, x_{150}$. We assume that the mean of the samples will be equal to 15 after $n = 150$. Try to calculate the fuzzy mean $\tilde{\mu}$.

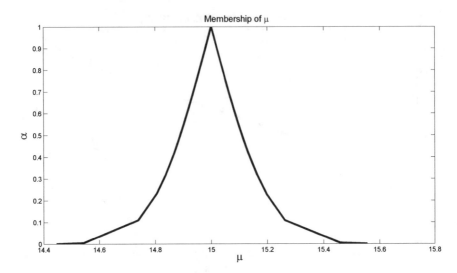

Fig. 6 Membership function of the mean for Example 5

We will adapt a similar approach which is represented in Eq. (35)

$$[\mu_1(\alpha), \mu_2(\alpha)] = \left[15 - 0.1633 z_{\alpha/2}, 15 + 0.1633 z_{\alpha/2}\right] \quad (36)$$

After the estimation of the mean, we need to formulate an approach to estimate the variance. In a similar manner we have to start with the definition of a normal process (Fig. 6).

Let x be a random variable having a probability density function $N(\mu, \sigma^2)$ where the mean μ and the variance σ^2 are unknown. In order to achieve σ^2 estimation, a random sample $x_1, x_2, \ldots, x_n$ is considered. In probability theory, $\frac{(n-1)s^2}{\sigma^2}$ is called *Chi-square distribution* with $n - 1$ degrees of freedom. Thus, fuzzy σ^2 could be estimated using a confidence interval as follows;

$$\left[\frac{(n-1)s^2}{\chi^2_{R,\beta/2}}, \frac{(n-1)s^2}{\chi^2_{L,\beta/2}}\right] \quad (37)$$

$\chi^2_{R,\beta/2}$ and $\chi^2_{L,\beta/2}$ denote the right and the left sides of χ^2 distribution, respectively. Furthermore, we can rewrite this expression to obtain an unbiased fuzzy estimator [4, 5, 9].

$$L(\lambda) = (1 - \lambda)\chi^2_{R,0.005} + \lambda(n - 1) \quad (38)$$

$$R(\lambda) = (1 - \lambda)\chi^2_{L,0.005} + \lambda(n - 1) \quad (39)$$

The unbiased $(1 - \beta)\ 100\ \%$ confidence interval is determined for σ^2 as follows;

$$\tilde{\sigma}^2 = \left[\frac{(n-1)s^2}{L(\lambda)}, \frac{(n-1)s^2}{R(\lambda)}\right] \tag{40}$$

In order to interpret the fuzzy case, we may substitute the λs with the α-cuts. Finally we may obtain the fuzzy estimator $\tilde{\sigma}^2$ as follows;

$$\tilde{\sigma}^2[\alpha] = \left[\frac{(n-1)s^2}{(1-\alpha)\chi^2_{R,0.005} + \alpha(n-1)}, \frac{(n-1)s^2}{(1-\alpha)\chi^2_{L,0.005} + \alpha(n-1)}\right] \tag{41}$$

Example 6 Assume that we have a random sample $x_1, x_2, \ldots, x_n$ having a normal probability density function $N(\mu, \sigma^2)$ where σ^2 is unknown. In order to estimate σ^2, let us consider that s^2 value is determined as 2.05 for 150 samples. Try to calculate the fuzzy variance $\tilde{\sigma}^2$.

We will adapt a similar approach which is represented in Eq. (41)

$$\tilde{\sigma}^2[\alpha] = \left[\frac{(149)(2.05)}{(1-\alpha)\chi^2_{R,0.005} + \alpha(149)}, \frac{(149)2.05}{(1-\alpha)\chi^2_{L,0.005} + \alpha(149)}\right] \tag{42}$$

Finally, we obtained $\tilde{\sigma}^2$ values by taking α-cuts. Figure 7 depicts this expression.

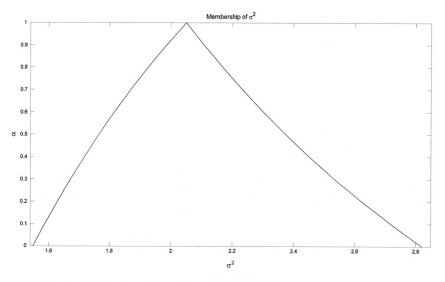

Fig. 7 Membership function of the variance for Example 6

Fuzzy Normal Approximations

• Approximation of Fuzzy Binomial Distribution

At this point, let us study the approximation between fuzzy normal and fuzzy binomial distribution. As a recall from the discrete case, let $p = 0.5$ and $m = 150$ which we will use to write fuzzy binomial distribution. Since p value is fuzzy, let $\tilde{p} = (0.3, 0.5, 0.55)$ we may write;

$$P_1[\alpha] = min\{150p^2(1-p)\} \quad p \in \tilde{p}[\alpha] \tag{43}$$

$$P_2[\alpha] = max\{150p^2(1-p)\} \quad p \in \tilde{p}[\alpha] \tag{44}$$

Then,

$$\tilde{P}(X)[\alpha] = \left[150(p_1(\alpha))^2(1-p_1(\alpha))150(p_2(\alpha))^2(1-p_2(\alpha))\right],$$
$$\tilde{p}[\alpha] = [p_1(\alpha), p_2(\alpha)] = [0.3 + 0.2\alpha, 0.55 - 0.05\alpha] \tag{45}$$

Therefore, we have to define the fuzzy mean and fuzzy variance of $B(m, \tilde{p})$; the fuzzy binomial distribution. Let $\tilde{\mu} = m\tilde{p}$ the fuzzy mean and we consider $B(150, \tilde{p})$ to find the fuzzy probability $\tilde{P}[30, 50]$. Thus, we may write;

$$\tilde{P}[30, 50][\alpha] = \sum_{i=30}^{50} \binom{150}{i} p^i (1-p)^{150-i} \tag{46}$$

Using the normal probability density function $f(x; 0, 1)$ we may approximate the same expression with $z_1 = \frac{29.5-\mu}{\sigma}$ and $z_2 = \frac{50.5-\mu}{\sigma}$ as follows;

$$\tilde{P}[30, 50][\alpha] \approx \int_{z_1}^{z_2} f(x; 0, 1)dx \quad \mu \in \tilde{\mu}[\alpha] \quad \sigma^2 \in \tilde{\sigma}^2[\alpha] \tag{47}$$

We should note that $\tilde{\mu}$ and $\tilde{\sigma}^2$ are the fuzzy binomial mean and the fuzzy binomial variance, respectively.

• Approximation of Fuzzy Poisson Distribution

The fuzzy Poisson distribution could be approximated using fuzzy normal distribution. Similar to fuzzy binomial case, let $\lambda = 15$ which we will use to express fuzzy Poisson distribution. Since λ value is fuzzy, let $\tilde{\lambda} = (10, 15, 18)$ we may write;

$$p_1(\alpha) = min\{m(\lambda)\} \tag{48}$$

$$p_2(\alpha) = max\{m(\lambda)\} \tag{49}$$

where $\lambda \in \tilde{\lambda}[\alpha]$ and $m(\lambda) = \lambda e^{-\lambda}/x!$.

Then, we may apply the same approximation technique to achieve a fuzzy probability value through fuzzy normal distribution. Let $11 \leq X \leq 17$ and we may note that;

$$\tilde{P}[11,17][\alpha] = \left\{ \sum_{x=11}^{17} \frac{\lambda^x e^{-\lambda}}{x!} \right\} \quad \lambda \in \tilde{\lambda}[\alpha] \tag{50}$$

Therefore, we may rewrite this expression using the fuzzy normal distribution by defining the boundaries, $z_1 = \frac{10.5-\mu}{\sigma}$ and $z_2 = \frac{17.5-\mu}{\sigma}$ as follows;

$$\tilde{P}[11,17][\alpha] \approx \int_{z_1}^{z_2} f(x;0,1)dx \quad \lambda \in \tilde{\lambda}[\alpha] \tag{51}$$

7 Fuzzy Lognormal Distribution

Lognormal distribution could be interpreted as an extended form of normal distribution. Let t be a random variable which has a normal distribution; mean μ and variance σ^2. $x = e^t$ is a lognormal random variable and the lognormal distribution $L(\tilde{\mu}, \tilde{\sigma}^2)$ is;

$$f(x) = \frac{1}{x\sigma\sqrt{2\pi}} e^{\left[-\frac{(ln(x)-\mu)^2}{2\sigma^2} \right]} \quad 0 < x < \infty \tag{52}$$

In order to develop the fuzzy lognormal distribution, we follow a similar approach as we developed in fuzzy normal distribution. Let $N(\tilde{\mu}, \tilde{\sigma}^2)$ be a fuzzy normal distribution with fuzzy numbers $\tilde{\mu}$ and $\tilde{\sigma}^2$. The fuzzy lognormal probability $\tilde{P}$ is calculated within an interval $[a,b]$. For $\alpha \in [0,1]$, $\mu \in \tilde{\mu}[\alpha]$ and $\sigma^2 \in \tilde{\sigma}^2[\alpha]$, $\tilde{P}[a,b]$ is calculated using the interval $[z_1, z_2]$ where $z_1 = (a-\mu)/\sigma$ and $z_2 = (b-\mu)/\sigma$ as follows;

$$\tilde{P}[a,b][\alpha] = \int_{z_1}^{z_2} f(x;0,1)dx \quad \alpha \in [0,1] \tag{53}$$

Example 7 Let the fuzzy mean be $\tilde{\mu} = (6, 6.5, 7.5)$ and the fuzzy variance be $\tilde{\sigma}^2 = (1.44, 1.69, 1.96)$. Compute the fuzzy probability value $\tilde{P}[500, 700]$ using fuzzy lognormal distribution.

In order to calculate this value we should use the following integration;

Table 4 Fuzzy probability values of Example 7

α	$min\{\tilde{P}[500, 700][\alpha]\}$	$max\{\tilde{P}[500, 700][\alpha]\}$
0	0.0868	0.1060
0.1	0.0887	0.1065
0.2	0.0904	0.1067
0.3	0.0922	0.1068
0.4	0.0939	0.1067
0.5	0.0955	0.1064
0.6	0.0971	0.1059
0.7	0.0986	0.1053
0.8	0.1000	0.1045
0.9	0.1013	0.1036
1	0.1026	0.1026

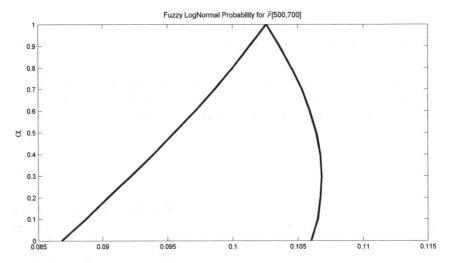

Fig. 8 Fuzzy probability values of Example 7

$$\tilde{P}[500, 700][\alpha] = \left\{ \frac{1}{\sqrt{2\pi}} \int_{z_1}^{z_2} e^{-z^2/2} dz \right\} \tag{54}$$

where $\mu \in \tilde{\mu}[\alpha]$, $\sigma^2 \in \tilde{\sigma}^2$. We set that $z_1 = \frac{ln(500)-\mu}{\sigma}$ and $z_2 = \frac{ln(700)-\mu}{\sigma}$. Thus, we generate Table 4 using the calculation of the integration in Eq. (54) and the Fig. 8.

Fuzzy Mean and Variance of Lognormal Distribution

The mean and the variance of fuzzy lognormal distribution are derived from the lognormal distribution. By using the α-cuts, the fuzzy mean; $\tilde{\mu}_L$ of fuzzy lognormal distribution could be calculated as follows;

$$\tilde{\mu}_L[\alpha] = e^{\tilde{\mu} + \tilde{\sigma}^2/2} \tag{55}$$

Similarly, the fuzzy variance; $\tilde{\sigma}_L^2$ of fuzzy lognormal distribution is calculated as follows;

$$\tilde{\sigma}_L^2[\alpha] = e^{\tilde{\sigma}^2 - 1} e^{2\tilde{\mu} + \tilde{\sigma}^2} \tag{56}$$

8 Conclusion

In fuzzy probability theory, continuous case serves frequently to better understand real applications and to develop realistic models. The theory is based on the use of fuzzy continuous random variables. Therefore, the similarity between discrete and continuous case facilitates the integration of fundamental definitions and to link the probability and possibility theory.

In this section, we applied the already developed theory of fuzzy random variables into the continuous case. Therefore, the expectations theory was derived and the fundamental parts were introduced to characterize the distributions. Variables could be derived using fuzzy continuous random variables. Fuzzy uniform, fuzzy exponential, fuzzy laplace, fuzzy normal and fuzzy lognormal distributions were explained starting with the crisp distributions. Therefore, the continuous case helped us to show the immigration into the fuzzy representation by adapting the α-cuts. The continuous expectation theory guided us to derive the fuzzy mean and the fuzzy variance of all distributions. The illustrative examples depicted the implementation of this theory and explained how to sketch the fuzzy probability in continuous case.

Moreover, fuzzy continuous theory is also utilized in fuzzy probability distributions while the occurrences of events appear by imprecise linguistic data. We tried to state more clearly the fuzzy probability theory in continuous form in a well-organized frame in this chapter.

For further research topics, construction of fuzzy uniform, fuzzy exponential, fuzzy laplace, fuzzy normal, and fuzzy lognormal probability distributions by type-2 fuzzy numbers can be accomplished by further studies. Also Markov models taking into account type-2 fuzzy numbers could be investigated in more developed form.

References

1. Zadeh, L.A.: Probability measures of fuzzy Events. J. Math. Anal. Appl. **23**(2), 421–427 (1968)
2. Zadeh, L.A.: Fuzzy set as the basis for the theory of possibility. Fuzzy Sets Syst. **1**, 3–28 (1978)
3. Buckley, J.J.: Fuzzy statistics. Springer-Verlag, Berlin (2004)
4. Buckley, J.J.: Fuzzy Probabilities. Springer, Berlin (2005)

5. Buckley, J.J.: Fuzzy Probability and Statistics. Springer, The Netherlands (2006)
6. Liu, B., Liu, Y.K.: Expected value of fuzzy variable and fuzzy expected value models. IEEE Trans. Fuzzy Syst. **10**, 445–450 (2002)
7. Liu, B., Liu, Y.K.: Fuzzy random variables: a scalar expected value operator. Fuzzy Optim. Decis. Making **2**, 143–160 (2003)
8. Nguyen, H.T., Wu, B.: Fundamentals of statistics with fuzzy data. Springer, The Netherlands (2010)
9. Kaya, I., Kahraman, C.: Fuzzy process capability analyses with fuzzy normal distribution. Expert Syst. Appl. **37**, 5390–5403 (2010)
10. Shapiro, A.F.: Fuzzy random variables. Insur. Math. Econ. **44**, 307–314 (2009)
11. Shapiro, A.F.: Modeling future lifetime as a fuzzy random variable. Insur. Math. Econ. **53**, 864–870 (2013)
12. Rakus-Andersson, E.: Continuous fuzzy sets as probabilities of continuous fuzzy events. In: Fuzzy Systems (FUZZ), 2010 IEEE International Conference Proceedings, pp. 1–7 (2010)
13. Oshmita, D., Debjani, C.: A fuzzy random continuous review inventory system. Int. J. Prod. Econ. **132**(1), 101–106 (2011)
14. Ignacio, M., Javier, H., Davide, M., Susana, M.: Characterization of continuous t-norms compatible with Zadeh's probability of fuzzy events. Fuzzy Sets Syst. **228**, 29–43 (2013)
15. Kwakernaak, H.: Fuzzy random variables-I. Definitions and theorems. Inf. Sci. **15**, 1–29 (1978)
16. Kwakernaak, H.: Fuzzy random variables-II. Algorithms and examples for the discrete case. Inf. Sci. **17**, 253–278 (1979)

On Fuzzy Bayesian Inference

Reinhard Viertl and Owat Sunanta

Abstract Bayesian inference deals with a-priori information in statistical analysis. However, usually Bayesians assume that all kind of uncertainty can be modeled by probability. Unfortunately, this is not always true due to how uncertainties are defined. The uncertainty of measurement results of continuous quantities differs from probabilistic uncertainty. Individual measurement results also contain another kind of uncertainty, which is called fuzziness. The combination of fuzziness and stochastic uncertainty calls for a generalization of Bayesian inference, i.e. fuzzy Bayesian inference. This chapter explains the generalized Bayes' theorem in handling fuzzy a-priori information and fuzzy data.

Keywords Bayesian inference · Characterizing functions · Fuzzy data · Fuzzy numbers · Fuzzy probability distributions · Fuzzy vectors · Generalized Bayes' theorem · Vector-characterizing functions

1 Introduction

In standard Bayesian inference, a-priori distributions are assumed to be standard probability distributions and the observations are assumed to be numbers or vectors. Bayes' theorem formulates the transition from an a-priori distribution of the stochastic quantity $\tilde{\theta}$, which describes the parameters of interest θ, to a corresponding a-posteriori distribution. For continuous stochastic model $X \sim f(\cdot|\theta); \theta \in \Theta$, based on observations $x_1, \ldots, x_n$ of X, the transition of an a-priori density with an updated information to the distribution of the stochastic quantity describing the parameter $\tilde{\theta}$ is given by the conditional density $\pi(\cdot|\ x_1, \ldots, x_n)$ of $\tilde{\theta}$, i.e.

R. Viertl (✉) · O. Sunanta
Institute of Statistics and Mathematical Methods in Economics,
Technische Universität Wien, Vienna, Austria
e-mail: r.viertl@tuwien.ac.at

© Springer International Publishing Switzerland 2016
C. Kahraman and Ö. Kabak (eds.), *Fuzzy Statistical Decision-Making*,
Studies in Fuzziness and Soft Computing 343,
DOI 10.1007/978-3-319-39014-7_4

$$\pi(\theta|x_1, \ldots, x_n) = \frac{\pi(\theta) \cdot l(\theta; x_1, \ldots, x_n)}{\int_\Theta \pi(\theta) \cdot l(\theta; x_1, \ldots, x_n) d\theta} \tag{1}$$

where $l(\theta; x_1, \ldots, x_n)$ is the likelihood function defined on the parameter space Θ.

However, the use of a-priori densities in form of standard probability densities is questionable in reality. Moreover, real observations from continuous quantities are not precise numbers, but rather fuzzy.

The first problem can be overcome by using a more general form of probability, i.e. fuzzy probability densities. The second problem can be solved by using fuzzy numbers and fuzzy vectors. The generalization of fuzzy Bayesian inference is, then, necessary (see related work in [1, 3, 7]).

In Sect. 2, the mathematical concepts for fuzzy numbers and vectors along with their characterizing functions are described. In Sect. 3, fuzzy probability densities are introduced through defining fuzzy a-priori distributions. In Sect. 4, the mathematical description of fuzzy observations and their corresponding characterizing functions is explained. In Sect. 5, the generalized Bayes' theorem is described for handling fuzzy a-priori densities by using the so-called δ-level functions. As a result, the fuzzy a-posteriori density is obtained. In Sect. 6, fuzzy predictive densities are described. The paper is, then, concluded with an example (Sect. 7) to show results of applying the streamlined concepts. Lastly, final remarks along with proposed future research are given in Sect. 8.

2   Fuzzy Numbers and Fuzzy Vectors

In order to describe observations or measurements of continuous quantities, the definition of general fuzzy numbers is useful.

Definition 1 A general fuzzy number x^* is defined by its characterizing function $\xi(\cdot)$, which is a real function of one real variable and possesses the following properties:

(1) $\xi: \mathbb{R} \to [0, 1]$
(2) The support of $\xi(\cdot)$, denoted by $\text{supp}[\xi(\cdot)]$ and defined by
     $\text{supp}[\xi(\cdot)] := \{x \in \mathbb{R}: \xi(x) > 0\}$,
     is a bounded subset of $\mathbb{R}$.
(3) For all $\delta \in (0, 1]$ the δ-cut $C_\delta[\xi(\cdot)]$, defined by
     $C_\delta[\xi(\cdot)] := \{x \in \mathbb{R}: \xi(x) \geq \delta\} = \bigcup_{j=1}^{k_\delta}[a_{\delta,j}, b_{\delta,j}]$,
     is non-empty and a finite union of compact intervals.

Along with general fuzzy numbers, a related critical question is how to obtain the characterizing function of a measurement result. For details, see [4, 5].

For multivariate continuous data, a standard measurement result is an n-dimensional real vector $(x_1, \ldots, x_n)$. In reality, there are two possibilities:

First, when the individual values of the variables x_i are fuzzy numbers x_i^*, a vector of fuzzy numbers $(x_1^*, \ldots, x_n^*)$ is obtained.

Second, a fuzzy version of a vector is obtained. For example, the position of a ship on a radar screen, which is, in the ideal case, a two-dimensional vector $(x, y) \in \mathbb{R}^2$. In real situation, such position is characterized by a light point on the screen, which is not a precise vector. Instead, the result is a fuzzy vector, denoted as $(x, y)^*$.

Definition 2 Using the notation $\mathbf{x} = (x_1, \ldots, x_n)$, an n-dimensional fuzzy vector $\mathbf{x}^*$ is determined by its so-called vector-characterizing function $\zeta(\cdot, \ldots, \cdot)$, which is a real function of n real variables $x_1, \ldots, x_n$ and possesses the following properties:

(1) $\zeta: \mathbb{R}^n \to [0, 1]$
(2) The support of $\zeta(\cdot, \ldots, \cdot)$ is a bounded set.
(3) For all $\delta \in (0, 1]$, the δ-cut $C_\delta[\mathbf{x}^*]$, defined by
$$C_\delta[\mathbf{x}^*] := \{ \mathbf{x} \in \mathbb{R}^n : \zeta(\mathbf{x}) \geq \delta \},$$

is non-empty, bounded, and a finite union of simply connected and closed sets.

Again, how to obtain the vector-characterizing function of a fuzzy vector is important (see [4] for details).

3 Fuzzy A-priori Distributions

Standard a-priori distributions in Bayesian inference are frequently not well justified in solving real-world problems. Therefore, a more general form of expressing a-priori information is more appropriate. These generalized a-priori distributions are called fuzzy a-priori densities. In order to define fuzzy densities, a special form of general fuzzy numbers is necessary.

Definition 3 A general fuzzy number whose δ-cuts are non-empty compact intervals $[a_\delta, b_\delta]$ is called fuzzy interval, $\mathcal{F}_I(\mathbb{R})$ identifying the set of all fuzzy intervals. For functions $f^*(\cdot)$ defined on a measure space $(M, \mathcal{A}, \mu)$, whose values $f^*(x)$ are fuzzy intervals, their so-called lower limit and upper limit δ-level functions $\underline{f}_\delta(\cdot)$ and $\bar{f}_\delta(\cdot)$ are defined in the following way:

Let $C_\delta[f^*(x)] = [a_\delta(x), b_\delta(x)] \; \forall \delta \in (0, 1]$, the lower limit and upper limit δ-level functions $a_\delta(\cdot)$ and $b_\delta(\cdot)$ are standard real-valued functions defined by their values $\underline{f}_\delta(x) := a_\delta(x) \; \forall x \in M$ and $\bar{f}_\delta(x) := b_\delta(x) \; \forall x \in M$.

Fuzzy densities are fuzzy valued function $f^*(\cdot)$ defined on a measure space $(M, \mathcal{A}, \mu)$ possessing the following properties:

(i) $f^*(\cdot)$ is fuzzy interval $\forall x \in M$
(ii) $\exists g : M \to [0, \infty)$ which is a standard probability density on $(M, \mathcal{A}, \mu)$ such that $\underline{f}_1(x) \leq g(x) \leq \bar{f}_1(x) \; \forall x \in M$
(iii) all δ-level functions $\underline{f}_\delta(\cdot)$ and $\bar{f}_\delta(\cdot)$ are integrable functions with finite integral.

Next, probabilities of events $A \in \mathcal{A}$ based on a fuzzy probability density $f^*(\cdot)$ are defined in the following way:

Let $\mathcal{D}_\delta$ be the set of all standard probability densities $h(\cdot)$ on $(M, \mathcal{A}, \mu)$ where $\underline{f}_\delta(x) \leq h(x) \leq \bar{f}_\delta(x) \; \forall x \in M$. The generalized probability $P^*(A)$ is a fuzzy interval, which is determined by a generating family of compact intervals $B_\delta = [a_\delta, b_\delta]$ where

$$b_\delta := \sup\left\{ \int_A h(x)d\mu(x) : h \in \mathcal{D}_\delta \right\}$$

$$\forall \delta \in (0, 1].$$

$$a_\delta := \inf\left\{ \int_A h(x)d\mu(x) : h \in \mathcal{D}_\delta \right\}$$

By applying the so-called construction lemma for general fuzzy numbers (see [4]), the characterizing function $\xi(\cdot)$ of $P^*(A)$ is given by its values
$\xi(x) = \sup\{\delta.\mathbb{1}_{[a_\delta, b_\delta]}(x) : \delta \in [0, 1]\} \quad \forall x \in \mathbb{R}$, where $\mathbb{1}_B(\cdot)$ denotes the indicator function of the set B, and $[a_0, b_0] := \mathbb{R}$.

As a result, fuzzy probability density is defined. Fuzzy probability densities are general forms of expressing a-priori information concerning parameters θ in stochastic models $X \sim f(\cdot|\theta); \; \theta \in \Theta$.

Remark 1 The concept of fuzzy probability differs from that of the lower/upper probabilities.

4 Fuzzy Data

Real observations of continuous stochastic quantities X are not precise numbers or vectors, whereas the measurement results are rather fuzzy. The best mathematical description (see also [2, 6]) of such observations is by means of general fuzzy numbers $x_1^*, \ldots, x_n^*$ with corresponding characterizing functions $\xi_1(\cdot), \ldots, \xi_n(\cdot)$. The fuzziness of an observation x_i^* resolves the problem in standard continuous stochastic models where observed data have zero probability.

There are different approaches to generalize Bayes' theorem to take care of fuzziness. A promising method is defining the likelihood function by the extension principle. For independent fuzzy observations $x_1^*, \ldots, x_n^*$ with characterizing functions $\xi_i(\cdot)$, where $i = 1, 2, \ldots, n$, the likelihood function can be defined based on the combined fuzzy sample element x^*.

x^* is the combined fuzzy sample whose vector-characterizing function $\zeta(\cdot,\ldots,\cdot)$ is defined by its values in the following way:

$$\zeta(x_1,\ldots,x_n) := \min\{\xi_1(x_1),\ldots,\xi_n(x_n)\} \quad \forall(x_1,\ldots,x_n) \in \mathbb{R}^n \qquad (3)$$

The generalized likelihood function $l^*(\theta;x^*)$ can, then, be represented by its lower limit δ-level function $\underline{l}_\delta(\theta;x^*)$ and the upper limit δ-level function $\bar{l}_\delta(\theta;x^*)$ for all $\delta \in (0, 1]$. For a δ-cut of the fuzzy likelihood function $l^*(\theta;x^*)$, the following fuzzy value is obtained:

$$C_\delta(l^*(\theta;x^*)) = \left[\underline{l}_\delta(\theta;x^*), \bar{l}_\delta(\theta;x^*)\right] \qquad (4)$$

Remark 2 The generalized likelihood function $l^*(\theta;x^*)$ is a fuzzy valued function, i.e. $l^*: \Theta \rightarrow \mathcal{F}_1([0; \infty))$.

5   Generalized Bayes' Theorem

The standard Bayes' theorem has to be generalized to handle fuzzy a-priori densities $\pi^*(\cdot)$ on the parameter space $\ominus$ and fuzzy data $x_1^*,\ldots,x_n^*$ of parametric stochastic models $X \sim f(\cdot|\theta); \theta \in \ominus$. This is possible by using the δ-level functions $\underline{\pi}_\delta(\cdot)$ and $\bar{\pi}_\delta(\cdot)$ of $\pi^*(\cdot)$ along with defining the corresponding δ-level functions $\underline{\pi}_\delta(\cdot|x_1^*,\ldots,x_n^*)$ and $\bar{\pi}_\delta(\cdot|x_1^*,\ldots,x_n^*)$ of the fuzzy a-posteriori density in the following way:

$$\bar{\pi}_\delta\left(\theta|x_1^*,\ldots,x_n^*\right) = \frac{\bar{\pi}_\delta(\theta) \cdot \bar{l}_\delta(\theta;x_1^*,\ldots,x_n^*)}{\int_\ominus \frac{1}{2}\left[\underline{\pi}_\delta(\theta) \cdot \underline{l}_\delta\left(\theta;x_1^*,\ldots,x_n^*\right) + \bar{\pi}_\delta(\theta) \cdot \bar{l}_\delta(\theta;x_1^*,\ldots,x_n^*)\right]d\theta}$$

$$(5)$$

$$\underline{\pi}_\delta\left(\theta|x_1^*,\ldots,x_n^*\right) = \frac{\underline{\pi}_\delta(\theta) \cdot \underline{l}_\delta(\theta;x_1^*,\ldots,x_n^*)}{\int_\ominus \frac{1}{2}\left[\underline{\pi}_\delta(\theta) \cdot \underline{l}_\delta\left(\theta;x_1^*,\ldots,x_n^*\right) + \bar{\pi}_\delta(\theta) \cdot \bar{l}_\delta(\theta;x_1^*,\ldots,x_n^*)\right]d\theta}$$

Remark 3 The averaging $\frac{1}{2}\int_\ominus \left[\underline{\pi}_\delta(\theta) \cdot \underline{l}_\delta\left(\theta;x_1^*,\ldots,x_n^*\right) + \bar{\pi}_\delta(\theta) \cdot \bar{l}_\delta(\theta;x_1^*,\ldots,x_n^*)\right]d\theta$ is necessary in order to keep the sequential updating of standard Bayes' theorem as shown in the following lemma.

Lemma 1 *For the given definition of the fuzzy a-posteriori density, the following holds*:

Let $x_1^*, \ldots, x_k^*$ *be a fuzzy sample and* $x_{k+1}^*, \ldots, x_n^*$ *a second fuzzy sample from a stochastic model* $X \sim f(\cdot|\theta)$; $\theta \in \Theta$. *Also, let* $\pi^*(\cdot)$ *be a fuzzy a-priori density on the parameter space* Θ.

Calculating the fuzzy a-posteriori density $\pi^*(\cdot \mid x_1^*, \ldots, x_k^*)$ *and taking this calculated density as new a-priori density for the second sample as well as further calculating the corresponding a-posteriori density* $\pi^*(\cdot \mid x_{k+1}^*, \ldots, x_n^*)$, *we obtain the same result as that from taking all the observations* $x_1^*, \ldots, x_n^*$ *and calculating the a-posteriori density* $\pi^*(\cdot \mid x_1^*, \ldots, x_n^*)$ *in one step.*

Proof The δ-level functions $\underline{\pi}_\delta(\cdot \mid x_1^*, \ldots, x_k^*)$ and $\bar{\pi}_\delta(\cdot \mid x_1^*, \ldots, x_k^*)$, $\underline{\pi}_\delta(\cdot \mid x_{k+1}^*, \ldots, x_n^*)$ and $\bar{\pi}_\delta(\cdot \mid x_{k+1}^*, \ldots, x_n^*)$ are considered.

For the δ-level functions, we obtain

$$C_\delta(\pi^*(\theta \mid x_1^*, \ldots, x_n^*)) = [\underline{\pi}_\delta(\theta \mid x_1^*, \ldots, x_n^*); \bar{\pi}_\delta(\theta \mid x_1^*, \ldots, x_n^*)].$$

Using the abbreviation

$$N(x_1^*, \ldots, x_n^*) = \frac{1}{2}\left[\int_\Theta \underline{\pi}_\delta(\theta) \cdot \underline{l}_\delta(\theta; \boldsymbol{x}^*)d\theta + \int_\Theta \bar{\pi}_\delta(\theta) \cdot \bar{l}_\delta(\theta; \boldsymbol{x}^*)d\theta \right]$$
$$= \int_\Theta \frac{1}{2}[\underline{\pi}_\delta(\theta) \cdot \underline{l}_\delta(\theta; \boldsymbol{x}^*) + \bar{\pi}_\delta(\theta) \cdot \bar{l}_\delta(\theta; \boldsymbol{x}^*)]d\theta,$$

and taking the fuzzy a-posteriori density $\pi^*(\cdot \mid x_1^*, \ldots, x_k^*)$ as new a-priori density and $x_{k+1}^*, \ldots, x_n^*$ as data, also defining the combined fuzzy sample element x_1^* of fuzzy sample $x_1^*, \ldots, x_k^*$ and x_2^* of fuzzy sample $x_{k+1}^*, \ldots, x_n^*$ in the same manner as $\boldsymbol{x}^*$ in Sect. 4, the lower limit δ-level function of the new a-posteriori density is defined by:

$$
\begin{aligned}
\underline{\pi}_\delta(\theta|x_1^*, \ldots, x_n^*) &= \frac{\underline{\pi}_\delta(\theta|x_1^*, \ldots, x_k^*) \cdot \underline{l}_\delta(\theta; x_2^*)}{\int_\Theta \frac{1}{2}[\underline{\pi}_\delta(\theta|x_1^*, \ldots, x_k^*) \cdot \underline{l}_\delta(\theta; x_2^*) + \bar{\pi}_\delta(\theta|x_1^*, \ldots, x_k^*) \cdot \bar{l}_\delta(\theta; x_2^*)]d\theta} \\
&= \frac{[N(x_1^*, \ldots, x_k^*)]^{-1}\underline{\pi}_\delta(\theta) \cdot \underline{l}_\delta(\theta; x_1^*) \cdot \underline{l}_\delta(\theta; x_2^*)}{\int_\Theta \frac{1}{2}N[(x_1^*, \ldots, x_k^*)]^{-1}[\underline{\pi}_\delta(\theta) \cdot \underline{l}_\delta(\theta; x_1^*) \cdot \underline{l}_\delta(\theta; x_2^*) + \bar{\pi}_\delta(\theta) \cdot \bar{l}_\delta(\theta; \underline{x}_1^*) \cdot \bar{l}_\delta(\theta; x_2^*)]d\theta} \\
&= \frac{\underline{\pi}_\delta(\theta) \cdot \underline{l}_\delta(\theta; x_1^*) \cdot \underline{l}_\delta(\theta; x_2^*)}{\int_\Theta \frac{1}{2}[\underline{\pi}_\delta(\theta) \cdot \underline{l}_\delta(\theta; x_1^*) \cdot \underline{l}_\delta(\theta; x_2^*) + \bar{\pi}_\delta(\theta) \cdot \bar{l}_\delta(\theta; x_1^*) \cdot \bar{l}_\delta(\theta; x_2^*)]d\theta} \\
&= \frac{\underline{\pi}_\delta(\theta) \cdot \underline{l}_\delta(\theta; \boldsymbol{x}^*)}{\int_\Theta \frac{1}{2}[\underline{\pi}_\delta(\theta) \cdot \underline{l}_\delta(\theta; \boldsymbol{x}^*) + \bar{\pi}_\delta(\theta) \cdot \bar{l}_\delta(\theta; \boldsymbol{x}^*)]d\theta} \\
&= \underline{\pi}_\delta(\theta|x_1^*, \ldots, x_n^*).
\end{aligned}
$$

The upper limit δ-level functions $\bar{\pi}_\delta(\theta|\boldsymbol{x}^*)$ are obtained analogously. Therefore, the sequential updating from standard Bayes' theorem also remains valid for fuzzy data.

6 Fuzzy Predictive Distributions

Standard predictive densities for stochastic model $X \sim f(\cdot|\theta)$; $\theta \in \Theta$ based on data D are defined as marginal density of the stochastic quantity X, i.e.

$$p(x|D) := \int_{\Theta} f(x|\theta)\pi(\theta|D)d\theta \quad \forall x \in M_X,$$

where M_X is the observation space of X.

In case of fuzzy a-posteriori density $\pi^*(\cdot|D^*)$ based on fuzzy data D^*, the integration has to be generalized. This generalized integration yields fuzzy intervals. Based on $\mathcal{D}_\delta$ as defined in Sect. 3, the generating family of intervals $[c_\delta, d_\delta]$, $\delta \in (0, 1]$ is defined by

$c_\delta := \inf\{\int_\Theta f(x|\theta)h(\theta)d\theta : h \in \mathcal{D}_\delta\}$

$d_\delta := \sup\{\int_\Theta f(x|\theta)h(\theta)d\theta : h \in \mathcal{D}_\delta\}.$

Definition 4 The fuzzy predictive density $p^*(\cdot|D^*)$ of X is defined by its values $p^*(x|D^*)$ $\forall x \in M_X$ whose characterizing function $\psi_x(\cdot)$ is given by the construction lemma, i.e.

$$\psi_x(y) = \sup\{\delta.\mathbb{1}_{[c_\delta, d_\delta]}(y) : \delta \in [0, 1]\} \quad \forall y \in \mathbb{R}, \text{ where } [c_0, d_0] := \mathbb{R}.$$

Remark 4 The standard case of precise a-priori density and precise data is a special case of the above construction. In this case, for standard a-posteriori densities $\pi(\cdot|D)$, the above definition yields as characterizing function $\psi_x(\cdot)$ of the value of the predictive density the indicator function of $p(x|D)$, i.e. $\psi_x(\cdot) = \mathbb{1}_{\{p(x|D)\}}(\cdot)$.

7 Example

Waiting times in a queuing system are modeled with an exponential distribution and with fuzzy gamma density as a-priori density for the parameter $\theta \in (0, \infty)$. Different δ-level functions (each with upper and lower limits) of the fuzzy a-priori density are generated (using the method described in Sect. 3) and shown in Fig. 1.

Using a fuzzy sample of waiting times $D^* = (x_1^*, \ldots, x_8^*)$ with characterizing functions as shown in Fig. 2, the application of the generalized Bayes' theorem (explained in Sect. 5) yields the corresponding fuzzy a-posteriori density $\pi^*(\cdot|D^*)$ with their δ-level functions. Some δ-level functions of the result are depicted in Fig. 3. Compared with the fuzzy a-priori density (Fig. 1), one notices the changes in both shape and magnitude of the distribution curves, i.e. the distribution is updated based on the fuzzy sample.

Some δ-level functions of the predictive density $p^*(\cdot|D^*)$ are approximated by simulation and the result is depicted in Fig. 4. Application of fuzzy a-posteriori density $\pi^*(\cdot|D^*)$ based on fuzzy data D^* along with the generalized integration as

$$\underline{\pi}_\delta(\theta), \overline{\pi}_\delta(\theta)$$

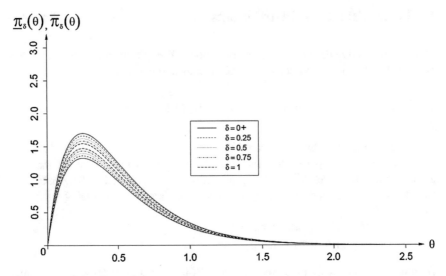

	$\delta = 0+$
	$\delta = 0.25$
	$\delta = 0.5$
	$\delta = 0.75$
	$\delta = 1$

Fig. 1 Fuzzy a-priori density

Fig. 2 Fuzzy sample

$$\xi_i(x)$$

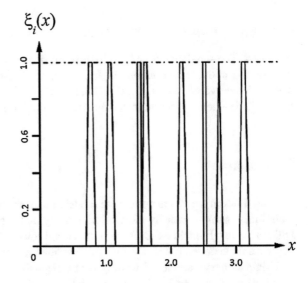

explained in Sect. 6, the predictive density (marginal density in standard case) is obtained.

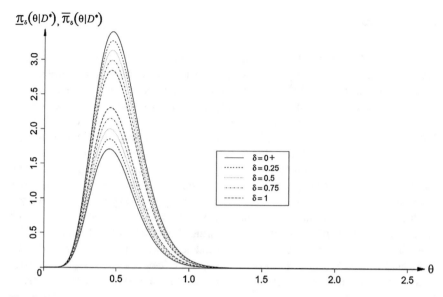

Fig. 3 Fuzzy a-posteriori density

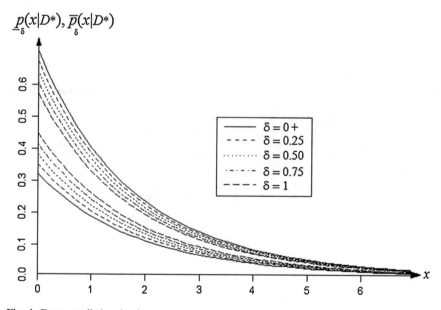

Fig. 4 Fuzzy predictive density

8   Final Remark

A more general concept of fuzzy a-priori densities is introduced to handle fuzzy a priori information and fuzzy data, as opposed to the use of standard probability. This concept is more suitable in modeling prior information, which is usually uncertain, i.e. fuzzy. Besides, concepts of general fuzzy numbers and fuzzy vectors along with their characterizing functions have been applied in capturing the imprecision of continuous quantities. The fuzzy probability densities with their δ-level functions are used to define fuzzy a-posteriori densities, where the a-priori densities are updated based on fuzzy samples. In addition, fuzzy a-posteriori densities can be used in order to generate fuzzy predictive distributions. In doing so, a generalized integration procedure for the construction of fuzzy predictive densities is introduced. As a result, Bayes' theorem is generalized for modeling data with explainable mathematical grounds in capturing the variability and imprecision of real observations. However, in many applications, the data to be analyzed are thought of as displayed in a matrix or in several matrices. A suggesting direction for future effort is, therefore, to explicitly consider the inference methods for multivariate fuzzy data.

References

1. Klir, G., Yuan, B.: Fuzzy Sets and Fuzzy Logic – Theory and Applications. Prentice Hall, Upper Saddle River (1995)
2. Stein, M., Beer, M., Kreinovich, V.: Bayesian approach for inconsistent information. Inf. Sci. **245**, 96–111 (2013)
3. Viertl, R.: Foundations of fuzzy Bayesian inference. J. Uncertain Syst. **2**(3), 187–191 (2008)
4. Viertl, R.: Statistical Methods for Fuzzy Data. Wiley, Chichester (2011)
5. Viertl, R., Sunanta, O.: Fuzzy Bayesian inference. Metron **71**(3), 207–216 (2013)
6. Wolkenhauer, O.: Data Engineering – Fuzzy Mathematics in Systems Theory and Data Analysis. Wiley, New York (2001)
7. Yang, C.C.: Fuzzy Bayesian inference. In: Proceeding of IEEE International Conference on Systems, Man, and Cybernetics, vol. 3, pp. 2707–2712. Florida, 12–15 Oct 1997

Fuzzy Central Tendency Measures

Cengiz Kahraman and İrem Uçal Sarı

Abstract This chapter converts the classical central tendency measures to their fuzzy cases. Fuzzy mean, fuzzy mode and fuzzy median are explained by numerical examples. Fuzzy frequency distribution is another subtitle of this chapter. Classical graphical illustrations are examined under fuzziness. A numerical example for each central tendency measure is given.

Keywords Fuzzy sets · Mode · Median · Geometric mean · Quadratic mean · Arithmetic mean · Harmonic mean

1 Introduction

A measure of central tendency is a single value that attempts to describe a set of data by identifying the central position within that set of data. The mean is most likely the measure of central tendency, but there are others, such as the median and the mode. The mean, median and mode are all valid measures of central tendency, but under different conditions, some measures of central tendency become more appropriate to use than others.

Statistical modeling is used to describe variability of quantities and errors in observations. But these models assume the observations to be numbers or vectors. This assumption is often not realistic because measurement results of continuous quantities are always not precise numbers but more or less non-precise. This kind of uncertainty is different from errors and variability. Whereas errors and variability can be modelled by stochastic variables and probability distributions, imprecision is another kind of uncertainty, called fuzziness. For a quantitative description of such data the most up-to-date method is to use fuzzy numbers and fuzzy vectors which are special fuzzy models [9].

C. Kahraman (✉) · İ.U. Sarı
Management Faculty, Department of Industrial Engineering, Istanbul Technical University, Istanbul, Turkey
e-mail: kahramanc@itu.edu.tr

© Springer International Publishing Switzerland 2016
C. Kahraman and Ö. Kabak (eds.), *Fuzzy Statistical Decision-Making*,
Studies in Fuzziness and Soft Computing 343,
DOI 10.1007/978-3-319-39014-7_5

65

This chapter briefly gives the definitions of classical central tendency measures, and then explains how to calculate the fuzzy central tendency measures when we have fuzzy data. Fuzzy mode, fuzzy median and fuzzy mean are explained by numerical examples. Fuzzy frequency distribution and graphical illustrations for fuzzy data are also given by numerical examples.

The remaining of this chapter is organized as follows. Section 2 briefly presents the literature review on fuzzy central tendency measures. Section 3 includes the classical measures of central tendency. Section 4 converts the classical definitions to their fuzzy cases. Section 5 gives the steps of fuzzy frequency distribution. Section 6 shows graphical illustrations for fuzzy data. Section 7 concludes the chapter.

2 Literature Review

There exist a few publications in the literature related to fuzzy central tendency measures. Goodman [1] determined the measures of central tendency of fuzzy sets by a new approach based on the characterization of homomorphic-like operators among fuzzy sets and related random sets. The measures of central tendency refer to the domain values of a given fuzzy set. A number of ad hoc approaches, such as the mean-of-maxima (MOM) of the fuzzy set membership function and the center-of-area (COA) approach for the measurement of central tendencies, were analyzed. Sun and Wu [8] proposed definitions of fuzzy mode, fuzzy median and fuzzy mean as well as investigation of their related properties and employ these techniques in the practical applications of real life. Empirical result shows that fuzzy statistics with soft computing is more realistic and reasonable for the statistical research. Teran (2009) presents an account of the notion of centrality which is based on fuzzy events and is valid for single distributions and for families of distributions. This unifying framework includes (a) univariate location estimators like the mean, the median and the mode, (b) the interquartile interval and the Lorenz curve of a random variable, (c) several generalized medians, trimmed regions and statistical depth functions from multivariate analysis, (d) most known location estimators for random sets, (e) the probability mass function of a discrete random variable and the coverage function of a random closed set, (f) the Choquet integral with respect to an infinitely alternating or infinitely monotone capacity. Lu and Jiao [3] used fuzzy average statistics and analyzed the questionnaire for university students. They found that fuzzy sample mean, fuzzy sample mode and fuzzy sample median could describe the fuzzy investing behavior of the university students and its average trends. Random fuzzy numbers has become a valuable tool to model and handle fuzzy-valued data generated through a random process. Recent studies have been devoted to introduce measures of the central tendency of random fuzzy numbers showing a more robust behaviour than the so-called Aumann-type mean value.

Sinova et al. [6] aimed to explore the extension of the median to random fuzzy numbers. This extension is based on the 1-norm distance and its adequacy is shown by analyzing its properties and comparing its robustness with that of the mean both theoretically and empirically. Parvathi and Atanassova [5] defined the theoretical aspects like sample mean, median, and mode of intuitionistic fuzzy data are defined. Sinova et al. [7] aimed to deepen in the analysis of these centrality measures and the Aumann-type mean by examining the situation of symmetric random fuzzy numbers.

3 Central Tendency Measures

A measure of central tendency is a single value that attempts to describe a whole set of data with a single value that describes the middle or center of a data set by identifying the central position within that set of data. In this section, the most used measures of central tendency and the graphical representation methods of data are detailed.

3.1 Mean

The mean is the arithmetic average for a set of data. To find the arithmetic average for a set of values the sum of all values in the set of data are divided by the number of the values in that data.

The mean of a sample is denoted by $\bar{x}$ and calculated by using Eq. 1 where $x_1, x_2, x_3, \ldots, x_n$ are the values in the set of data and n is the number of the sample size.

$$\bar{x} = \frac{x_1 + x_2 + x_3, + \cdots + x_n}{n} = \frac{\sum_{i=1}^{n} x_i}{n} \tag{1}$$

The mean of a population is denoted by μ and calculated by using Eq. 2 where N is the number of the population.

$$\mu = \frac{x_1 + x_2 + x_3, + \cdots + x_N}{N} = \frac{\sum_{i=1}^{N} x_i}{N} \tag{2}$$

Example 1 There are 50 observations of execution times of an experiment are given in Table 1.

Sum of all the values in Table 1 is calculated as 6723. By using Eq. 2 the mean is calculated as 6723/50 = 134.46.

Table 1 The observations

141	137	144	128	137	127	125	143	128	132
125	125	139	129	137	136	139	123	127	141
135	130	140	123	140	135	131	139	128	139
139	138	140	143	137	137	139	140	142	134

3.2 Median

The median of a data set is the middle value which divides the data set into two equal groups, after the values are ordered from lowest to highest. If there is an even number of values in the set, the median is calculated by taking the mean of the two values in the center of the data.

Example 2 To calculate median of the data set given in Table 1, first the data are ordered from the least to the largest:

122, 123, 123, 123, 124, 125, 125, 125, 127, 127, 128, 128, 128, 129, 130, 131, 131, 131, 132, 134, 135, 135, 136, 136, 137, 137, 137, 137, 137, 138, 138, 139, 139, 139, 139, 139, 139, 140, 140, 140, 140, 141, 141, 141, 142, 142, 143, 143, 143, 144

The middle values of the data are 137 and 137. Median of the data is calculated by taking the arithmetic mean of these two data which is equal to 137.

3.3 Mode

The mode of a data set is the most commonly occurring value in the set. Mode can also be defined as the element with the largest frequency in a given data set. The frequency of a particular data value is the number of times the data value occurs.

Example 3 To calculate mode of the data set given in Table 1, first the frequencies of the values which are written in brackets below are determined:

122 (1), 123 (3), 124 (1), 125 (3), 127 (2), 128 (3), 129 (1), 130 (1), 131 (3), 132 (1), 134 (1), 135 (2), 136 (2), 137 (5), 138 (2), 139 (6), 140 (4), 141 (3), 142 (2), 143 (3), 144 (1).

The mode of the data set is determined as 139 which is repeated 6 times in the data set.

3.4 Geometric Mean

Geometric mean indicates the typical value of a data set by using the product of their values. Geometric mean is defined as the nth root of the product of n numbers. The geometric mean is preferred when working with percentages (which are derived from values), whereas the standard arithmetic mean is preferred when working with the values themselves.

Geometric mean of a sample is denoted by GM and calculated by using Eq. 3:

$$GM = \sqrt[n]{x_1 \times x_2 \times x_3 \times \cdots \times x_n} \tag{3}$$

Example 4 Geometric mean of the data set given in Table 1 is calculated as 134.3 as shown below:

$$GM = \sqrt[50]{\begin{array}{l} 141 \times 125 \times 135 \times 139 \times 123 \times 137 \\ \times 125 \times 130 \times 138 \times 142 \times \cdots \times 122 \end{array}} = 134.3$$

3.5 Harmonic Mean

Harmonic mean indicates the typical value of a data set by dividing the number of observations by the reciprocal of each number in the series. Harmonic mean of a sample is denoted by HM and calculated by using Eq. 4:

$$HM = \frac{1}{\frac{1}{n}\left(\frac{1}{x_1} + \frac{1}{x_2} + \frac{1}{x_3} + \ldots + \frac{1}{x_n}\right)} = \frac{n}{\sum_{i=1}^{n} \frac{1}{x_i}} \tag{4}$$

Example 5 Harmonic mean of the data set given in Table 1 is calculated as 134.137 as shown below:

$$HM = \frac{50}{\left(\frac{1}{141} + \frac{1}{125} + \frac{1}{135} + \frac{1}{139} + \frac{1}{123} + \frac{1}{137} + \frac{1}{125} + \frac{1}{130} + \cdots + \frac{1}{122}\right)}$$
$$= 134.137$$

3.6 Quadratic Mean

Quadratic mean (QM) is a statistical measure of the magnitude of a varying quantity. It can be calculated for a series of discrete values or for a continuously varying function. It is calculated by taking the square root of the mean of the squares of the values. It is a special case of the generalized mean with the exponent p = 2.

Quadratic mean of a sample is denoted by QM and calculated by using Eq. 5:

$$QM = \sqrt{\frac{x_1^2 + x_2^2 + x_3^2 + \cdots + x_n^2}{n}} = \sqrt{\frac{\sum_{i=1}^{n} x_i^2}{n}} \tag{5}$$

Example 6 Quadratic mean of the data set given in Table 1 is calculated as 134.61 as shown below:

$$QM = \sqrt{\frac{141^2 + 125^2 + 135^2 + 139^2 + 123^2 + 137^2 + 125^2 + 130^2 + 138^2 + \cdots + 122^2}{50}}$$
$$= 134.61$$

3.7 Graphical Representation of Data

3.7.1 Stem-and-Leaf Plots

A stem and leaf plot is a graphical method of displaying data where each data value is split into a stem which is the first digit or digits of the number and a leaf which is usually the last digit of the number. It is particularly useful when the data are not too numerous.

On a standard stem and leaf plot the stem is on the left and in this column the first digits of the numbers in the data set are ordered from smallest to largest. A vertical line is drawn between stem and leaf columns. The leaves are on the left side of the plot. Each number on the leaf represents one single value from the data set. The numbers in the leaf are organized from smallest to largest and separated by commas.

Example 7 The stem and leaf plot of the data set given in Table 1 is shown in Fig. 1:

Fig. 1 Stem and leaf plot

12	2,3,3,3,4,5,5,5,7,7,8,8,8,9
13	0,1,1,1,2,4,5,5,6,6,7,7,7,7,7,8,8,9,9,9,9,9,9
14	0,0,0,0,1,1,1,2,2,3,3,3,4

3.7.2 Frequency Distribution Tables

Frequency distributions are visual displays that organize and present frequency counts so that the information can be interpreted more easily. Frequency distributions can show either the actual number of observations falling in each range or the percentage of observations. Frequency distributions are particularly useful in summarizing large data sets and assigning probabilities.

There are four steps for constructing a frequency distribution from a data set.

1. The number of the classes is decided which is mostly preferred between 5 and 20.
2. The class width is defined by rounding up the value which is found by dividing the range of the data by the number of classes.
3. The class limits are determined which are mostly defined by taking the minimum data entry as the lower limit of the first class. To get the lower limit of the next class, simply the class width is added to the lower limit of a class. After determining lower limits for all classes, upper limits for each class is determined considering that the classes cannot overlap.
4. The number of data entries for each class are counted and in the row of the table for that class.

Example 8 To construct a frequency distribution table from the data set given in Table 1 we decide the number of classes as 8. The range of the data set is 144–122 = 22. The class width is defined as 3 by rounding up 22/8 = 2.75. The class limits are determined as 122–124, 125–127, 128–130, 131–133, 134–136, 137–138, 139–141 and 142–145. The frequency table is constructed in Table 2 by counting the number of data which belongs to a class.

3.7.3 Histograms

A histogram is a graphical display of data using bars of different heights. It shows a count of the data points falling in various ranges. It is particularly useful when there

Table 2 Frequency distribution table

Class limits	Frequency
122–124	5
125–127	5
128–130	5
131–133	4
134–136	5
137–139	13
140–142	9
143–145	4

Fig. 2 Histogram

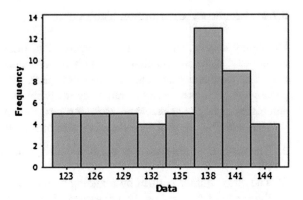

are a large number of observations. In order to make a histogram, first the frequency distributions are calculated. Then histogram is constructed by using the groups and frequencies in the frequency distribution on the horizontal and vertical axes, respectively.

Example 9 Histogram of the data set given in Table 1 is constructed by using the frequency distribution table given in Table 2 and shown in Fig. 2.

3.7.4 Frequency Polygons

Frequency polygons are a graphical device for understanding the shapes of distributions. To construct a frequency polygon, midpoints of the interval of corresponding rectangle in a histogram are joined together by straight lines. They serve the same purpose as histograms, but are especially helpful for comparing sets of data.

Example 10 Frequency polygon of the data set given in Table 1 is shown in Fig. 3.

Fig. 3 Frequency polygon

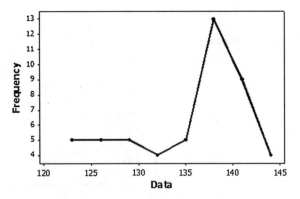

3.7.5 Bar Charts

Bar charts are used to display and compare the number, frequency or other measure for different discrete categories or groups. One axis of the chart shows the specific categories being compared, and the other axis represents a discrete value.

Example 11 A manufacturer produces 5 different types of items. The production amounts of the items are given in Table 3.

Bar chart of the data given in Table 3 is shown in Fig. 4:

3.7.6 Pie Charts

A pie chart is a circular statistical graphic, which shows the relative contribution that different categories contribute to an overall total. In a pie chart, the arc length of each category (and consequently its central angle and area), is proportional to the quantity it represents.

Example 12 Pie chart of the data set given in Table 3 is shown in Fig. 5:

Table 3 Production amounts of the items

	Item A	Item B	Item C	Item D	Item E
Production amounts	5000	8000	3000	17,000	12,000

Fig. 4 Bar chart

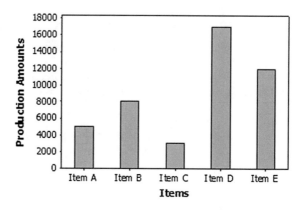

Fig. 5 Pie chart

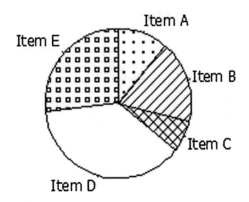

4 Central Tendency Measures Under Fuzziness

4.1 Fuzzy Arithmetic Mean

In case of fuzziness, the data are vague and are represented by fuzzy sets. Table 4 represents a group of fuzzy data selected from a population randomly:

Fuzzy arithmetic mean for the sample can be calculated by Eq. (6):

$$\tilde{\tilde{x}} = \frac{\sum_{i=1}^{n} \tilde{x}_i}{n} \tag{6}$$

Assume $\tilde{x}_i$ values are represented by triangular fuzzy numbers in the form of (x_l, x_m, x_u) where $x_l > 0$. Then Eq. (6) becomes

$$\tilde{\tilde{x}} = \frac{\sum_{i=1}^{n} \tilde{x}_i}{n} = \left(\sum_{i=1}^{n} x_{li}/n, \sum_{i=1}^{n} x_{mi}/n, \sum_{i=1}^{n} x_{ui}/n \right) \tag{7}$$

Now assume $\tilde{x}_i$ values are represented by trapezoidal fuzzy numbers in the form of $(x_l, x_{m1}, x_{m2}, x_u)$ where $x_l > 0$. Then Eq. (6) becomes

$$\tilde{\tilde{x}} = \frac{\sum_{i=1}^{n} \tilde{x}_i}{n} = \left(\sum_{i=1}^{n} x_{li}/n, \sum_{i=1}^{n} x_{m1i}/n, \sum_{i=1}^{n} x_{m2i}/n, \sum_{i=1}^{n} x_{ui}/n \right) \tag{8}$$

Table 4 Fuzzy data

$\tilde{x}_1$	$\tilde{x}_2$	...	$\tilde{x}_k$
$\tilde{x}_{k+1}$	$\tilde{x}_{k+2}$	...	$\tilde{x}_l$
$\tilde{x}_{l+1}$	$\tilde{x}_{l+2}$	...	$\tilde{x}_n$

Fuzzy arithmetic mean for the population can be calculated by Eq. (9):

$$\tilde{\mu} = \frac{\sum_{i=1}^{N} \tilde{x}_i}{N} \qquad (9)$$

Assume $\tilde{x}_i$ values are represented by triangular fuzzy numbers in the form of (x_l, x_m, x_u) where $x_l > 0$. Then Eq. (9) becomes

$$\tilde{\mu} = \frac{\sum_{i=1}^{N} \tilde{x}_i}{N} = \left(\sum_{i=1}^{N} x_{li}/N, \sum_{i=1}^{N} x_{mi}/N, \sum_{i=1}^{N} x_{ui}/N \right) \qquad (10)$$

Now assume $\tilde{x}_i$ values are represented by trapezoidal fuzzy numbers in the form of $(x_l, x_{m1}, x_{m2}, x_u)$ where $x_l > 0$. Then Eq. (9) becomes

$$\tilde{\mu} = \frac{\sum_{i=1}^{N} \tilde{x}_i}{N} = \left(\sum_{i=1}^{N} x_{li}/N, \sum_{i=1}^{N} x_{m1i}/N, \sum_{i=1}^{N} x_{m2i}/N, \sum_{i=1}^{N} x_{ui}/N \right) \qquad (11)$$

Example 13 Consider the fuzzy data belonging to a sample in Table 5 and calculate the arithmetic mean.

We can apply a fuzzification percentage to obtain their corresponding fuzzy numbers in Table 5. For instance, if a 10 % fuzzification percentage is used, the triangular fuzzy numbers given in Table 6 are obtained.

$$\tilde{\tilde{x}} = \left(\sum_{i=1}^{n} x_{li}/n, \sum_{i=1}^{n} x_{mi}/n, \sum_{i=1}^{n} x_{ui}/n \right) = (20.7, 23, 25.3)$$

Nguyen and Wu [4] give the definition of fuzzy sample mean as follows:

Let U be the universal set, $L = \{L_1, L_2, \ldots, L_k\}$ be a set of k-linguistic variables on U, and $\left\{ Fx_i = \frac{m_{i1}}{L_1} + \frac{m_{i2}}{L_2} + \cdots + \frac{m_{ik}}{L_k}, i = 1, 2, \ldots, n \right\}$ be a sequence of random

Table 5 Fuzzy data

Around 20	Around 23	Around 20	Around 21
Around 28	Around 19	Around 22	Around 32
Around 14	Around 29	Around 21	Around 27

Table 6 10 % fuzzification

(18, 20, 22)	(20.7, 23, 25.3)	(18, 20, 22)	(18.9, 21, 23.1)
(25.2, 28, 30.8)	(17.1, 19, 20.9)	(19.8, 22, 24.2)	(28.8, 32, 35.2)
(12.6, 14, 15.4)	(26.1, 29, 31.9)	(18.9, 21, 23.1)	(24.3, 27, 29.7)

fuzzy sample on U. m_{ij} $\left(\sum_{j=1}^{k} m_{ij} = 1\right)$ is the membership with respect to L_j. Then, the fuzzy sample mean is defined as

$$F\bar{x} = \frac{\frac{1}{n}\sum_{i=1}^{n} m_{i1}}{L_1} + \frac{\frac{1}{n}\sum_{i=1}^{n} m_{i2}}{L_2} + \cdots + \frac{\frac{1}{n}\sum_{i=1}^{n} m_{ik}}{L_k} \qquad (12)$$

Nguyen and Wu [4] give the definition of fuzzy sample mean for the interval-valued data as follows:

Let U be the universe set, and $\{Fx_i = [a_i, b_i], a_i, b_i \in R, i = 1, 2, \ldots, n\}$ be a sequence of random fuzzy sample on U. Then the fuzzy sample mean value is defined as

$$F\bar{x} = \left[\frac{1}{n}\sum_{i=1}^{n} a_i, \frac{1}{n}\sum_{i=1}^{n} b_i\right] \qquad (13)$$

Example 14 Let's consider the data in Table 7.

Using Eq. (13), the fuzzy sample mean is calculated as follows:

$$F\bar{x} = \left[\frac{3+4+2+2+4+5+\cdots+5}{16}, \frac{7+6+5+4+5+7+\cdots+6}{16}\right]$$
$$= [3.5625, 6.3125]$$

4.2 Fuzzy Median

To be able to find the median when the values are fuzzy numbers, it is necessary to rank the data from the least value to the largest value. Hence, a ranking method for fuzzy numbers is needed. In the literature, there are many ranking methods developed by different researchers, which may give different ranking results. This means that you may not obtain the same rank for the fuzzy data if you use a few ranking methods and you may find different medians for each of these methods.

Example 15 Consider the data in Table 6. Let us use Lee and Li's [2] method to rank these numbers. Lee and Li's [2] method for triangular fuzzy numbers is defined as in Eq. (14):

Table 7 The interval valued data

[3, 7]	[4, 6]	[2, 5]	[2, 4]
[4, 5]	[5, 7]	[3, 8]	[4, 7]
[3, 6]	[3, 5]	[4, 7]	[6, 8]
[2, 7]	[4, 7]	[3, 6]	[5, 6]

$$\bar{x}_P(\tilde{x}_i) = \frac{x_{il} + 2x_{im} + xi_u}{4} \tag{14}$$

Since we applied 10 % fuzzification of the middle point, the result of Eq. (12) will be equal to the middle point of each fuzzy number. Thus we have 14, 19, 20, 20, 21, 21, 22, 23, 27, 28, 29, and 32. The median is (21 + 22)/2 = 21.5. We can write it as a fuzzy number by 10 % fuzzification again to be (19.35, 21.5, 23.65). You would find the same result if you had used the fuzzy TFNs (18.9, 21, 23.1) and (19.8, 22, 24.2) by summing them and dividing by 2.

Nguyen and Wu [4] give the definition of fuzzy sample median as follows:

Let U be the universe set and $L = \{L_1, L_2, \ldots, L_k\}$ be a ordered set of k-linguistic variables on U, and $\left\{x_i = \frac{m_{i1}}{L_1} + \frac{m_{i2}}{L_2} + \cdots + \frac{m_{ik}}{L_k}, i = 1, 2, \ldots, n\right\}$ be a sequence of random fuzzy sample on U. Let $S_j = \sum_{i=1}^{n} m_{ij}$, j = 1,2,...,k and $T = 1.S_1 + 2.S_2 + \cdots + k.S_k$. Then, the minimum L_j such that $\sum_{i=1}^{j} S_i \geq \left[\frac{T}{2}\right]$ is called the fuzzy median of this sample. Here, $\left[\frac{T}{2}\right]$ means the largest integral that equal or less than $\frac{T}{2}$:

$$F\ median(x_i) = \left\{ L_j : minimum\ j\ such\ that \sum_{i=1}^{j} S_i \geq \left[\frac{T}{2}\right] \right\} \tag{15}$$

Nguyen and Wu [4] give the definition of fuzzy sample median for the interval-valued data as follows:

Let U be the universe set, and $\{Fx_i = [a_i, b_i], a_i, b_i \in R, i = 1, 2, \ldots, n\}$ be a sequence of random fuzzy sample on U. Let c_j be center of the interval of $[a_i, b_i]$ and l_j be the length of $[a_i, b_i]$. Then the fuzzy sample mean is defined by Eq. (16):

$$F\ median = (c; r), c = median\{c_j\}, r = \frac{median\{l_i\}}{2} \tag{16}$$

4.3 Fuzzy Mode

To be able to find the mode when the values are fuzzy numbers, one way is first to defuzzify the data and then find the element having the largest frequency. However, this approach may generally result in "no mode" for the data since defuzzification of fuzzy numbers will possibly give a unique value different from the others. Another approach for fuzzy mode may be to examine the intersections of intervals, if any, of fuzzy numbers since the same values on the x-axis may be determined as a mode or modes.

Example 16 Consider the data in Table 6. Let us use Lee and Li's [2] method to rank these numbers. The method is given by Eq. (12). Since we applied 10 %

Fig. 6 Determination of mode

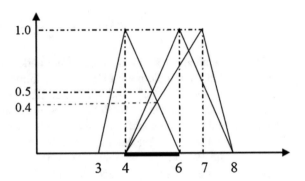

fuzzification of the middle point, the result of Eq. (12) will be equal to the middle point of each fuzzy number. Thus we have 14, 19, 20, 20, 21, 21, 22, 23, 27, 28, 29, and 32. The data are bimodal and they are (18, 20, 22) and (18.9, 21, 23.1). Now consider the following fuzzy numbers and determine the mode or modes: (4, 6, 8), (4, 7, 8), (3, 4, 6), (9, 10, 12), (1, 2, 3).

The fuzzy numbers (9, 10, 12) and (1, 2, 3) have no intersection with the other fuzzy numbers. Hence we do not consider these numbers while calculating the mode. The mode is somewhere in the interval [4, 6] and the most possible value in this interval is found in the following way (see Fig. 6).

$$\alpha_{Mode} = max \left\{ \begin{array}{l} \min_{x=4}(1,0,0), \min_{x=5}(0.5, 0.5, 0.33), \\ \min_{x=5.5}(0.75, 0.4, 0.4), \min_{x=6}(1, 0.67, 0) \end{array} \right\}$$

$$\alpha_{Mode} = max\{0, 0.33, 0.4, 0\} = 0.4$$

Thus, the mode is 5.5 with a membership of 0.4.

Nguyen and Wu [4] give the definition of fuzzy sample mode as follows:

Let U be the universal set, $L = \{L_1, L_2, \ldots, L_k\}$ a set of k-linguistic variables on U, and $\{FS_i, i = 1, 2, \ldots, n\}$ a sequence of random fuzzy sample on U. For each sample FS_i, ASSİGN a linguistic variable L_j a normalized membership $m_{ij}\left(\sum_{j=1}^{k} m_{ij} = 1\right)$, let $S_j = \sum_{i=1}^{n} m_{ij}$, $j = 1, 2, \ldots, k$. Then, the maximum value of S_j with respect to L_j is called the fuzzy mode (FM) of this sample. That is $FM = \left\{L_j | S_j = \max_{1 \le i \le k} S_i\right\}$.

Nguyen and Wu [4] give the definition of fuzzy sample mode for interval-valued data as follows:

Let U be the universal set, $L = \{L_1, L_2, \ldots, L_k\}$ a set of k-linguistic variables on U, and $\{FS_i = [a_i, b_i], a_i, b_i \in R, i = 1, 2, \ldots, n\}$ be a sequence of random fuzzy sample on U. For each sample FS_i, if there is an interval $[c, d]$ which is covered by certain samples, we call these samples as a cluster. Let MS be the set of clusters which contains the maximum number of sample, then the fuzzy mode FM is defined by Eq. (17):

$$FM = [a, b] = \left\{ \bigcap [a_i, b_i] \mid [a_i, b_i] \subset MS \right\} \qquad (17)$$

If $[a, b]$ does not exist (i.e. $[a, b]$ is an empty set), it is said that the sample data do not have fuzzy mode.

5 Fuzzy Frequency Distribution

Obtaining a fuzzy frequency distribution will be explained by solving a numerical example. First, assume we have the data in Table 8.

Step 1. Determine the number of values in the data set and let be N.

Let it be N = 36 in Table 8.

Step 2. Determine the least and largest values in the data set to find the range.

Using a ranking method, the values in Table 8 are ranked. Assume that in Table 8, the least value is (15, 17, 19) and the largest value is (53, 56, 58).

Step 3. Calculate the range of the data.

R = (53, 56, 58) − (15, 17, 19) = (34, 39, 43)

Step 4. Calculate the number of classes, k. It can be calculated by using the equation $k = \sqrt{N} = \sqrt{36} = 6$.

Step 5. Calculate the class width, h using Eq. (18).

$$h \geq \frac{R}{k} \qquad (18)$$

$$h \geq \frac{(34, 39, 43)}{6} = (5.67, 6.5, 7.17)$$

Step 6. Establish the class limits.

To establish the class limits, Table 9 is used. We start by adding "h-1" to the least value in order to find the upper limit of the first class.

If you made it correctly, the largest value of the data must be within the class limits of the last class. In our case, the interval (43.35, 49.5, 54.85) and (48.02, 55, 61.02) involves (53, 56, 58), which indicates that the process is correct.

Step 7. Establish the class boundaries.

The class boundaries can be obtained by subtracting 0.5 from the class limits as in Table 10.

Step 8. Determine the class frequencies.

Table 8 The fuzzy data for frequency distribution

(34, 36, 39)	(41, 45, 52)	(21, 23, 26)
(30, 34, 35)	(15, 17, 19)	(53, 56, 58)
(36, 41, 44)	...	(43, 46, 48)

Table 9 Class limits

Class number	Lower limit	Upper limit
1	(15, 17, 19)	(19.67, 22.5, 25.17)
2	(20.67, 23.5, 26.17)	(25.34, 29, 32.34)
3	(26.34, 30, 33.34)	(31.01, 35.5, 39.51)
4	(32.01, 36.5, 40.51)	(36.68, 42, 46.68)
5	(37.68, 43, 47.68)	(42.35, 48.5, 53.85)
6	(43.35, 49.5, 54.85)	(48.02, 55, 61.02)

Table 10 Class boundaries

Class number	Lower class boundary (LCB)	Upper class boundary (UCB)
1	(14.5, 16.5, 18.5)	(20.17, 23, 25.67)
2	(20.17, 23.0, 25.67)	(25.84, 29.5, 32.84)
3	(25.84, 29.5, 32.84)	(31.51, 36, 40.01)
4	(31.51, 36, 40.01)	(37.18, 42.5, 47.18)
5	(37.18, 42.5, 47.18)	(42.85, 49, 54.35)
6	(42.85, 49, 54.35)	(48.52, 55.5, 61.52)

The values in Table 8 must be defuzzified in order to determine to which class they belong. For instance, (34, 36, 39) can be defuzzified by the formula $(a + 2b + c)/4$, and the result becomes 36.25. Applying the same formula to the class boundaries, we find that 36.25 belongs to the fourth class. Assuming that we made it for all of the values in Table 8, we obtained Table 11.

Table 11 Defuzzified boundaries and class frequencies

Class number	Defuzzified lower boundary	Defuzzified upper boundary	Class frequency
1	16.5	22.96	3
2	22.96	29.42	7
3	29.42	35.88	11
4	35.88	42.34	8
5	42.34	48.8	5
6	48.8	55.26	2

6   Fuzzy Graphics

6.1   *Fuzzy Stem-and-Leaf Display*

Classical stem and leaf diagram is converted its fuzzy case for the data in Table 8. Figure 7 illustrates the stem and leaf diagram for only 8 fuzzy numbers in Table 8. This figure shows us a left-skewed distribution.

6.2   *Fuzzy Histograms*

Figure 8 shows the fuzzy lower and upper class boundaries of a class. As it can be seen from Fig. 8, some data may belong to the previous or next class at the same time.

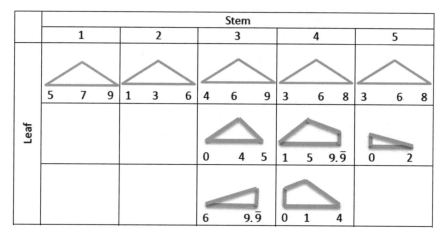

Fig. 7 Fuzzy stem and leaf diagram

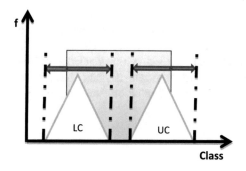

Fig. 8 Fuzzy class boundaries of a class

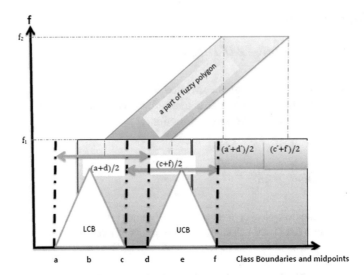

Fig. 9 A part of fuzzy polygon

6.3   *Fuzzy Frequency Polygons*

Fuzzy frequency polygon is illustrated for two classes in Fig. 9. Let LCB = (a, b, c) and UCB = (d, e, f). Since the class marks (class midpoints) may be between [(a + d)/2, (c + f)/2] for each class, the frequency polygon for two classes will be as in Fig. 9.

7   Conclusion

Central tendency measures are summary measures that attempt to describe a whole set of data with a single value that represents the middle or center of its distribution. When we have fuzzy data, classical central tendency measures should be transformed to their fuzzy cases. We proposed these fuzzy measures and gave an example for each measure. We also proposed fuzzy graphs for histogram, polygon, and stem and leaf diagram. For further research, the new extensions of fuzzy sets can be used to convert the classical central tendency measures to obtain hesitant fuzzy mean, intuitionistic fuzzy mean, hesitant fuzzy mode, intuitionistic fuzzy mode, hesitant fuzzy median, and intuitionistic fuzzy median.

References

1. Goodman, I.R.: Applications of random set representations of fuzzy sets to determining measures of central tendency. In: Proceedings of the Joint Conference on Information Sciences, Pinehurst, NC, United States, pp. 196–199 (1994)
2. Lee, E.S., Li, R.L.: Comparison of fuzzy numbers based on the probability measure of fuzzy events. Comput. Math Appl. **15**, 887–896 (1988)
3. Lu, W., Jiao, P.: Fuzzy statistical description and fuzzy nonparametric test for investing decision. In: Proceedings, 6th International Conference on Natural Computation, ICNC 2010, 10–12 August, pp. 308–311
4. Nguyen, H.T., Wu, B.: Fundamentals of statistics with fuzzy data, Studies in Fuzziness and Soft Computing, vol. 198. Springer, Berlin Heidelberg (2006)
5. Parvathi, R., Atanassova, V.: Intuitionistic fuzzy statistical tools for filters in image processing. In: 6th IEEE International Conference Intelligent Systems, IS 2012; Sofia; Bulgaria; 6–8 Sept, pp. 150–152 (2012)
7. Sinova, B., Gil, M.Á., Colubi, A., Van Aelst, S.: The median of a random fuzzy number. the 1-norm distance approach. Fuzzy Sets Syst. **200**, 99–115 (2012)
6. Sinova, B., Casals, M.R., Gil, M.Á.: Central tendency for symmetric random fuzzy numbers. Inf. Sci. **278**, 599–613 (2014)
8. Sun, C.-M., Wu, B.: New statistical approaches for fuzzy data. Int. J. Uncertainty Fuzz. Knowl. Based Syst. **15**(2), 89–106 (2007)
9. Viertl, R.: Statistical Methods for Fuzzy Data. Wiley (2011)

Fuzzy Dispersion Measures

İrem Uçal Sarı, Cengiz Kahraman and Özgür Kabak

Abstract Dispersion measures are very useful tools to measure the variability of data. Under uncertainty, the fuzzy set theory can be used to capture the vagueness in the data. This chapter develops the fuzzy versions of classical dispersion measures namely, standard deviation and variance, mean absolute deviation, coefficient of variation, range, and quartiles. Initially, we summarize the classical dispersion measures and then we develop their fuzzy versions for triangular fuzzy data. A numerical example for each fuzzy dispersion measure is given.

Keywords Dispersion measures · Fuzzy sets · Standard deviation · Variance · Absolute deviation · Coefficient of variation · Range · Quartiles

1 Introduction

Dispersion measures are used to describe the spread of the data, or its variation around a central value. Dispersion measures are important to understand two data sets which have the same mean or median, but entirely different levels of variability. There are various methods that can be used to measure the dispersion of a dataset, each with its own set of advantages and disadvantages.

The classical dispersion measures need exact data whereas we sometimes meet imprecise and vague data in real life problems. Especially some problems include categorical data which are represented by linguistic terms. For instance, classification of the people for whom we do not have their age data can be made by observing them and using linguistic terms such as very old, old, middle-aged, young, and very young. In such cases, the fuzzy set theory can capture the uncertainty included in these linguistic data.

İ.U. Sarı · C. Kahraman (✉) · Ö. Kabak
Industrial Engineering Department, Management Faculty, Istanbul Technical University, Istanbul, Turkey
e-mail: kahramanc@itu.edu.tr

© Springer International Publishing Switzerland 2016
C. Kahraman and Ö. Kabak (eds.), *Fuzzy Statistical Decision-Making*, Studies in Fuzziness and Soft Computing 343, DOI 10.1007/978-3-319-39014-7_6

85

In the literature there is not much research on fuzzy dispersion measures. DiCesare et al. [1] presented a new approach to the summarization of linguistic data and defined a measure of variation of the data, a fuzzy variance. They constructed a normalized measure of dispersion based on the fuzzy variance and the range of the fuzzy data. The concepts of variance and range are extended to the fuzzy sets theory and the values obtained are interpreted in a linguistic fashion. Spadoni and Stefanini [8] proposed an algorithm for the computation of the interval and fuzzy variance. Pizzi [6] described dispersion-adjusted fuzzy quartile encoding based preprocessing method for the classification of patterns. They defined fuzzy quantile and interquartile range to determine reasonable values for the fuzzy set boundaries. Kaya and Kahraman [3, 4] used fuzzy process mean, and fuzzy variance, which are obtained by using the fuzzy extension principle to add more information and flexibility to process capability indices. Tsao [9] proposed novel equations for computing the fuzzy variance and standard deviation by applying the fuzzy arithmetic with requisite constraints.

In this paper the classical dispersion measures are fuzzified by using triangular fuzzy data. The proposed equations can also be used for trapezoidal fuzzy data as well. The rest of the chapter is organized as follows: Sect. 2 presents the classical dispersion measures. Section 3 fuzzifies the dispersion measures and gives numerical examples. Section 4 concludes the chapter.

2   Classical Dispersion Measures

There are six frequently used measures of variability: standard deviation, variance, the range, mean absolute deviation, coefficient of variance and interquartile range.

2.1   Standard Deviation and Variance

The standard deviation and variance are more complete measures of dispersion which take into account every score in a distribution. Variance is the average squared difference of scores from the mean score of a distribution. Standard deviation is the square root of the variance. Two different calculations are used to determine standard population or form a sample of a larger population.

Variance of a population is the average squared distance of all measurements from the population mean. It is determined by σ^2 and calculated by using Eq. 1 (Soong [7]):

$$\sigma^2 = \frac{\sum_{i=1}^{N} (x_i - \mu)^2}{N} \tag{1}$$

where x_i represents the ith value in the data set, μ is the mean of the data set and N is the number of the number of the data in the set.

Standard deviation of a population is determined by σ and calculated by using Eq. 2:

$$\sigma = \sqrt{\frac{\sum_{i=1}^{N} (x_i - \mu)^2}{N}} \tag{2}$$

Variance of a sample is determined by s^2 and calculated by using Eq. 3:

$$s^2 = \frac{\sum_{i=1}^{n} (x_i - \bar{x})^2}{n - 1} \tag{3}$$

where x_i is the ith value in the sample set, $\bar{x}$ is the mean of the sample set and n is the number of the data in the sample set.

Standard deviation of a sample is determined by s and calculated by using Eq. 4:

$$s = \sqrt{\frac{\sum_{i=1}^{n} (x_i - \bar{x})^2}{n - 1}} \tag{4}$$

Example 1 There are 50 observations of execution times of an experiment are given in Table 1.

Sum of all the values in Table 1 is equal to 6723. The mean is calculated as 6723/50 = 134.46. By using Eq. 1 variance of the data set is calculated as 42.32 and by using Eq. 2 standard deviation of the data set is calculated as 6.50.

$$\sigma^2 = \frac{(141 - 134.46)^2 + (125 - 134.46)^2 + (135 - 134.46)^2 + \cdots + (122 - 134.46)^2}{50} = 42.32$$

$$\sigma = \sqrt{\frac{(141 - 134.46)^2 + (125 - 134.46)^2 + (135 - 134.46)^2 + \cdots + (122 - 134.46)^2}{50}} = 6.50$$

Table 1 The observations

141	137	144	128	137	127	125	143	128	132
125	125	139	129	137	136	139	123	127	141
135	130	140	123	140	135	131	139	128	139
139	138	140	143	137	137	139	140	142	134
123	142	124	141	131	143	138	131	136	122

2.2 *Mean Absolute Deviation*

The mean absolute deviation (MAD) of a data set is the average distance between each data value and the mean. It helps to understand whether the mean of the data set is useful or not. When the mean absolute deviation is large, the mean is not relevant mostly because of an outlier. MAD of a population is determined by Eq. 5 (Montgomery and Runger [5]):

$$MAD = \frac{1}{N} \sum_{i=1}^{N} |x_i - \mu| \qquad (5)$$

MAD of a sample is determined by Eq. 6:

$$MAD = \frac{1}{n} \sum_{i=1}^{n} |x_i - \bar{x}| \qquad (6)$$

Example 2 To calculate the mean absolute deviation of the data set given in Table 1, Eq. 5 is applied and MAD of the data set is calculated as 5.728.

$$MAD = \frac{1}{50} \left((|141 - 134.46|) + (|125 - 134.46|) + (|135 - 134.46|) \right.$$
$$\left. + \cdots + (|122 - 134.46|) \right) = 5.728$$

2.3 *Coefficient of Variation*

Coefficient of variation is the ratio of the standard deviation to the mean which shows the extent of variability in relation to the mean of the population. The coefficient of variation should be computed only for data which can take non-negative values. Coefficient of variation is denoted by *CV* and is determined by Eq. 6 for a population data set as follows:

$$CV = \frac{\sigma}{\mu} \qquad (7)$$

Coefficient of variation of a sample data set is determined by Eq. 8:

$$CV = \frac{s}{\bar{x}} \qquad (8)$$

Example 3 Coefficient of variation is calculated as 0.048 for the data set given in Table 1 by using Eq. 7.

$$CV = \frac{6.50}{134.46} = 0.048$$

2.4 Range

The range is the simplest measure of dispersion. It is the difference between the maximum and minimum values in the data set. Due to the fact that range is determined by the furthest outliers (extreme scores) it does not give information about the typical values of the data set. Range is denoted by R and determined by Eq. 9:

$$R = \max(x_i) - \min(x_i) \tag{9}$$

Example 4 Range of the data set given in Table 1 is found as 22.

$$R = 144 - 122 = 22.$$

2.5 The Interquartile Range

The interquartile range (IQR) is a measure of variability that indicates the extent to which the central 50 % of values within the data set are dispersed. It is based on dividing a data set into quartiles which are the values that divide a rank-ordered data set into four equal parts. First quartile (Q_1) which is also known as lower quartile is the middle value in the first half of the rank-ordered data set, second quartile (Q_2) is the median value in the data set and the third quartile (Q_3) which is also known as higher quartile is the middle value in the second half of the rank ordered data set. The interquartile range is the difference between the first and third quartiles which is obtained by using Eq. 10 [2]:

$$IQR = Q_3 - Q_1 \tag{10}$$

Since only the middle 50 % of the data affects this measure, it is robust to outliers.

Example 5 Interquartile range of the data set given in Table 1 is calculated as 11.5.

$$IQR = 139.75 - 128.25 = 11.5.$$

3 Dispersion Measures Under Fuzziness

3.1 Fuzzy Variance and Fuzzy Standard Deviation

When the data are ambiguous and they could be represented by fuzzy sets. In Table 2 a fuzzy data is represented:

Fuzzy variance for the sample can be calculated by Eq. 11:

$$\tilde{s}^2 = \frac{\sum_{i=1}^{n} (\tilde{x}_i - \tilde{\bar{x}})^2}{n-1} \tag{11}$$

Fuzzy standard deviation for the sample can be calculated by Eq. 12:

$$\tilde{s} = \sqrt{\frac{\sum_{i=1}^{n} (\tilde{x}_i - \tilde{\bar{x}})^2}{n-1}} \tag{12}$$

Assume $\tilde{x}_i$ and $\tilde{\bar{x}}$ values are represented by triangular fuzzy numbers in the form of $\tilde{x}_i = (x_{li}, x_{mi}, x_{ui})$ and $\tilde{\bar{x}} = (\bar{x}_l, \bar{x}_m, \bar{x}_u)$ where $x_{li} > 0$ and $\bar{x}_l > 0$. Then fuzzy variance formula for the sample data represented by triangular fuzzy numbers can be calculated by Eq. 13:

$$\tilde{s}^2 = \left(\frac{\sum_{i=1}^{n} (x_{li} - \bar{x}_l)^2}{n-1}, \frac{\sum_{i=1}^{n} (x_{mi} - \bar{x}_m)^2}{n-1}, \frac{\sum_{i=1}^{n} (x_{ui} - \bar{x}_u)^2}{n-1} \right) \tag{13}$$

For the independent any two fuzzy numbers the largest possible value of the second number is subtracted from the least possible value of the first number to get the first term of the subtraction, and similar operations for the other two terms. In Eq. 13, as it is seen the least possible values are subtracted in the first term; the most possible values in the second term and the largest possible values in the third term are subtracted since $\tilde{x}_i$ values produce $\tilde{\bar{x}}$ value; in other words $\tilde{\bar{x}}$ is dependent on $\tilde{x}_i$ values.

Fuzzy standard deviation formula for the sample data represented by triangular fuzzy numbers is given in Eq. 14:

$$\tilde{s} = \left(\sqrt{\frac{\sum_{i=1}^{n} (x_{li} - \bar{x}_l)^2}{n-1}}, \sqrt{\frac{\sum_{i=1}^{n} (x_{mi} - \bar{x}_m)^2}{n-1}}, \sqrt{\frac{\sum_{i=1}^{n} (x_{ui} - \bar{x}_u)^2}{n-1}} \right) \tag{14}$$

Table 2 Fuzzy data

$\tilde{x}_1$	$\tilde{x}_2$	$\ldots$	$\tilde{x}_k$
$\tilde{x}_{k+1}$	$\tilde{x}_{k+2}$	$\ldots$	$\tilde{x}_l$
$\tilde{x}_{l+1}$	$\tilde{x}_{l+2}$	$\ldots$	$\tilde{x}_n$

Now assume $\tilde{x}_i$ and $\tilde{\bar{x}}$ values are represented by trapezoidal fuzzy numbers in the form of $\tilde{x}_i = (x_{li}, x_{m1i}, x_{m2i}, x_{ui})$ and $\tilde{\bar{x}} = (\bar{x}_l, \bar{x}_{m1}, \bar{x}_{m2}, \bar{x}_u)$ where $x_{li} > 0$ and $\bar{x}_l > 0$. Then Eq. 11 becomes Eq. 15:

$$\tilde{s}^2 = \left(\frac{\sum_{i=1}^{n} (x_{li} - \bar{x}_l)^2}{n-1}, \frac{\sum_{i=1}^{n} (x_{m1} - \bar{x}_{m1})^2}{n-1}, \frac{\sum_{i=1}^{n} (x_{m2} - \bar{x}_{m2})^2}{n-1}, \frac{\sum_{i=1}^{n} (x_u - \bar{x}_u)^2}{n-1} \right)$$

(15)

Fuzzy standard deviation formula for the sample data represented by trapezoidal fuzzy numbers is given in Eq. 16:

$$\tilde{s} = \left(\sqrt{\frac{\sum_{i=1}^{n} (x_{li} - \bar{x}_l)^2}{n-1}}, \sqrt{\frac{\sum_{i=1}^{n} (x_{m1i} - \bar{x}_{m1})^2}{n-1}}, \sqrt{\frac{\sum_{i=1}^{n} (x_{m2i} - \bar{x}_{m2})^2}{n-1}}, \sqrt{\frac{\sum_{i=1}^{n} (x_{ui} - \bar{x}_u)^2}{n-1}} \right)$$

(16)

Fuzzy variance for the population can be calculated by Eq. 17:

$$\tilde{\sigma}^2 = \frac{\sum_{i=1}^{N} (\tilde{x}_i - \tilde{\mu})^2}{N}$$

(17)

Fuzzy standard deviation for the population can be calculated by Eq. 18:

$$\tilde{\sigma} = \sqrt{\frac{\sum_{i=1}^{N} (\tilde{x}_i - \tilde{\mu})^2}{N}}$$

(18)

Assume $\tilde{x}_i$ and $\tilde{\mu}$ values are represented by triangular fuzzy numbers in the form of $\tilde{x}_i = (x_{li}, x_{mi}, x_{ui})$ and $\tilde{\mu} = (\mu_l, \mu_m, \mu_u)$ where $x_{li} > 0$ and $\mu_l > 0 \, x_1 > 0$. Then Eq. 17 becomes

$$\tilde{\sigma}^2 = \frac{\sum_{i=1}^{N} (\tilde{x}_i - \tilde{\mu})^2}{N} = \left(\frac{\sum_{i=1}^{N} (x_{li} - \mu_l)^2}{N}, \frac{\sum_{i=1}^{N} (x_{mi} - \mu_m)^2}{N}, \frac{\sum_{i=1}^{N} (x_{ui} - \mu_u)^2}{N} \right)$$

(19)

Fuzzy standard deviation formula for the population data represented by triangular fuzzy numbers is given in Eq. 20:

$$\tilde{s} = \left(\sqrt{\frac{\sum_{i=1}^{n} (y_{li} - \mu_l)^2}{N}}, \sqrt{\frac{\sum_{i=1}^{n} (y_{mi} - \mu_m)^2}{N}}, \sqrt{\frac{\sum_{i=1}^{n} (y_{ui} - \mu_u)^2}{N}} \right)$$

(20)

Now assume $\tilde{x}_i$ and $\tilde{\mu}$ values are represented by trapezoidal fuzzy numbers in the form of $\tilde{x}_i = (x_{li}, x_{m1i}, x_{m2i}, x_{ui})$ and $\tilde{\mu} = (\mu_l, \mu_{m1}, \mu_{m2}, \mu_u)$ where $x_{li} > 0$ and $\mu_l > 0$. Then Eq. 17 becomes

$$\tilde{\sigma}^2 = \frac{\sum_{i=1}^{N} (\tilde{x}_i - \tilde{\mu})^2}{N}$$

$$= \left(\frac{\sum_{i=1}^{N} (x_{li} - \mu_l)^2}{N}, \frac{\sum_{i=1}^{N} (x_{m1i} - \mu_{m1})^2}{N}, \frac{\sum_{i=1}^{N} (x_{m2i} - \mu_{m2})^2}{N}, \frac{\sum_{i=1}^{N} (x_{ui} - \mu_u)^2}{N} \right)$$

$$(21)$$

Fuzzy standard deviation formula for the population data represented by trapezoidal fuzzy numbers is given in Eq. 22:

$$\tilde{\sigma} = \sqrt{\frac{\sum_{i=1}^{N} (\tilde{x}_i - \tilde{\mu})^2}{N}}$$

$$= \left(\sqrt{\frac{\sum_{i=1}^{N} (x_{li} - \mu_l)^2}{N}}, \sqrt{\frac{\sum_{i=1}^{N} (x_{li} - \mu_l)^2}{N}}, \sqrt{\frac{\sum_{i=1}^{N} (x_{li} - \mu_l)^2}{N}}, \sqrt{\frac{\sum_{i=1}^{N} (x_{li} - \mu_l)^2}{N}} \right)$$

$$(22)$$

Example 6 Consider the fuzzy data belonging to a sample in Table 3 and calculate the arithmetic mean.

We can apply a fuzzification percentage to obtain their corresponding fuzzy numbers in Table 3. For instance, if a 10 % fuzzification percentage is used, the triangular fuzzy numbers given in Table 4 are obtained:

Mean of the sample data given in Table 4 is calculated as (20.7, 23, 25.3). Variance of the sample data set is calculated by using Eq. 13 as (20.765, 25.636, 31.02). Standard deviation of the sample data set given in Table 4 is calculated by using Eq. 14 as (4.557, 5.063, 5.569).

Table 3 Fuzzy data

Around 20	Around 23	Around 20	Around 21
Around 28	Around 19	Around 22	Around 32
Around 14	Around 29	Around 21	Around 27

Table 4 10% fuzzification

(18, 20, 22)	(20.7, 23, 25.3)	(18, 20, 22)	(18.9, 21, 23.1)
(25.2, 28, 30.8)	(17.1, 19, 20.9)	(19.8, 22, 24.2)	(28.8, 32, 35.2)
(12.6, 14, 15.4)	(26.1, 29, 31.9)	(18.9, 21, 23.1)	(24.3, 27, 29.7)

3.2 Fuzzy Mean Absolute Deviation

Fuzzy mean absolute deviation of a population data is represented by $M\tilde{A}D$ and can be calculated by using Eq. 23:

$$M\tilde{A}D = \frac{1}{N}\sum_{i=1}^{N}|\tilde{x}_i - \tilde{\mu}| \tag{23}$$

Assume $\tilde{x}_i$ and $\tilde{\mu}$ values are represented by triangular fuzzy numbers in the form of $\tilde{x}_i = (x_{li}, x_{mi}, x_{ui})$ and $\tilde{\mu} = (\mu_l, \mu_m, \mu_u)$ where $x_{li} > 0$ and $\mu_l > 0$. Then Eq. 23 becomes

$$M\tilde{A}D = \frac{1}{N}\sum_{i=1}^{N}|\tilde{x}_i - \tilde{\mu}| = \left(\frac{\sum_{i=1}^{N}|x_{li} - \mu_l|}{N}, \frac{\sum_{i=1}^{N}|x_{mi} - \mu_m|}{N}, \frac{\sum_{i=1}^{N}|x_{ui} - \mu_u|}{N}\right) \tag{24}$$

Now assume $\tilde{x}_i$ and $\tilde{\mu}$ values are represented by trapezoidal fuzzy numbers in the form of $\tilde{x}_i = (x_{li}, x_{m1i}, x_{m2i}, x_{ui})$ and $\tilde{\mu} = (\mu_l, \mu_{m1}, \mu_{m2}, \mu_u)$ where $y_{li} > 0$ and $\mu_l > 0$. Then Eq. 23 becomes

$$\begin{aligned} M\tilde{A}D &= \frac{1}{N}\sum_{i=1}^{N}|\tilde{x}_i - \tilde{\mu}| \\ &= \left(\frac{\sum_{i=1}^{N}|x_{li} - \mu_l|}{N}, \frac{\sum_{i=1}^{N}|x_{m1i} - \mu_{m1}|}{N}, \frac{\sum_{i=1}^{N}|x_{m2i} - \mu_{m2}|}{N}, \frac{\sum_{i=1}^{N}|x_{ui} - \mu_u|}{N}\right) \end{aligned} \tag{25}$$

$M\tilde{A}D$ of a sample is determined by Eq. 26:

$$M\tilde{A}D = \frac{1}{n}\sum_{i=1}^{n}|\tilde{x}_i - \tilde{\bar{x}}| \tag{26}$$

Assume $\tilde{x}_i$ and $\bar{x}$ values are represented by triangular fuzzy numbers in the form of $\tilde{x}_i = (x_{li}, x_{mi}, x_{ui})$ and $\tilde{\bar{x}} = (\bar{x}_l, \bar{x}_m, \bar{x}_u)$ where $x_{li} > 0$ and $\bar{x}_l > 0$. Then fuzzy mean absolute devision formula for the sample data represented by triangular fuzzy numbers can be calculated by Eq. 27:

$$M\tilde{A}D = \frac{1}{n}\sum_{i=1}^{n}|\tilde{x}_i - \tilde{\bar{x}}| = \left(\frac{\sum_{i=1}^{n}|x_{li} - \bar{x}_l|}{n}, \frac{\sum_{i=1}^{n}|x_{mi} - \bar{x}_m|}{n}, \frac{\sum_{i=1}^{n}|x_{ui} - \bar{x}_u|}{n}\right) \tag{27}$$

Now assume $\tilde{x}_i$ and $\bar{x}$ values are represented by trapezoidal fuzzy numbers in the form of $\tilde{x}_i = (x_{li}, x_{m1i}, x_{m2i}, x_{ui})$ and $\bar{x} = (\bar{x}_l, \bar{x}_{m1}, \bar{x}_{m2}, \bar{x}_u)$ where $x_{li} > 0$ and $\bar{x}_l > 0$. Then Eq. 26 becomes Eq. 28:

$$\tilde{MAD} = \frac{1}{n}\sum_{i=1}^{n}|\tilde{x}_i - \tilde{\bar{x}}|$$
$$= \left(\frac{\sum_{i=1}^{n}|x_{li} - \bar{x}_l|}{n}, \frac{\sum_{i=1}^{n}|x_{m1i} - \bar{x}_{m1}|}{n}, \frac{\sum_{i=1}^{n}|x_{m2i} - \bar{x}_{m2}|}{n}, \frac{\sum_{i=1}^{n}|x_{ui} - \bar{x}_u|}{n}\right)$$
(28)

Example 7 To calculate the fuzzy mean absolute deviation of the sample data set given in Table 4, Eq. 27 is applied and $\tilde{MAD}$ of the data set is calculated as (3.6,4,4.4).

$$\tilde{MAD} = \left(\frac{43.2}{12}, \frac{48}{12}, \frac{52.8}{12}\right) = (3.6, 4, 4.4)$$

3.3 Fuzzy Coefficient of Variation

Fuzzy coefficient of variation is the ratio of the fuzzy standard deviation to the fuzzy mean which shows the extent of variability in relation to the fuzzy mean of the population. Fuzzy coefficient of variation is denoted by $\tilde{CV}$ and is determined by Eq. 29 for a population data set:

$$\tilde{CV} = \frac{\tilde{\sigma}}{\tilde{\mu}}$$
(29)

Assume $\tilde{\sigma}$ and $\tilde{\mu}$ values are represented by triangular fuzzy numbers in the form of $\tilde{\sigma} = (\sigma_l, \sigma_m, \sigma_u)$ and $\tilde{\mu} = (\mu_l, \mu_m, \mu_u)$ where $\sigma_l > 0$ and $\mu_l > 0$. $\tilde{CV}$ of a population data represented by triangular fuzzy numbers is given in Eq. 30:

$$\tilde{CV} = \frac{\tilde{\sigma}}{\tilde{\mu}} = \left(\frac{\sigma_l}{\mu_l}, \frac{\sigma_m}{\mu_m}, \frac{\sigma_u}{\mu_u}\right)$$
(30)

Now assume $\tilde{\sigma}$ and $\tilde{\mu}$ values are represented by trapezoidal fuzzy numbers in the form of $\tilde{\sigma} = (\sigma_l, \sigma_{m1}, \sigma_{m2}, \sigma_u)$ and $\tilde{\mu} = (\mu_l, \mu_{m1}, \mu_{m2}, \mu_u)$ where $\sigma_l > 0$ and $\mu_l > 0$. $\tilde{CV}$ of a population data represented by trapezoidal fuzzy numbers is given in Eq. 31:

$$\tilde{CV} = \frac{\tilde{\sigma}}{\tilde{\mu}} = \left(\frac{\sigma_l}{\mu_l}, \frac{\sigma_{m1}}{\mu_{m1}}, \frac{\sigma_{m2}}{\mu_{m2}}, \frac{\sigma_u}{\mu_u}\right)$$
(31)

$C\tilde{V}$ is determined by Eq. 32 for a sample data set:

$$C\tilde{V} = \frac{\tilde{s}}{\tilde{\bar{x}}} \tag{32}$$

Assume $\tilde{s}$ and $\tilde{\bar{x}}$ values are represented by triangular fuzzy numbers in the form of $\tilde{s} = (s_l, s_m, s_u)$ and $\tilde{\bar{x}} = (\bar{x}_l, \bar{x}_m, \bar{x}_u)$ where $s_l > 0$ and $\bar{x}_l > 0$. $C\tilde{V}$ of a sample data represented by triangular fuzzy numbers is given in Eq. 33:

$$C\tilde{V} = \frac{\tilde{s}}{\tilde{\bar{x}}} = \left(\frac{s_l}{\bar{x}_l}, \frac{s_m}{\bar{x}_m}, \frac{s_u}{\bar{x}_u}\right) \tag{33}$$

Now assume $\tilde{s}$ and $\tilde{\bar{x}}$ values are represented by trapezoidal fuzzy numbers in the form of $\tilde{s} = (s_l, s_{m1}, s_{m2}, s_u)$ an $\bar{x} = (\bar{x}_l, \bar{x}_{m1}, \bar{x}_{m2}, \bar{x}_u)$ where $s_l > 0$ and $\bar{x}_l > 0$. $C\tilde{V}$ of a sample data represented by trapezoidal fuzzy numbers is given in Eq. 34:

$$C\tilde{V} = \frac{\tilde{s}}{\tilde{\bar{x}}} = \left(\frac{s_l}{\bar{x}_l}, \frac{s_{m1}}{\bar{x}_{m2}}, \frac{s_{m1}}{\bar{x}_{m2}}, \frac{s_u}{\bar{x}_u}\right) \tag{34}$$

Example 8 Coefficient of variation is calculated as (4.5424, 4.5427, 4.5430) for the data set given in Table 4 by using Eq. 33:

$$C\tilde{V} = \frac{\tilde{s}}{\tilde{\bar{x}}} = \left(\frac{20.7}{4.556}, \frac{23}{5.063}, \frac{25.3}{5.569}\right) = (4.5424, 4.5427, 4.5430)$$

3.4 Fuzzy Range

In a fuzzy data set, sometimes it is hard to define the maximum and the minimum values of the data. When the maximum and the minimum fuzzy numbers of a data set cannot be determined easily, first ranking methods should be used to find out the maximum and the minimum values. Then by applying Eq. 35, fuzzy range which is denoted by $\tilde{R}$ can be determined:

$$\tilde{R} = \max(\tilde{x}_i) - \min(\tilde{x}_i) \tag{35}$$

Assume $\tilde{x}_i$ values are represented by triangular fuzzy numbers in the form of $\tilde{x}_i = (x_{il}, x_{im}, x_{iu})$ where $x_{il} > 0$. $\tilde{R}$ of a data set represented by triangular fuzzy numbers is given in Eq. 36:

$$\tilde{R} = \left((x_{\max, l} - x_{\min, u}), (x_{\max, m} - x_{\min, m}), (x_{\max, u} - x_{\min, l})\right) \tag{36}$$

where $\max(\tilde{x}_i) = (x_{\max, l}, x_{\max, m}, x_{\max, u})$ and $\min(\tilde{x}_i) = (x_{\min, l}, x_{\min, m}, x_{\min, u})$ providing that $x_{\max, l} - x_{\min, u} \geq 0$. Otherwise; it is not certain that there are clear maximum and minimum values in the data set. In that case range could be calculated by using defuzzified values of the data set. Then maximum and minimum values can be easily determined.

Assume $\tilde{x}_i$ values are represented by trapezoidal fuzzy numbers in the form of $\tilde{x}_i = (x_{il}, x_{im1}, x_{im2}, x_{iu})$ where $x_{il} > 0$. $\tilde{R}$ of a data set represented by trapezoidal fuzzy numbers is given in Eq. 37:

$$\tilde{R} = ((x_{\max, l} - x_{\min, u}), (x_{\max, m1} - x_{\min, m2}), (x_{\max, m2} - x_{\min, m1}), (x_{\max, u} - x_{\min, l}))$$
(37)

where $\max(\tilde{x}_i) = (x_{\max, l}, x_{\max, m1}, x_{\max, m2}, x_{\max, u})$ and $\min(\tilde{x}_i) = (x_{\min, l}, x_{\min, m1}, x_{\min, m2}, x_{\min, u})$ providing that $x_{\max, l} - x_{\min, u} \geq 0$.

Example 9 In the data set given in Table 4, it is obvious that the maximum value of the data is (28.8, 32, 35.2) and the minimum value of the data is (12.6, 14, 15.4). By using Eq. 36 fuzzy range of the data set is calculated as (13.4, 18, 22.6).

$$\tilde{R} = (28.8, 32, 35.2) - (12.6, 14, 15.4) = (13.4, 18, 22.6)$$

3.5 Fuzzy Interquartile Range

The formula of fuzzy interquartile range $(I\tilde{Q}R)$ is given in Eq. 38

$$I\tilde{Q}R = \tilde{Q}_3 - \tilde{Q}_1$$
(38)

where $\tilde{Q}_1$ is the middle value in the first half of the rank ordered fuzzy data set which is the $((n + 1)/4)$th element of the data set and $\tilde{Q}_3$ is the middle value in the second half of the rank ordered fuzzy data set which is the $(3(n + 1)/4)$th element of the data set.

Assume $\tilde{Q}_i$ values are represented by triangular fuzzy numbers in the form of $\tilde{Q}_i = (Q_{il}, Q_{im}, Q_{iu})$ where $Q_{il} > 0$. $I\tilde{Q}R$ of a data set represented by triangular fuzzy numbers is given in Eq. 39 where $Q_{3l} - Q_{1u} \geq 0$:

$$I\tilde{Q}R = ((Q_{3l} - Q_{1u}), (Q_{3m} - Q_{1m}), (Q_{3u} - Q_{1l}))$$
(39)

Now assume $\tilde{Q}_i$ values are represented by trapezoidal fuzzy numbers in the form of $\tilde{Q}_i = (Q_{il}, Q_{im1}, Q_{im2}, Q_{iu})$ where $Q_{il} > 0$. $I\tilde{Q}R$ of a data set represented by trapezoidal fuzzy numbers is given in Eq. 40 where $Q_{3l} - Q_{1u} \geq 0$:

$$I\tilde{Q}R = ((Q_{3l} - Q_{1u}), (Q_{3m1} - Q_{1m2}), (Q_{3m2} - Q_{1m1}), (Q_{3u} - Q_{1l})) \qquad (40)$$

Example 10 There are lots of ranking methods which could be used to rank fuzzy numbers. In this example it is preferred to rank fuzzy numbers by using their most expected values. The fuzzy numbers in the data set given in Table 4 are ranked as follows:

(12.6,14,15.4), (17.1, 19, 20.9), (18, 20, 22), (18, 20, 22), (18.9, 21, 23.1), (18.9, 21, 23.1), (19.8, 22, 24.2), (20.7, 23, 25.3), (24.3, 27, 29.7), (25.2, 28, 30.8), (26.1, 29, 31.9), (28.8, 32, 35.2)

There are 12 numbers in the data set which means the first quartile is the $(12 + 1)/4 = 3.25$th number in the ranked ordered fuzzy data set. The first quartile is obtained as follows:

$$\tilde{Q}_1 = (18, 20, 22) + (3.25 - 3)((18, 20, 22) - (18, 20, 22)) = (18, 20, 22)$$

The third quartile is obtained as follows:

$$\tilde{Q}_3 = (24.3, 27, 29.7) + (3.25x3 - 9)((25.2, 28, 30.8) - (24.3, 27, 29.7))$$
$$= (20.925, 27.75, 34.575)$$

By applying Eq. 39 $I\tilde{Q}R$ is calculated as $(-1.075, 7.75, 16.575)$ which means there is no clear third quartile and first quartile in the data set.

$$I\tilde{Q}R = \tilde{Q}_3 - \tilde{Q}_1 = (20.925, 27.75, 34.575) - (18, 20, 22)$$
$$= (-1.075, 7.75, 16.575)$$

In this case, $I\tilde{Q}R$ can be calculated by defuzzifying $\tilde{Q}_1$ and $\tilde{Q}_3$. In this example, center of area defuzzification method is used. This technique can be expressed as shown in Eq. 41:

$$x^* = \frac{\int \mu_i(x)x\,dx}{\int \mu_i(x)\,dx} \qquad (41)$$

By using center of area defuzzification method *IQR* is calculated as 7.75:

$$Q_1^* = \frac{\int_{18}^{20} \left(\frac{x-18}{2}\right) x dx + \int_{20}^{22} \left(\frac{22-x}{2}\right) x dx}{\int_{18}^{20} \left(\frac{x-18}{2}\right) dx + \int_{20}^{22} \left(\frac{22-x}{2}\right) dx} = \frac{40}{2} = 20$$

$$Q_3^* = \frac{\int_{20.925}^{27.75} \left(\frac{x-20.925}{2}\right) x dx + \int_{27.75}^{34.575} \left(\frac{34.575-x}{2}\right) x dx}{\int_{20.925}^{27.75} \left(\frac{x-20.925}{2}\right) dx + \int_{27.75}^{34.575} \left(\frac{34.575-x}{2}\right) dx} = \frac{646.306}{23.290} = 27.75$$

$$IQR^* = Q_3^* - Q_1^* = 27.75 - 20 = 7.75$$

4 Conclusion

Dispersion measures are important concepts to measure the variability of data. The classical approaches cannot deal with vague and imprecise data. Therefore; in this chapter, we have introduced corresponding fuzzy concepts of variance, standard deviation, mean absolute deviation, coefficient of variation, range, and quartiles.

The resulting fuzzy dispersion measures should be defuzzified for comparison purposes. Since different defuzzification methods might give different defuzzified values, interpretation of the fuzzy result should be based on more than one defuzzification method. For instance; fuzzy coefficient of variations of two samples from two populations can be compared based on the defuzzified values which may depend on the selected defuzzification method.

For further research the new extensions of fuzzy sets such as intuitionistic, type-2, or hesitant fuzzy sets can be used in developing the fuzzy dispersion measures. Intuitionistic fuzzy variance, hesitant fuzzy variance, and type-2 fuzzy variance have not yet been introduced in the literature.

References

1. DiCesare, F., Sahnoun, Z., Bonissone, P.P.: Linguistic summarization of fuzzy data. Inf. Sci. **52** (2), 141–152 (1990)
2. Howell, D.: Fundamental Statistics for the Behavioral Sciences. Cengage Learning (2013)
3. Kaya, İ., Kahraman, C.: Fuzzy process capability indices with asymmetric tolerances. Expert Syst. Appl. **38**(12), 14882–14890 (2011)
4. Kaya, İ., Kahraman, C.: Process capability analyses with fuzzy parameters. Expert Syst. Appl. **38**(9), 11918–11927 (2011)
5. Montgomery, D.C., Runger, G.C.: Applied Statistics and Probability for Engineers. Wiley (2010)
6. Pizzi, N.J.: Fuzzy quartile encoding as a preprocessing method for biomedical pattern classification. Theoret. Comput. Sci. **412**(42), 5909–5925 (2011)
7. Soong, T.T.: Fundamentals of Probability and Statistics for Engineers. Wiley (2004)

8. Spadoni, M., Stefanini, L.: Computing the variance of interval and fuzzy data. Fuzzy Sets Syst. **165**(1), 24–36 (2011)
9. Tsao, C.-T.: Fuzzy net present values for capital investments in an uncertain environment. Comput. Oper. Res. **39**(8), 1885–1892 (2012)

Sufficiency, Completeness, and Unbiasedness Based on Fuzzy Sample Space

Mohsen Arefi and S. Mahmoud Taheri

Abstract A new approach is introduced to the estimation of a parameter in the statistical models, based on fuzzy sample space. Two basic concepts of the point estimation theory, i.e. sufficiency and completeness, are extended to the fuzzy data case. Then, the unbiased estimator and the UMVU estimator are defined for such situations. The properties of these estimators are investigated, and some procedures are provided to obtain the UMVU estimators, based on fuzzy data.

Keywords Completeness · Fuzzy random sample · Point estimation · Sufficiency · Unbiasedness · Uniformly minimum variance unbiased estimator (UMVUE)

1 Introduction and Motivation

An important topic in parametric statistical inference is the point estimation of an unknown parameter of the model of interest. In classical approaches to the problem of point estimation, some essential assumptions are imposed on the underlying model. Specially, it is often assumed that the available data derived from the model are observed to be exact quantities. But, in real world problems, there are many situations in which the available data are imprecise (fuzzy) rather than crisp.

For instance, in survival analysis, in the study to estimate the proportion of a population who are infected with a certain virus, we may not able to identify the presence or absence of the virus in each member of the population, exactly. In this case, we may instead derive some vague inference about the situation, represented

M. Arefi
Department of Statistics, Faculty of Mathematical Sciences and Statistics,
University of Birjand, Birjand, Iran
e-mail: Arefi@Birjand.ac.ir

S. Mahmoud Taheri (✉)
Faculty of Engineering Science, College of Engineering,
University of Tehran, Tehran, Iran
e-mail: sm_taheri@ut.ac.ir

© Springer International Publishing Switzerland 2016
C. Kahraman and Ö. Kabak (eds.), *Fuzzy Statistical Decision-Making*,
Studies in Fuzziness and Soft Computing 343,
DOI 10.1007/978-3-319-39014-7_7

101

by linguistic terms such as: *it is quite certain that he/she has virus, we are 0.50 certain that he/she has the virus*, and so on. As another example, in social studies we may not able to measure the public opinion on a certain subject, precisely. People usually reflect their opinions on a social subject in non-exact (fuzzy) terms, such as: *I more or less agree, I completely agree, I cannot fully agree*, and the like. We can provide a list of such real world examples in which it is actually non-precise data that must be handled.

Since such non-precise data typically do not lend themselves to analysis by the classical statistical methods, we, therefore, need to develop new methods of statistical analysis based on non-precise data. The fuzzy set theory provides the necessary tools to extend the classical statistical methods to the analysis of non-precise (fuzzy) data.

The main contribution of this research is to investigate the problem of point estimation when the data available are fuzzy rather than crisp. In this regard, we extend the basic concepts of point estimation such as: sufficiency, completeness, and unbiasedness to the case the data available are fuzzy. Then, to investigate the properties of the introduced concepts, we develop the concept of UMVU estimators based on fuzzy data.

The organization of this paper is as follows. In Sect. 2, a literature survey related to sampling theory under fuzziness is presented. In Sect. 3, we recall some preliminaries concerning fuzzy sample space and fuzzy random sample. In Sect. 4, we extend the concepts of sufficiency and completeness based on fuzzy data. In Sect. 5, the unbiased estimators based on fuzzy data are studied and, as a special case, the UMVU estimators for such a data are investigated. Some examples to find the proposed estimators are presented in Sect. 6. Section 7 concludes the paper.

2 Literature Survey

Over the past decades, there have been a lot of attempts to investigate the statistical methods with fuzzy data. But, as far as the authors know, there have been only a few works on the topic of sampling theory and point estimation under fuzziness. In this section, we briefly review some important works in these topics.

Let us, first, review some works related to the sampling theory in fuzzy environment. Kruse [34] and Miyakoshi and Shimbo [40] investigated the strong law of large numbers for the sequences of independent and identically distributed fuzzy random variables. Kelement et al. [31] proved a strong law of large numbers and a central limit theorem for independent and identically distributed fuzzy random variables, whose values are fuzzy sets with compact levels. Boswell and Taylor [8] stated and proved a central limit theorem for fuzzy random variables. Guangyuan and Zhong [24] defined two types of generalized metrics in the set of all fuzzy random variables on a probability space. Then, they studied several different kinds of convergence for sequences of fuzzy random variables with respect to the introduced metrics. They also investigated some relations of these kinds of convergence. Using some limit properties of fuzzy numbers, Wu [51] extended the weak and strong convergence of fuzzy distribution functions. He also investigated a central limit theorem for fuzzy random variables based on a notion of fuzzy normal distribution. He studied the law of large numbers for fuzzy random variables, too [52]. Joo and Kim [28] obtained the Kolmogorov's strong law of large numbers for sums of independent and level-wise identically distributed fuzzy random variables, (see also Kim [30] and Joo [27]). Feng [18] proposed a formulation of strong and weak laws of large numbers for fuzzy random variables based on the concept of variance of fuzzy random variables. A strong law of large numbers, a central limit theorem and a Gliwenko-Cantelli theorem for fuzzy random variables are proved by Kratschmer [32] based on the L_p-metrics on the fuzzy sample spaces (see also [33]). Li and Ogura [37] presented the strong laws of large numbers for independent (not necessary identically distributed) fuzzy set-valued random variables whose base space is a separable Banach space or an Euclidean space, in the sense of an extended Hausdorff metric. Strong laws of large numbers for t-norm-based addition of fuzzy random variables are studied by Teran [47] and Hong [26]. Guan and Li [23] proved a strong law of large numbers for exchangeable random variables with respect to nonadditive measures and based on the relationship between set-valued random variables. See also Colubi [15], Joo et al. [29], and Fu and Zhang [19] for some limit theorems for fuzzy random sets. It should be mentioned that, beside the above works, a few works have been done by researchers on sampling methods under fuzziness. For instance, Garcia et al. [20] considered the problem of estimating the expected value of a fuzzy-valued random element in the stratified random sampling from finite populations. Lin and Lee [38, 39] introduced a fuzzy sense of sampling to express the degree of interviewee's feelings in sampling survey via questionnaire.

Now, we review briefly some important works on the point estimation in fuzzy environment. Kruse [35] studied a method to obtain a statistical estimation with linguistic data. Kruse and Meyer [36] investigated both consistent estimators and unbiased estimators using set representation techniques. Gil et al. [21] and Schnatter [43] studied some fuzzy Bayes estimators for real-valued parameters. Gil et al. [21] developed their methods by borrowing the concept of probability measures of fuzzy events from Zadeh [56], but Schnatter [43] focused on the different combination rules for fuzzy data. Cai [11] studied parameter estimation based on normal fuzzy variables. Yao and Hwang [55] proposed the concept of the sufficient statistic and the unbiased estimator based on a sample of fuzzy random variables. Hong [25] extended Cai's approach to the study of parameter estimations of non-normal fuzzy variables. Viertl [49], using the extension principle, investigated some methods to obtain a point estimation based on fuzzy data. Wu [53] developed a procedure for fuzzy estimating of fuzzy parameters based on fuzzy random variables. Buckley [9, 10] considered the confidence interval as the α-cuts of a triangular shaped fuzzy estimation for the parameter of interest (see also, Falsafain and Taheri [16] for an improved method on the Buckley's approach). An approach for obtaining the UMVU estimators based on the α-cuts of fuzzy observations is studied by Akbari and Rezaei [1]. Parchami and Mashinchi [42] applied the Buckley's approach to find fuzzy estimates of the process capability indices. Also, Falsafain et al. [17] studied a method to find the explicit formula for membership functions of the fuzzy estimations in statistical models by developing Buckley's approach. Taheri and Arefi [46] used the estimation introduced by Buckley to extend an approach for testing fuzzy hypotheses. Arefi and Taheri [5] investigated some aspects of a Bayesian approach to the estimation problem for fuzzy data. For more on statistics in imprecise environments, the reader is referred to the relevant literature, e.g. Taheri [45], Viertl [50], and Blanco-Fernandez et al. [7].

In sequel, we introduce and investigate a new approach to the theory of point estimation for fuzzy data, in which, for the first time, some basic concepts of the point estimation are investigated.

3 Preliminaries

Assume that $(\Omega, \mathcal{A})$ is a measurable space, in which Ω is a sample space, and $(\Omega, \mathcal{A}, \mathcal{P})$ is a probability space, where P is a probability measure on $(\Omega, \mathcal{A})$. In the following, we recall two definitions from Casals et al. [12] and Torabi et al. [48], but in a slightly different way.

Definition 1 A fuzzy sample space (associated with the probability space $(\Omega, \mathcal{A}, \mathcal{P})$), denoted by $\tilde{\chi} = \{\tilde{x}_1, \ldots, \tilde{x}_k\}$, is a collection of fuzzy sets $\tilde{x}_i, i = 1, 2, \ldots, k$ of Ω whose membership functions are Borel measurable and satisfy the orthogonality constraint: $\sum_{i=1}^{k} \tilde{x}_i(x) = 1, \forall x \in \Omega$.

Remark 1 The orthogonality constraint in Definition 1 is not a strong condition. If $\tilde{x}_1, \ldots, \tilde{x}_k$ are some fuzzy numbers such that they do not satisfy the orthogonality constraint, we can render a translation on $\tilde{x}_i$'s, so that they obey the orthogonality constraint. See the following example.

Example 1 Let $(\Omega, \mathcal{A}, \mathcal{P})$ be a probability space with $\Omega = [0, 1]$. Consider the membership functions of fuzzy sets as follows:

$$\tilde{x}_1(x) = 1 - x, \quad 0 \leq x \leq 1,$$

$$\tilde{x}_2(x) = \begin{cases} 2x & 0 \leq x < 0.5, \\ 2(1 - x) & 0.5 \leq x \leq 1, \end{cases}$$

$$\tilde{x}_3(x) = x, \quad 0 \leq x \leq 1.$$

The above fuzzy numbers do not obey the orthogonality constraint, since

$$\sum_{i=1}^{3} \tilde{x}_i(x) = \begin{cases} 1 + 2x & 0 \leq x < 0.5, \\ 3 - 2x & 0.5 \leq x \leq 1. \end{cases}$$

But, we can obtain a fuzzy sample space based on $\tilde{x}_1$, $\tilde{x}_2$, and $\tilde{x}_3$, denoted by $\tilde{\chi} = \{\tilde{x}_1^*, \tilde{x}_2^*, \tilde{x}_3^*\}$, as follows

$$\tilde{x}_1^*(x) = \begin{cases} \frac{1-x}{1+2x} & 0 \le x < 0.5, \\ \frac{1-x}{3-2x} & 0.5 \le x \le 1, \end{cases}$$

$$\tilde{x}_2^*(x) = \begin{cases} \frac{2x}{1+2x} & 0 \le x < 0.5, \\ \frac{2(1-x)}{3-2x} & 0.5 \le x \le 1, \end{cases}$$

$$\tilde{x}_3^*(x) = \begin{cases} \frac{x}{1+2x} & 0 \le x < 0.5, \\ \frac{x}{3-2x} & 0.5 \le x \le 1. \end{cases}$$

Remark 2 In this paper, we assume that $(\Omega, \mathcal{A})$ and $(\tilde{\chi}, \mathcal{F})$ are two measurable spaces, where $\tilde{\chi}$ is the fuzzy sample space on Ω, and $\mathcal{F}$ is a σ-field on $\tilde{\chi}$. Also, $\tilde{\chi}^n = \tilde{\chi} \times \cdots \times \tilde{\chi}$ is the Cartesian product on the fuzzy sample spaces, and $\mathcal{F}^n = \mathcal{F} \otimes \cdots \otimes \mathcal{F}$ *is a σ-field on $\tilde{\chi}^n$.*

Definition 2 Let $(\Omega, \mathcal{A}, \mathcal{P})$ be a probability space. A fuzzy random sample (of size n) based on the original random variable X with PDF $f(.)$, denoted by $\tilde{\mathbf{X}} = (\tilde{X}_1, \ldots, \tilde{X}_n)$, is a measurable function from $(\Omega, \mathcal{A})$ to $(\tilde{\chi}^n, \mathcal{F}^n)$, whose probability (following Zadeh's probability 1968) is given by

$$h(\tilde{\mathbf{x}}) = h(\tilde{x}_{r_1}, \ldots, \tilde{x}_{r_n}) = P(\tilde{\mathbf{X}} = \tilde{\mathbf{x}})$$

$$= \int_{\chi} \cdots \int_{\chi} \prod_{i=1}^{n} \tilde{x}_{r_i}(x_i) f(x_i) dv(x_1, \ldots, x_n),$$

where $\tilde{\mathbf{x}} = (\tilde{x}_{r_1}, \ldots, \tilde{x}_{r_n})$, $\tilde{x}_{r_i} \in \tilde{\chi}$ and $f(.)$ is the Radon-Nikodym derivative of P with respect to v (a σ-finite measure). The measure v usually is "counting measure" or "Lebesgue measure", and $\chi = \{x \in R \,|\, f(x) > 0\}$ is the support of X.

Note that, using Fubini's theorem ([6], pp. 233–234), we obtain

$$h(\tilde{x}_{r_1}, \ldots, \tilde{x}_{r_n}) = h(\tilde{x}_{r_1}) \cdots h(\tilde{x}_{r_n}), \quad \forall \tilde{x}_{r_i} \in \tilde{\chi},$$

where

$$h(\tilde{x}_{r_i}) = \int_{\chi} \tilde{x}_{r_i}(x_i) f(x_i) dv(x_i), \quad i = 1, \ldots, n.$$

The $h(\tilde{x}_{r_i})$ forms a PDF on $\tilde{\chi}$, since based on the orthogonality of the $\tilde{x}_{r_i}$, we have

$$
\sum_{\tilde{x}_{r_i} \in \tilde{\chi}} h(\tilde{x}_{r_i}) = \sum_{\tilde{x}_{r_i} \in \tilde{\chi}} \int_{\chi} \tilde{x}_{r_i}(x_i) f(x_i) dv(x_i)
$$

$$
= \int_{\chi} f(x_i) \sum_{\tilde{x}_{r_i} \in \tilde{\chi}} \tilde{x}_{r_i}(x_i) dv(x_i)
$$

$$
= \int_{\chi} f(x_i dv)(x_i) = 1.
$$

Theorem 1 [48] *Let $(\Omega, \mathcal{A}, \mathcal{P})$ be a probability space and $\tilde{\mathbf{X}}$ be a fuzzy random sample related to the fuzzy sample space $\tilde{\chi}$. If g is a measurable function from $\tilde{\chi}^n$ to R, then $Y = g(\tilde{\mathbf{X}})$ is a crisp (non-fuzzy) random variable.*

Suppose that we have a fuzzy sample space $\tilde{\chi}$ based on the original random variable X (with the probability density function or probability mass function $f_X(x; \theta)$, $\theta \in \Theta$), and that by taking a random sample of size n, we obtain the fuzzy random sample $\tilde{\mathbf{X}} = (\tilde{X}_1, \ldots, \tilde{X}_n)$. In the following, using such a sample data, we provide an approach to point estimation for the unknown parameter θ.

4 Sufficiency and Completeness

In this section, we extend the concepts of sufficiency and completeness to the case the available data are fuzzy rather than crisp.

Definition 3 Let $\tilde{\mathbf{X}}$ be a fuzzy random sample, and g be a measurable function from $\tilde{\chi}^n$ to R. The function $g(\tilde{\mathbf{X}})$ is called a **statistic** if $g(\tilde{\mathbf{X}})$ does not depend on θ.

Note that, when $g(\tilde{\mathbf{X}})$ is applied to estimating the unknown parameter θ, it is commonly called the **estimator** of θ.

Definition 4 Let $\tilde{\mathbf{X}} = (\tilde{X}_1, \ldots, \tilde{X}_n)$ be a fuzzy random sample related to the fuzzy sample space $\tilde{\chi}$. Suppose that

$$
\tilde{A}_1 = (\tilde{X}_{r_1}, \ldots, \tilde{X}_{r_{m_1}}) \subset \tilde{\mathbf{X}}
$$

and

$$
\tilde{A}_2 = (\tilde{X}_{r_{m_2}}, \ldots, \tilde{X}_{r_{m_3}}) \subset \tilde{\mathbf{X}}.
$$

Based on Definition 2, we define the joint probability of $\tilde{A}_1$ and $\tilde{A}_2$ as follows

(i) If $m_1 < m_2$, then

$$P(\tilde{A}_1 = \tilde{\mathbf{x}}, \tilde{A}_2 = \tilde{\mathbf{y}}) = P(\tilde{X}_{r_1} = \tilde{x}_{r_1}, \ldots, \tilde{X}_{r_{m_1}}$$
$$= \tilde{x}_{r_{m_1}}, \tilde{X}_{r_{m_2}} = \tilde{y}_{r_{m_2}}, \ldots, \tilde{X}_{r_{m_3}} = \tilde{y}_{r_{m_3}}),$$

(ii) If $r_1 \leq m_2 \leq m_1 < m_3$, and for each $i = m_2, \ldots, m_1$, $\tilde{x}_i = \tilde{y}_i$, then

$$P(\tilde{A}_1 = \tilde{\mathbf{x}}, \tilde{A}_2 = \tilde{\mathbf{y}}) = P(\tilde{X}_{r_1} = \tilde{x}_{r_1}, \ldots, \tilde{X}_{r_{m_1}}$$
$$= \tilde{x}_{r_{m_1}}, \tilde{X}_{r_{m_1}+1} = \tilde{y}_{r_{m_1}+1}, \ldots, \tilde{X}_{r_{m_3}} = \tilde{y}_{r_{m_3}}),$$

(iii) If $r_1 \leq m_2 \leq m_1 < m_3$, and $\exists i \in \{m_2, \ldots, m_1\}$, such that $\tilde{x}_i \neq \tilde{y}_i$, then

$$P(\tilde{A}_1 = \tilde{\mathbf{x}}, \tilde{A}_2 = \tilde{\mathbf{y}}) = 0.$$

Remark 3 Based on Definition 4, the conditional probability of $\tilde{A}_1$ given $\tilde{A}_2$ is defined as follows

$$P(\tilde{A}_1 = \tilde{\mathbf{x}} | \tilde{A}_2 = \tilde{\mathbf{y}}) = \frac{P(\tilde{A}_1 = \tilde{\mathbf{x}}, \tilde{A}_2 = \tilde{\mathbf{y}})}{P(\tilde{A}_2 = \tilde{\mathbf{y}})}, \quad P(\tilde{A}_2 = \tilde{\mathbf{y}}) \neq 0.$$

Definition 5 Let $\tilde{\mathbf{X}} = (\tilde{X}_1, \ldots, \tilde{X}_n)$ be a fuzzy random sample related to the fuzzy sample space $\tilde{\chi}$, and g be a measurable function from $\tilde{\chi}^n$ to R. Then, $T = g(\tilde{\mathbf{X}})$ is a sufficient statistic for θ (parameter of the statistical model $f(x; \theta)$ if and only if the conditional distribution $P(\tilde{\mathbf{X}} = \tilde{\mathbf{x}} | T = t)$ does not depend on θ.

Definition 6 Let $T = g(\tilde{\mathbf{X}})$ be a statistic based on fuzzy data. $T = g(\tilde{\mathbf{X}})$ is called a complete statistic if and only if for every statistic $d(T)$, if $E(d(T)) = 0$, then ($\forall \theta \in \Theta$)

$$P_\theta[d(T) = 0] = P_\theta[\tilde{\mathbf{X}} = (\tilde{x}_1, \ldots, \tilde{x}_n) | d(\tilde{\mathbf{X}}) = 0] = 1,$$

where $P_\theta(.)$ is Zadeh's probability based on Definition 2.

Remark 4 When the available data are crisp numbers $x_1, \ldots, x_n$, then Definition 5 and Definition 6 reduce to the ordinary definitions of sufficiency and completeness.

Example 2 Consider a study for estimating the proportion θ of a kind of trees in a forest, which are infected with a plague. We take a sample of n trees and examine each tree for the presence of plague. Suppose that we do not have any precise mechanism to make an exact distinction between the presence and absence of plague, but rather they can inform us whether (a) with much certainty the tree presents infection; or whether (b) with much certainty the tree does not present infection.

The usual model for this problem starts from the Bernoulli experiment X associated with the presence of plague $(Bin(1,\theta), 0 < \theta < 1)$. Thereafter, the model gathers the available information in a fuzzy partition $\tilde{\chi} = \{\tilde{x}_1, \tilde{x}_2\}$. We identify the information with much certainty the tree presents infection and with much certainty the tree does not present infection with the fuzzy sets $\tilde{x}_1$ and $\tilde{x}_2$ on $x = 0, 1$, respectively. The membership functions of $\tilde{x}_1$ and $\tilde{x}_2$ are given by

$$\tilde{x}_1 = \left\{\frac{0.9}{0}, \frac{0.3}{1}\right\}, \quad \tilde{x}_2 = \left\{\frac{0.1}{0}, \frac{0.7}{1}\right\},$$

whose PDF's are as

$$
\begin{aligned}
h(\tilde{x}) &= \sum_{x=0,1} \tilde{x}(x) f(x) \\
&= \begin{cases} 0.9(1-\theta) + 0.3\theta & \tilde{x} = \tilde{x}_1 \\ 0.1(1-\theta) + 0.7\theta & \tilde{x} = \tilde{x}_2 \end{cases} \\
&= \begin{cases} 0.9 - 0.6\theta & \tilde{x} = \tilde{x}_1, \\ 0.1 + 0.6\theta & \tilde{x} = \tilde{x}_2. \end{cases}
\end{aligned}
$$

Suppose that, we have taken a fuzzy random sample of size $n = 2$ as $\tilde{\mathbf{X}} = (\tilde{X}_1, \tilde{X}_2)$. Let the statistic $\tilde{T} = \tilde{X}_1 \oplus \tilde{X}_2$ be as follows

$$
\begin{aligned}
\tilde{t}_1 &= \tilde{x}_1 \oplus \tilde{x}_1 = \left\{\frac{0.81}{0}, \frac{0.27}{1}, \frac{0.09}{2}\right\}, \\
\tilde{t}_2 &= \tilde{x}_1 \oplus \tilde{x}_2 = \tilde{x}_2 \oplus \tilde{x}_1 = \left\{\frac{0.18}{0}, \frac{0.66}{1}, \frac{0.42}{2}\right\}, \\
\tilde{t}_3 &= \tilde{x}_2 \oplus \tilde{x}_2 = \left\{\frac{0.01}{0}, \frac{0.07}{1}, \frac{0.49}{2}\right\},
\end{aligned}
$$

where the PDF of $\tilde{T}$ are as

$$
\begin{aligned}
p_1 &= P(\tilde{T} = \tilde{t}_1) = 0.81(1-\theta)^2 + 0.54\theta(1-\theta) + 0.09\theta^2, \\
p_2 &= P(\tilde{T} = \tilde{t}_2) = 0.18(1-\theta)^2 + 1.32\theta(1-\theta) + 0.42\theta^2, \\
p_3 &= P(\tilde{T} = \tilde{t}_3) = 0.01(1-\theta)^2 + 0.14\theta(1-\theta) + 0.49\theta^2.
\end{aligned}
$$

Based on Definition 4 and Remark 3, we obtain

$$
\begin{aligned}
P(\tilde{X}_1 = \tilde{x}, \tilde{X}_2 = \tilde{y} | \tilde{T} = \tilde{t}_1) &= \begin{cases} \frac{P(\tilde{X}_1 = \tilde{x}_1, \tilde{X}_2 = \tilde{x}_1)}{P(\tilde{T} = \tilde{t}_1)} & \tilde{x} = \tilde{x}_1 \text{ and } \tilde{y} = \tilde{x}_1 \\ 0 & otherwise \end{cases} \\
&= \begin{cases} \frac{P(\tilde{X}_1 = \tilde{x}_1, \tilde{X}_2 = \tilde{x}_1)}{P(\tilde{X}_1 = \tilde{x}_1, \tilde{X}_2 = \tilde{x}_1)} & \tilde{x} = \tilde{x}_1 \text{ and } \tilde{y} = \tilde{x}_1 \\ 0 & otherwise \end{cases} \\
&= \begin{cases} 1 & \tilde{x} = \tilde{x}_1 \text{ and } \tilde{y} = \tilde{x}_1, \\ 0 & otherwise, \end{cases}
\end{aligned}
$$

$$P(\tilde{X}_1 = \tilde{x}, \tilde{X}_2 = \tilde{y} | \tilde{T} = \tilde{t}_2) = \begin{cases} \frac{P(\tilde{X}_1 = \tilde{x}_1, \tilde{X}_2 = \tilde{x}_2)}{P(\tilde{T} = \tilde{t}_2)} & \tilde{x} = \tilde{x}_1 \text{ and } \tilde{y} = \tilde{x}_2 \\ \frac{P(\tilde{X}_1 = \tilde{x}_2, \tilde{X}_2 = \tilde{x}_1)}{P(\tilde{T} = \tilde{t}_2)} & \tilde{x} = \tilde{x}_2 \text{ and } \tilde{y} = \tilde{x}_1 \\ 0 & \text{otherwise} \end{cases}$$

$$= \begin{cases} \frac{P(\tilde{X}_1 = \tilde{x}_1, \tilde{X}_2 = \tilde{x}_2)}{P(\tilde{X}_1 = \tilde{x}_1, \tilde{X}_2 = \tilde{x}_2) + P(\tilde{X}_1 = \tilde{x}_2, \tilde{X}_2 = \tilde{x}_1)} & \tilde{x} = \tilde{x}_1 \text{ and } \tilde{y} = \tilde{x}_2 \\ \frac{P(\tilde{X}_1 = \tilde{x}_2, \tilde{X}_2 = \tilde{x}_1)}{P(\tilde{X}_1 = \tilde{x}_1, \tilde{X}_2 = \tilde{x}_2) + P(\tilde{X}_1 = \tilde{x}_2, \tilde{X}_2 = \tilde{x}_1)} & \tilde{x} = \tilde{x}_2 \text{ and } \tilde{y} = \tilde{x}_1 \\ 0 & \text{otherwise} \end{cases}$$

$$= \begin{cases} \frac{1}{2} & \tilde{x} = \tilde{x}_1 \text{ and } \tilde{y} = \tilde{x}_2, \\ \frac{1}{2} & \tilde{x} = \tilde{x}_2 \text{ and } \tilde{y} = \tilde{x}_1, \\ 0 & \text{otherwise,} \end{cases}$$

$$P(\tilde{X}_1 = \tilde{x}, \tilde{X}_2 = \tilde{y} | \tilde{T} = \tilde{t}_3) = \begin{cases} \frac{P(\tilde{X}_1 = \tilde{x}_2, \tilde{X}_2 = \tilde{x}_2)}{P(\tilde{T} = \tilde{t}_3)} & \tilde{x} = \tilde{x}_2 \text{ and } \tilde{y} = \tilde{x}_2 \\ 0 & \text{otherwise} \end{cases}$$

$$= \begin{cases} \frac{P(\tilde{X}_1 = \tilde{x}_2, \tilde{X}_2 = \tilde{x}_2)}{P(\tilde{X}_1 = \tilde{x}_2, \tilde{X}_2 = \tilde{x}_2)} & \tilde{x} = \tilde{x}_2 \text{ and } \tilde{y} = \tilde{x}_2 \\ 0 & \text{otherwise} \end{cases}$$

$$= \begin{cases} 1 & \tilde{x} = \tilde{x}_2 \text{ and } \tilde{y} = \tilde{x}_2, \\ 0 & \text{otherwise.} \end{cases}$$

Therefore, $\tilde{T}$ is a sufficient statistics for θ. For completeness, suppose that $g(\tilde{T})$ is defined as

$$g(\tilde{T}) = \begin{cases} m_1 & \tilde{T} = \tilde{t}_1, \\ m_2 & \tilde{T} = \tilde{t}_2, \\ m_3 & \tilde{T} = \tilde{t}_3. \end{cases}$$

If $E(g(\tilde{T})) = 0$, then $m_1 p_1 + m_2 p_2 + m_3 p_3 = 0$, and we have

$$\begin{cases} 0.81 m_1 + 0.18 m_2 + 0.01 m_3 = 0, \\ -1.08 m_1 + 0.96 m_2 + 0.12 m_3 = 0, \Rightarrow m_1 = m_2 = m_3 = 0. \\ 0.36 m_1 - 0.72 m_2 + 0.36 m_3 = 0. \end{cases}$$

Hence, $\forall \theta \in (0, 1)$, $g(\tilde{T}) = 0$, and so $\tilde{T}$ is a complete statistic.

5 Unbiasedness

In this section, we define the concepts of unbiased estimator and UMVUE based on fuzzy data.

Definition 7 Let $\tilde{\mathbf{X}}$ be a fuzzy random sample related to the fuzzy sample space $\tilde{\chi}$, and g be a measurable function from $\tilde{\chi}^n$ to R. The expectation and the mean squared error (MSE) of the estimator $Y = g(\tilde{\mathbf{X}})$ are defined as follows (see also [48]):

$$E(Y) = E(g(\tilde{\mathbf{X}})) = \sum_{\tilde{x} \in \tilde{\chi}^n} g(\tilde{\mathbf{x}}) h(\tilde{\mathbf{x}}),$$

and

$$MSE_Y(\theta) = E(Y - \theta)^2 = Var(g(\tilde{\mathbf{X}})) + \left(E(g(\tilde{\mathbf{X}})) - \theta \right)^2.$$

Definition 8 Let $\tilde{\mathbf{X}}$ be a fuzzy random sample related to the fuzzy sample space $\tilde{\chi}$ and g be a measurable function from $\tilde{\chi}^n$ to R. Then, $U = g(\tilde{\mathbf{X}})$ is called an unbiased estimator for θ if

$$E(U) = E(g(\tilde{\mathbf{X}})) = \theta, \quad \forall \theta \in \Theta.$$

Example 3 Consider Example 2. To obtain an unbiased estimator for θ, consider the estimator $g(\tilde{\mathbf{X}})$ as follows

$$g(\tilde{\mathbf{X}}) = \begin{cases} m_1 & \tilde{\mathbf{x}} = (\tilde{x}_1, \tilde{x}_1), \\ m_2 & \tilde{\mathbf{x}} = (\tilde{x}_1, \tilde{x}_2) \, or \, (\tilde{x}_2, \tilde{x}_1), \\ m_3 & \tilde{\mathbf{x}} = (\tilde{x}_2, \tilde{x}_2), \end{cases}$$

where the PDF of $\tilde{\mathbf{X}} = (\tilde{X}_1, \tilde{X}_2)$ is given by

$$P(\tilde{\mathbf{X}} = \tilde{\mathbf{x}}) = \begin{cases} 0.36\theta^2 - 1.08\theta + 0.81 & \tilde{\mathbf{x}} = (\tilde{x}_1, \tilde{x}_1), \\ -0.36\theta^2 + 0.48\theta + 0.09 & \tilde{\mathbf{x}} = (\tilde{x}_1, \tilde{x}_2), \\ -0.36\theta^2 + 0.48\theta + 0.09 & \tilde{\mathbf{x}} = (\tilde{x}_2, \tilde{x}_1), \\ 0.36\theta^2 + 0.12\theta + 0.01 & \tilde{\mathbf{x}} = (\tilde{x}_2, \tilde{x}_2). \end{cases}$$

Then, $g(\tilde{\mathbf{X}})$ is an unbiased estimator for θ, if $m_1 = -\frac{1}{6}$, $m_2 = \frac{2}{3}$, and $m_3 = \frac{3}{2}$. If we observe $(\tilde{x}_1, \tilde{x}_2)$ or $(\tilde{x}_2, \tilde{x}_1)$, then the unbiased estimator $g(\tilde{\mathbf{X}})$ is in the range $(0, 1)$ and would be an acceptable estimator.

Definition 9 Let $\tilde{\mathbf{X}}$ be a fuzzy random sample related to the fuzzy sample space $\tilde{\chi}$. Let g_1 and g_2 be two measurable functions from $\tilde{\chi}^n$ to R. The estimator $Y_1 = g_1(\tilde{\mathbf{X}})$ is called more efficient than the estimator $Y_2 = g_2(\tilde{\mathbf{X}})$ if for each $\theta \in \Theta$, $MSE_{Y_1}(\theta) \leq MSE_{Y_2}(\theta)$.

Example 4 Suppose that in Example 2, we want to estimate the parameter θ based on a fuzzy random sample of size $n = 1$. Consider two estimators as follows:

$$Y_1 = \begin{cases} \frac{1}{3} & \tilde{x} = \tilde{x}_1, \\ \frac{1}{2} & \tilde{x} = \tilde{x}_2, \end{cases} \qquad Y_2 = \begin{cases} \frac{2}{3} & \tilde{x} = \tilde{x}_1, \\ \frac{1}{7} & \tilde{x} = \tilde{x}_2, \end{cases}$$

$MSE_{Y_1}(\theta)$ and $MSE_{Y_2}(\theta)$ are calculated as

$$MSE_{Y_1}(\theta) = \frac{1}{8} - \frac{37}{60}\theta + \frac{4}{5}\theta^2,$$

$$MSE_{Y_2}(\theta) = \frac{197}{490} - \frac{218}{147}\theta + \frac{57}{35}\theta^2,$$

hence, the estimator Y_1 is more efficient than Y_2, since $MSE_{Y_1}(\theta) < MSE_{Y_2}(\theta)$, $\forall \theta \in (0,1)$.

Definition 10 Let $\tilde{\mathbf{X}}$ be a fuzzy random sample related to the fuzzy sample space $\tilde{\chi}$, and g_1 and g_2 be two measurable functions from $\tilde{\chi}^n$ to R. Then, the unbiased estimator $g_1(\tilde{\mathbf{X}})$ of θ is called the "uniformly minimum variance unbiased estimator" (UMVUE) if and only if $Var(g_1(\tilde{\mathbf{X}})) \leq Var(g_2(\tilde{\mathbf{X}}))$ for all θ and for any other unbiased estimator $g_2(\tilde{\mathbf{X}})$ of θ.

Theorem 2 *Let $T = g(\tilde{\mathbf{X}})$ be a sufficient and complete statistic for θ based on a fuzzy random sample $\tilde{\mathbf{X}}$. Let $q(T)$ be any unbiased estimator for the parameter θ with a Borel function q. Then, $q(T)$ is the unique UMVUE for θ.*

Proof T is a sufficient and complete statistic, and is also an ordinary random variable based on Theorem 1. Hence, based on Lehmann-Scheffe theorem ([44], Theorem 3.1), $q(T)$ is the unique UMVUE for θ. □

Corollary 1 *Let $T = g(\tilde{\mathbf{X}})$ be a complete sufficient statistic. If the distribution of $T(\tilde{\mathbf{X}})$ is available, then the UMVUE is a function q of $T(\tilde{\mathbf{X}})$ such that*

$$E\big[q(T(\tilde{\mathbf{X}}))\big] = \theta.$$

Theorem 3 *Let $g(\tilde{\mathbf{X}})$ be any unbiased estimator of θ based on Definition 8, and $T(\tilde{\mathbf{X}})$ be a sufficient and complete statistic. Then $q(T(\tilde{\mathbf{X}})) = E(g(\tilde{\mathbf{X}})|T(\tilde{\mathbf{X}}))$ is the UMVUE of θ.*

Proof We have

$$\begin{aligned} Var\big(g(\tilde{\mathbf{X}})\big) &= Var\big(E(g(\tilde{\mathbf{X}})|T(\tilde{\mathbf{X}}))\big) + E\big(Var(g(\tilde{\mathbf{X}})|T(\tilde{\mathbf{X}}))\big) \\ &= Var\big(q(T(\tilde{\mathbf{X}}))\big) + E\big(Var(g(\tilde{\mathbf{X}})|T(\tilde{\mathbf{X}}))\big) \\ &\geq Var\big(q(T(\tilde{\mathbf{X}}))\big), \end{aligned}$$

hence, $q(T(\tilde{\mathbf{X}}))$ is the UMVUE of θ. □

6 Numerical Examples

Example 5 In Example 2, we show that $\tilde{T} = \tilde{X}_1 \oplus \tilde{X}_2$ is a complete sufficient statistic. Now, consider the estimator $q(\tilde{T})$ as follows:

$$q(\tilde{T}) = \begin{cases} w_1 & \tilde{T} = \tilde{t}_1, \\ w_2 & \tilde{T} = \tilde{t}_2, \\ w_3 & \tilde{T} = \tilde{t}_3. \end{cases}$$

If we take $w_1 = -\frac{1}{6}, w_2 = \frac{2}{3}, w_3 = \frac{3}{2}$, then $q(\tilde{T})$ would be an unbiased estimator. Hence, the UMVUE for θ is obtained as follows:

$$q(\tilde{T}) = \begin{cases} -\frac{1}{6} & \tilde{T} = \tilde{t}_1, \\ \frac{2}{3} & \tilde{T} = \tilde{t}_2, \\ \frac{3}{2} & \tilde{T} = \tilde{t}_3. \end{cases}$$

In the general case, if we take a fuzzy random sample of size n, then $q(\tilde{T}) = \frac{3}{2} - \frac{5}{3n} T(\tilde{\mathbf{X}})$ is the UMVUE for θ, where $T(\tilde{\mathbf{X}})$ is given by

$$T(\tilde{\mathbf{X}}) = \begin{cases} n & \tilde{\mathbf{x}} = (\tilde{x}_1, \tilde{x}_1, \ldots, \tilde{x}_1, \tilde{x}_1), \\ n-1 & \tilde{\mathbf{x}} = (\tilde{x}_2, \tilde{x}_1, \ldots, \tilde{x}_1, \tilde{x}_1) \, or \ldots (\tilde{x}_1, \tilde{x}_1, \ldots, \tilde{x}_1, \tilde{x}_2), \\ n-2 & \tilde{\mathbf{x}} = (\tilde{x}_2, \tilde{x}_2, \ldots, \tilde{x}_1, \tilde{x}_1) \, or \ldots (\tilde{x}_1, \tilde{x}_1, \ldots, \tilde{x}_2, \tilde{x}_2), \\ \vdots & \vdots \\ 0 & \tilde{\mathbf{x}} = (\tilde{x}_2, \tilde{x}_2, \ldots, \tilde{x}_2, \tilde{x}_2). \end{cases}$$

For $T(\tilde{\mathbf{X}}) = 0$ and $T(\tilde{\mathbf{X}}) = n$, the unbiased estimator $q(\tilde{T}) = \frac{3}{2} - \frac{5}{3n} T(\tilde{\mathbf{X}})$ is not in the range $(0, 1)$. Hence, the unbiased estimator $q(\tilde{T}) = \frac{3}{2} - \frac{5}{3n} T(\tilde{\mathbf{X}})$ is acceptable for $T(\tilde{\mathbf{X}}) = 1, 2, \ldots, n-1$.

Example 6 Let X be a random variable from a beta distribution $Beta(\theta, 1)$ with a density of

$$f(x; \theta) = \theta x^{\theta-1}, \quad \theta > 0, \quad 0 < x < 1.$$

Suppose that, the fuzzy sample space $\tilde{\chi}$ contains the following fuzzy events

$$\tilde{x}_1(x) = \begin{cases} 1 - 2x & 0 \le x < 0.5, \\ 2x - 1 & 0.5 \le x < 1, \end{cases}$$

$$\tilde{x}_2(x) = \begin{cases} 2x & 0 \le x < 0.5, \\ 2 - 2x & 0.5 \le x < 1. \end{cases}$$

The PDFs of the fuzzy events $\tilde{x}_1$ and $\tilde{x}_2$ are as follows

$$p_1 = h(\tilde{x}_1) = \frac{\theta - 1}{\theta + 1}, \quad p_2 = h(\tilde{x}_2) = \frac{2}{\theta + 1}.$$

Let the parameter of interest be $\beta = \frac{a+b\theta}{\theta+1}$, where $a \geq 0$ and $b > 0$ are known constants and $a \neq b$. Suppose that we have a fuzzy random sample of size n. Consider the sufficient and complete statistic as follows

$$T(\tilde{\mathbf{X}}) = \begin{cases} z_0 & \tilde{\mathbf{X}} = (\tilde{x}_1, \tilde{x}_1, \ldots, \tilde{x}_1, \tilde{x}_1), \\ z_1 & \tilde{\mathbf{X}} = (\tilde{x}_2, \tilde{x}_1, \ldots, \tilde{x}_1, \tilde{x}_1) \, or \ldots (\tilde{x}_1, \tilde{x}_1, \ldots, \tilde{x}_1, \tilde{x}_2), \\ z_2 & \tilde{\mathbf{X}} = (\tilde{x}_2, \tilde{x}_2, \ldots, \tilde{x}_1, \tilde{x}_1) \, or \ldots (\tilde{x}_1, \tilde{x}_1, \ldots, \tilde{x}_2, \tilde{x}_2), \\ \vdots & \vdots \\ z_n & \tilde{\mathbf{X}} = (\tilde{x}_2, \tilde{x}_2, \ldots, \tilde{x}_2, \tilde{x}_2). \end{cases}$$

On the other hand, $T(\tilde{\mathbf{X}})$ is an unbiased estimator if $E(T(\tilde{\mathbf{X}})) = \theta$. Hence, we obtain

$$\sum_{i=0}^{n} z_i \binom{n}{i} 2^i (\theta - 1)^{n-i} = (a + b\theta)(\theta + 1)^{n-1}.$$

The above relation can rewrite as follows

$$z_0 \lambda^n + \sum_{i=0}^{n-2} z_{i+1} \binom{n}{i+1} 2^{i+1} \lambda^{n-i-1} + z_n 2^n$$

$$= b\lambda^n + \sum_{i=0}^{n-2} \left[(a+b)\binom{n-1}{i} 2^i + b\binom{n-1}{i+1} 2^{i+1} \right] \lambda^{n-i-1} + (a+b)2^{n-1}$$

where $\lambda = \theta - 1$. Hence, $T(\tilde{\mathbf{X}})$ is the UMVUE for $\beta = \frac{a+b\theta}{\theta+1}$ if and only if

$$z_0 = b, \quad z_i = \frac{ia + (2n - i)b}{2n}, \quad i = 1, 2, \ldots, n.$$

Example 7 Suppose that we have taken a fuzzy random sample of $n = 2$ from a uniform distribution $U(0, \theta)$. Assume further that the fuzzy sample space $\tilde{\chi}$ contains the following fuzzy events

$$\tilde{x}_1(x) = \begin{cases} 1 & 0 \leq x < 0.5, \\ 2 - 2x & 0.5 \leq x < 1, \end{cases}$$

$$\tilde{x}_2(x) = \begin{cases} 2x - 1 & 0.5 \leq x < 1, \\ 1 & 1 \leq x < 1.5 \\ 4 - 2x & 1.5 \leq x < 2, \end{cases}$$

$$\tilde{x}_3(x) = \begin{cases} 2x - 3 & 1.5 \leq x < 2, \\ 1 & 2 \leq x. \end{cases}$$

The probabilities of these fuzzy events are obtained as

$$p_1 = h(\tilde{x}_1) = \frac{3}{4\theta},$$
$$p_2 = h(\tilde{x}_2) = \frac{1}{\theta},$$
$$p_3 = h(\tilde{x}_3) = \frac{4\theta - 7}{4\theta}.$$

We can obtain two sufficient statistics as follows

$$T_1(\tilde{\mathbf{X}}) = \begin{cases} m_1 & \tilde{\mathbf{x}} = (\tilde{x}_1, \tilde{x}_1), \\ m_2 & \tilde{\mathbf{x}} = (\tilde{x}_2, \tilde{x}_2), \\ m_3 & \tilde{\mathbf{x}} = (\tilde{x}_3, \tilde{x}_3), \\ m_4 & \tilde{\mathbf{x}} = (\tilde{x}_1, \tilde{x}_2), (\tilde{x}_2, \tilde{x}_1), \\ m_5 & \tilde{\mathbf{x}} = (\tilde{x}_1, \tilde{x}_3), (\tilde{x}_3, \tilde{x}_1), \\ m_6 & \tilde{\mathbf{x}} = (\tilde{x}_2, \tilde{x}_3), (\tilde{x}_3, \tilde{x}_2), \end{cases}$$

and

$$T_2(\tilde{\mathbf{X}}) = \begin{cases} w_1 & \tilde{\mathbf{x}} = (\tilde{x}_1, \tilde{x}_1), (\tilde{x}_2, \tilde{x}_2), (\tilde{x}_1, \tilde{x}_2), (\tilde{x}_2, \tilde{x}_1), \\ w_2 & \tilde{\mathbf{x}} = (\tilde{x}_3, \tilde{x}_1), (\tilde{x}_1, \tilde{x}_3), (\tilde{x}_2, \tilde{x}_3), (\tilde{x}_3, \tilde{x}_2), \\ w_3 & \tilde{\mathbf{x}} = (\tilde{x}_3, \tilde{x}_3). \end{cases}$$

Note that, $T_2(\tilde{\mathbf{X}})$ is the only complete statistic. Also, the UMVUE for θ does not exist, because if $E(T_2(\tilde{\mathbf{X}})) = \theta$, then we cannot find w_1, w_2, w_3 satisfy the following relation

$$(49w_1 - 98w_2 + 49w_3) + (56w_2 - 56w_3)\theta + 16w_3\theta^2 = 16\theta^3.$$

Now, let the parameter of interest be $\beta = \frac{1}{\theta}$. We will obtain below the UMVUE for β based on the two methods proposed in Corollary 1 and Theorem 3, respectively.

A1: Based on Corollary 1, we obtain w_1, w_2, and w_3 such that $E(T_2(\tilde{\mathbf{X}})) = \beta$. If $E(T_2(\tilde{\mathbf{X}})) = \beta$, then $(49w_1 - 98w_2 + 49w_3) + (56w_2 - 56w_3)\theta + 16w_3\theta^2 = 16\theta$. Therefore, the UMVUE for $\beta = \frac{1}{\theta}$ is obtained as

$$h(T_2) = \begin{cases} w_1 = \frac{4}{7} & \tilde{\mathbf{x}} = (\tilde{x}_1, \tilde{x}_1), (\tilde{x}_2, \tilde{x}_2), (\tilde{x}_1, \tilde{x}_2), (\tilde{x}_2, \tilde{x}_1), \\ w_2 = \frac{2}{7} & \tilde{\mathbf{x}} = (\tilde{x}_3, \tilde{x}_1), (\tilde{x}_1, \tilde{x}_3), (\tilde{x}_2, \tilde{x}_3), (\tilde{x}_3, \tilde{x}_2), \\ w_3 = 0 & \tilde{\mathbf{x}} = (\tilde{x}_3, \tilde{x}_3). \end{cases}$$

A2: Based on Theorem 3, first we obtain an unbiased estimator H for β, then it is concluded that $h(T_2) = E(H|T_2(\tilde{\mathbf{X}}))$ is the UMVUE. For example, let H be as follows

$$H = H(\tilde{X}_1) = \begin{cases} 1 & \tilde{X}_1 = \tilde{x}_1, \\ \frac{1}{4} & \tilde{X}_1 = \tilde{x}_2, \\ 0 & \tilde{X}_1 = \tilde{x}_3. \end{cases}$$

The conditional distribution of H given T_2 is obtained as

$$P(H = y|T_2 = w_1) = \begin{cases} \frac{3}{7} & y = 1, \\ \frac{4}{7} & y = \frac{1}{4}, \\ 0 & y = 0, \end{cases}$$

$$P(H = y|T_2 = w_2) = \begin{cases} \frac{3}{14} & y = 1, \\ \frac{2}{7} & y = \frac{1}{4}, \\ \frac{1}{2} & y = 0, \end{cases}$$

$$P(H = y|T_2 = w_3) = \begin{cases} 0 & y = 1, \frac{1}{4}, \\ 1 & y = 0. \end{cases}$$

Hence, the UMVUE for $\beta = \frac{1}{\theta}$ is given by

$$\begin{aligned} h(T_2) &= E(H(\tilde{X}_1)|T_2(\tilde{\mathbf{X}})) \\ &= \begin{cases} \frac{4}{7} & T_2 = w_1 \\ \frac{2}{7} & T_2 = w_2 \\ 0 & T_2 = w_3 \end{cases} \\ &= \begin{cases} \frac{4}{7} & \tilde{\mathbf{x}} = (\tilde{x}_1, \tilde{x}_1), (\tilde{x}_2, \tilde{x}_2), (\tilde{x}_1, \tilde{x}_2), (\tilde{x}_2, \tilde{x}_1), \\ \frac{2}{7} & \tilde{\mathbf{x}} = (\tilde{x}_3, \tilde{x}_1), (\tilde{x}_1, \tilde{x}_3), (\tilde{x}_2, \tilde{x}_3), (\tilde{x}_3, \tilde{x}_2), \\ 0 & \tilde{\mathbf{x}} = (\tilde{x}_3, \tilde{x}_3), \end{cases} \end{aligned}$$

which is exactly the same as obtained in Part **A1**.

7 Conclusion

In practical real world problems, we usually confront with non-precise observations. Conventional statistical methods are not appropriate for analyzing such situations. We, therefore, need some new statistical methods to deal with non-precise data.

In this regard, a new approach was introduced to the estimation of a parameter when the available data were reported as non-precise (fuzzy) quantities. In the proposed approach, the concepts of sufficient, complete, unbiased, and UMVU estimators are extended to the case the data are fuzzy. Some procedures to find such estimators, based on fuzzy data, are presented.

Applicability of the theory of point estimation provided in this research for constructing and evaluating confidence intervals based on fuzzy data (Chachi and Taheri [13] and Chachi et al. [14]), testing hypothesis using fuzzy data (Grzegorzewski [22], Wu [54], and Arefi and Taheri [2, 3]), and fuzzy regression analysis (Namdari et al. [41] and Arefi and Taheri [4]) are some topics for more research. An effective concept to obtain unbiased estimators is U-statistic (see [44], Definition 3.2). The U-statistic presents a large class of unbiased estimators in parametric and nonparametric problems. The study of U-statistic based on different kernel functions and under fuzzy data may be a subject of interest for future research. Moreover, investigation of the proposed approach in a decision theoretic framework may also be a subject of interest for future research.

References

1. Akbari, MGh, Rezaei, A.: An uniformly minimum variance unbiased point estimator using fuzzy observations. Austrian J. Stat. **36**, 307–317 (2007)
2. Arefi, M., Taheri, S.M.: Testing fuzzy hypotheses using fuzzy data based on fuzzy test statistic. J. Uncertain Syst. **5**, 45–61 (2011)
3. Arefi, M., Taheri, S.M.: A new approach for testing fuzzy hypotheses based on fuzzy data. Int. J. Comput. Intell. Syst. **6**, 318–327 (2013)
4. Arefi, M., Taheri, S.M.: Least-squares regression based on Atanassov's intuitionistic fuzzy inputs-outputs and Atanassov's intuitionistic fuzzy parameters. IEEE Trans. Fuzzy Syst. **23**, 1142–1154 (2015)
5. Arefi, M., Taheri, S.M.: Possibilistic Bayesian inference based on fuzzy data. Int. J. Mach. Learn. & Cyber. doi:10.1007/s13042-014-0291-8, (to appear) (2016)
6. Billingsley, P.: Probability and Measure, 3rd edn. Wiley, New York (1995)
7. Blanco-Fernandez, A., Casals, M.R., Colubi, A., Corral, N., Garcia-Barzana, M., Gil, M.A., Gonzalez-Rodriguez, G., Lopez, M.T., Lubiano, M.A., Montenegro, M., Ramos-Guajardo, A. B., De La Rosa De Saa, S., Silnova, B.: Random fuzzy sets: a mathematical tool to develop statistical fuzzy data analysis. Iran. J. Fuzzy Syst. **10**, 1–28 (2013)
8. Boswell, S.B., Taylor, M.S.: A central limit theorem for fuzzy random variables. Fuzzy Sets Syst. **24**, 331–344 (1987)
9. Buckley, J.J.: Fuzzy Statistics. Springer, Heidelberg (2005)
10. Buckley, J.J.: Fuzzy statistics: hypothesis testing. Soft. Comput. **9**, 512–518 (2005)
11. Cai, K.Y.: Parameter estimations of normal fuzzy variables. Fuzzy Sets Syst. **55**, 179–185 (1993)
12. Casals, M.R., Gil, M.A., Gil, P.: On the use of Zadeh's probabilistic definition for testing statistical hypotheses from fuzzy information. Fuzzy Sets Syst. **20**, 175–190 (1986)
'13. Chachi, J., Taheri, S.M.: Fuzzy confidence intervals for mean of Gaussian fuzzy random variables. Expert Syst. Appl. **38**, 5240–5244 (2011)
14. Chachi, J., Taheri, S.M., Viertl, R.: Testing statistical hypotheses based on fuzzy confidence intervals. Austrian J. Stat. **41**, 267–286 (2012)
15. Colubi, A., López-Díaz, M., Domínguez-Menchero, J.S., Gil, M.A.: A generalized strong law of large numbers. Probab. Theory Relat. Fields **114**, 401–417 (1999)
16. Falsafain, A., Taheri, S.M.: On Buckley's approach to fuzzy estimation. Soft. Comput. **15**, 345–349 (2011)
17. Falsafain, A., Taheri, S.M., Mashinchi, M.: Fuzzy estimation of parameters in statistical models. Int. J. Comput. Math. Sci. **2**, 79–85 (2008)

18. Feng, Y.: An approach to generalize laws of large numbers for fuzzy random variables. Fuzzy Sets Syst. **128**, 237–245 (2002)
19. Fu, K., Zhang, L.: Strong limit theorems for random sets and fuzzy random sets with slowly varying weights. Inf. Sci. **178**, 2648–2660 (2008)
20. Garcia, D., Lubiano, M.A., Alonso, C.: Estimating the expected value of fuzzy random variables in the stratified random sampling from finite populations. Inf. Sci. **138**, 165–184 (2001)
21. Gil, M.A., Corral, N., Gil, P.: The fuzzy decision problem: An approach to the point estimation problem with fuzzy information. Eur. J. Oper. Res. **22**, 26–34 (1985)
22. Grzegorzewski, P.: Testing fuzzy hypotheses with vague data. In: Bertoluzza, C., et al. (eds.) Statistical Modeling, Analysis and Management of Fuzzy Data, pp. 213–225. Springer, Heidelberg (2002)
23. Guan, L., Li, S.: A law of large numbers for exchangeable random variables on nonadditive measures. In: Kruse, R. et al. (eds.) Synergies of Soft Computing and Statistics for Intelligent Data Analysis, pp. 145–152. Springer (2013)
24. Guangyuan, W., Zhong, Q.: Convergence of sequences of fuzzy random variables and its application. Fuzzy Sets Syst. **63**, 187–199 (1994)
25. Hong, D.H.: Parameter estimations of mutually T-related fuzzy variables. Fuzzy Sets Syst. **123**, 63–71 (2001)
26. Hong, D.H.: Strong laws of large numbers for t-norm-based addition of fuzzy set-valued random variables. Fuzzy Sets Syst. **223**, 26–38 (2013)
27. Joo, S.Y.: Weak laws of large numbers for fuzzy random variables. Fuzzy Sets Syst. **147**, 453–464 (2004)
28. Joo, S.Y., Kim, Y.K.: Kolmogorov's strong law of large numbers for fuzzy random variables. Fuzzy Sets Syst. **120**, 499–503 (2001)
29. Joo, S.Y., Kim, Y.K., Kwon, J.S.: Strong convergence for weighted sums of fuzzy random sets. Inf. Sci. **176**, 1086–1099 (2006)
30. Kim, Y.K.: Strong law of large numbers for fuzzy random variables. Fuzzy Sets Syst. **111**, 319–323 (2000)
31. Klement, E.P., Puri, M.L., Ralescu, D.A.: Limit theorems for fuzzy random variables. Proc. Roy. Soc. London Ser. A **407**, 171–182 (1986)
32. Kratschmer, V.: Limit theorems for fuzzy-random variables. Fuzzy Sets Syst. **126**, 253–263 (2002)
33. Kratschmer, V.: Probability theory in fuzzy sample spaces. Metrika **60**, 167–189 (2004)
34. Kruse, R.: The strong law of large numbers for fuzzy random variables. Inf. Sci. **28**, 233–241 (1982)
35. Kruse, R.: Statistical estimation with linguistic data. Inf. Sci. **33**, 197–207 (1984)
36. Kruse, R., Meyer, K.D.: Statistics with Vague Data. Reidel, Dordrecht (1987)
37. Li, Sh, Ogura, Y.: Strong laws of large numbers for independent fuzzy set-valued random variables. Fuzzy Sets Syst. **157**, 2569–2578 (2006)
38. Lin, L., Lee, H.M.: Fuzzy assessment method on sampling survey analysis. Expert Syst. Appl. **36**, 5955–5961 (2009)
39. Lin, L., Lee, H.M.: Fuzzy assessment for sampling survey defuzzification by signed distance method. Expert Syst. Appl. **37**, 7852–7857 (2010)
40. Miyakoshi, M., Shimbo, M.: A strong law of large numbers for fuzzy random variables. Fuzzy Sets Syst. **12**, 133–142 (1984)
41. Namdari, M., Yoon, J.H., Abadi, A., Taheri, S.M., Choi, S.H.: Fuzzy logistic regression with least absolute deviations estimators. Soft. Comput. **19**, 909–917 (2015)
42. Parchami, A., Mashinchi, M.: Fuzzy estimation for process capability indices. Inf. Sci. **177**, 1452–1462 (2007)
43. Schnatter, S.: On fuzzy Bayesian inference. Fuzzy Sets Syst. **60**, 41–58 (1993)
44. Shao, J.: Mathematical Statistics, 2nd edn. Springer, New York (2003)
45. Taheri, S.M.: Trends in Fuzzy Statistics. Austrian J. Stat. **32**, 239–257 (2003)

46. Taheri, S.M., Arefi, M.: Testing fuzzy hypotheses based on fuzzy test statistic. Soft. Comput. **13**, 617–625 (2009)
47. Teran, P.: Strong law of large numbers for t-normed arithmetics. Fuzzy Sets Syst. **159**, 343–360 (2008)
48. Torabi, H., Behboodian, J., Taheri, S.M.: Neyman-Pearson Lemma for fuzzy hypotheses testing with vague data. Metrika **64**, 289–304 (2006)
49. Viertl, R.: Statistics with one-dimensional fuzzy data. In: Bertoluzza, C., et al. (eds.) Statistical Modeling, Analysis and Management of Fuzzy Data, pp. 199–212. Physica-Verlag, Heidelberg (2002)
50. Viertl, R.: Statistical Methods for Fuzzy Data. Wiley, Chichester (2011)
51. Wu, HCh.: The central limit theorems for fuzzy random variables. Inf. Sci. **120**, 239–256 (1999)
52. Wu, HCh.: The law of large numbers for fuzzy random variables. Fuzzy Sets Syst. **116**, 245–262 (2000)
53. Wu, HCh.: The fuzzy estimators of fuzzy parameters based on fuzzy random variables. Eur. J. Oper. Res. **146**, 101–114 (2003)
54. Wu, HCh.: Statistical hypotheses testing for fuzzy data. Inf. Sci. **175**, 30–56 (2005)
55. Yao, J.S., Hwang, C.M.: Point estimation for the n sizes of random sample with one vague data. Fuzzy Sets Syst. **80**, 205–215 (1996)
56. Zadeh, L.A.: Probability measures of fuzzy events. J. Math. Anal. Appl. **23**, 421–427 (1968)

Fuzzy Confidence Regions

Reinhard Viertl and Shohreh Mirzaei Yeganeh

Abstract Confidence regions are usually based on exact data. However, continuous data are always more or less non-precise, also called fuzzy. For fuzzy data the concept of confidence regions has to be generalized. This is possible and the resulting confidence regions are fuzzy subsets of the parameter space. The construction is explained for classical statistics as well as for Bayesian analysis. An example is given in the last section.

Keywords Confidence regions · Fuzzy confidence regions · Fuzzy data · Fuzzy Highest Posterior Density (HPD)-regions · Fuzzy sets

1 Introduction

Classical confidence regions for parameters θ in stochastic models $X \sim f(.|\theta), \theta \in \Theta$ based on standard data $x_1, \ldots, x_n$ of X are classical subsets $\Theta_{1-\alpha}$ of Θ, with $0 < \alpha < 1$, such that $\Pr[\theta \in \Theta_{1-\alpha}] = 1 - \alpha$, where $1 - \alpha$ is the so-called *confidence level*. Let M_X denote the set of possible values which X can assume. Such confidence sets are based on *confidence functions* $\kappa : M_X^n \to \wp(\Theta)$ which are functions from the sample space M_X^n of X to the system of subsets of Θ. For a mathematical sample $X_1, \ldots, X_n$ from X, i.e. a finite sequence of independent and identical as X distributed random variables X_i, a confidence function has to obey

$$\Pr_\theta[\kappa(X_1, \ldots, X_n) \in \Theta_{1-\alpha}] = 1 - \alpha \quad \forall \theta \in \Theta.$$

For observed sample $x_1, \ldots, x_n$ with $x_i \in M_X$, $\kappa(x_1, \ldots, x_n)$ is a classical subset of Θ, called *confidence set* with coverage probability $1 - \alpha$.

R. Viertl (✉) · S.M. Yeganeh
Vienna University of Technology, 1040 Vienna, Austria
e-mail: r.viertl@tuwien.ac.at

S.M. Yeganeh
e-mail: shohreh_my@yahoo.com

© Springer International Publishing Switzerland 2016
C. Kahraman and Ö. Kabak (eds.), *Fuzzy Statistical Decision-Making*,
Studies in Fuzziness and Soft Computing 343,
DOI 10.1007/978-3-319-39014-7_8

2   Fuzzy Data

Since all observations of continuous quantities are more or less fuzzy, a realistic sample in this case consists of n fuzzy numbers $x_1^*, \ldots, x_n^*$ with corresponding characterizing functions $\xi_1(.), \ldots, \xi_n(.)$. Therefore the concept of confidence sets has to be adapted to this situation. This generalization is based on the so-called *combined fuzzy sample* $\underline{x}^*$, which is a fuzzy element of the sample space, characterized by its membership function $\zeta(., \ldots, .)$, defined on the sample space M_X^n (see [1]). This membership function $\zeta(., \ldots, .)$ is obtained from the characterizing functions $\xi_i(.), i = 1(1)n$ of x_i^* by the so-called *minimum-t-norm*, i.e.

$$\zeta(x_1, \ldots, x_n) := \min\{\xi_1(x_1), \ldots, \xi_n(x_n)\} \quad \forall (x_1, \ldots, x_n) \in M_X^n.$$

3   Generalized Confidence Sets

In case of fuzzy data it is natural that related confidence sets become fuzzy subsets of the parameter space Θ. Therefore it is necessary to determine the membership function $\varphi(.)$ of the generalized (fuzzy) subsets $\Theta_{1-\alpha}^*$. Using the notation $\underline{x} = (x_1, \ldots, x_n) \in M_X^n$, and based on a classical confidence function $\kappa : M_X^n \to \wp(\Theta)$ with confidence level $1 - \alpha$, the construction of the membership function $\varphi(.)$ is possible in the following way.

Based on the combined fuzzy sample $\underline{x}^*$ whose membership function $\zeta(., \ldots, .)$ is given, the membership function $\varphi(.)$ of a so-called *fuzzy confidence region* is defined by its values

$$\varphi(\theta) := \begin{cases} \sup\{\zeta(\underline{x}) : \theta \in \kappa(\underline{x})\}, & \text{if } \exists \ \underline{x} \ \in M_X^n : \theta \in \kappa(\underline{x}) \\ 0, & \text{if } \nexists \ \underline{x} \ \in M_X^n : \theta \in \kappa(\underline{x}) \end{cases} \quad \forall \theta \in \Theta.$$

Remark In case of classical samples $\underline{x} = (x_1, \ldots, x_n) \in M_X^n$ the resulting membership function is the indicator function $\mathbf{1}_{\kappa(\underline{x})}(.)$ of the classical confidence set $\kappa(\underline{x})$.

In case of fuzzy data, the membership function $\varphi(.)$ of the fuzzy confidence set above is fulfilling the following inequality:

$$\mathbf{1}_{\bigcup_{\underline{x} : \zeta(\underline{x}) = 1} \kappa(\underline{x})}(\theta) \leqq \varphi(\theta) \quad \forall \theta \in \Theta$$

This can be seen by considering $\sup\{\zeta(\underline{x}) : \theta \in \kappa(\underline{x})\}$.

Example Let X have exponential distribution with density $f(x|\theta) = \frac{1}{\theta}\exp\{-\frac{x}{\theta}\}$ $\mathbf{1}_{(0,\infty)}(x)$ with parameter space $\Theta = (0, \infty)$. A classical confidence function $\kappa(X_1, \ldots, X_n)$ based on sample $X_1, \ldots, X_n$ of X with confidence level $1 - \alpha$ is given by

$$\kappa(X_1,\ldots,X_n) = \left[\frac{2\sum_{i=1}^n X_i}{\chi^2_{2n;1-\frac{\alpha}{2}}}, \frac{2\sum_{i=1}^n X_i}{\chi^2_{2n;\frac{\alpha}{2}}}\right].$$

Hereby $\chi^2_{2n;p}$ is denoting the p-fractile of the chi-square distribution with $2n$ degrees of freedom.

In Fig. 1, the membership functions of a fuzzy sample of X are depicted.

In Fig. 2, the membership functions of related fuzzy confidence intervals based on the fuzzy data from Fig. 1 for different coverage probabilities $1 - \alpha$ are illustrated.

For one-dimensional parameters and classical confidence functions which result in confidence intervals, the following proposition holds:

Proposition 1 *Let $\underline{x}^*$ be a combined fuzzy sample of size n with vector-characterizing function $\zeta : \mathbb{R}^n \to [0,1]$ whose δ-cuts are simply connected, and $\kappa(.,\ldots,.)$ be a level $1 - \alpha$ confidence function for the one-dimensional parameter θ of a stochastic model $X \sim f(.|\theta), \theta \in \Theta \subseteq \mathbb{R}$ with*

$$\kappa(X_1,\ldots,X_n) = [\underline{\kappa}(X_1,\ldots,X_n), \bar{\kappa}(X_1,\ldots,X_n)]$$

Fig. 1 Fuzzy sample

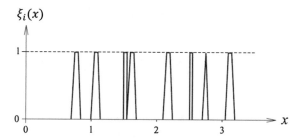

Fig. 2 Fuzzy confidence intervals

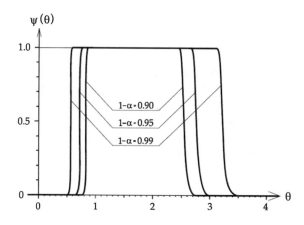

where $\underline{\kappa}(.,\ldots,.)$ and $\bar{\kappa}(.,\ldots,.)$ are continuous functions. Then the membership function $\varphi(.)$ of the fuzzy confidence region defined at the beginning of this section is the characterizing function of a fuzzy interval $\Theta^*_{1-\alpha}$ whose δ-cuts obey the following:

$$C_\delta(\Theta^*_{1-\alpha}) = [\min\{\underline{\kappa}(\underline{x}) : \underline{x} \in C_\delta(\underline{x}^*)\}, \quad \max\{\bar{\kappa}(\underline{x}) : \underline{x} \in C_\delta(\underline{x}^*)\}] \quad \forall \delta \in (0,1]$$

Proof By the continuity of $\underline{\kappa}(.,\ldots,.)$ and $\bar{\kappa}(.,\ldots,.)$ the set $\kappa^{-1}(\{\theta\}) := \{\underline{x} : \theta \in \kappa(\underline{x})\} = \{\underline{x} : \underline{\kappa}(\underline{x}) \le \theta \le \bar{\kappa}(\underline{x})\}$ is closed. The next step is to prove

$$C_\delta(\Theta^*_{1-\alpha}) = \bigcup_{\underline{x} \in C_\delta(\underline{x}^*)} \kappa(\underline{x}) = \bigcup_{\underline{x} \in C_\delta(\underline{x}^*)} [\underline{\kappa}(\underline{x}), \bar{\kappa}(\underline{x})].$$

Firstly, let $\theta \in \bigcup_{\underline{x} \in C_\delta(\underline{x}^*)} \kappa(\underline{x}) \Rightarrow \exists \underline{x}_o \in C_\delta(\underline{x}^*) : \theta \in \kappa(\underline{x}_o) \Rightarrow \zeta(\underline{x}_o) \ge \delta \Rightarrow \sup \{\zeta(\underline{x}) : \theta \in \kappa(\underline{x})\} \ge \delta \Rightarrow \theta \in C_\delta(\Theta^*_{1-\alpha})$.

On the other hand for $\theta \in C_\delta(\Theta^*_{1-\alpha})$ we have $\varphi(\theta) = \sup\{\zeta(\underline{x}) : \theta \in \kappa(\underline{x})\} \ge \delta$, and by the compactness of $\kappa(\underline{x})$ we have $\sup\{\zeta(\underline{x}) : \theta \in \kappa(\underline{x})\} \ge \delta = \max\{\xi(\underline{x}) : \theta \in k(\underline{x})\} \ge \delta \Rightarrow \exists \underline{x}_o : \zeta(\underline{x}_o) \ge \delta$, and therefore $\theta \in \bigcup_{\underline{x} \in C_\delta(\underline{x}^*)} \kappa(\underline{x})$.

In order to prove

$$\bigcup_{\underline{x} \in C_\delta(\underline{x}^*)} [\underline{\kappa}(\underline{x}), \bar{\kappa}(\underline{x})] = [\min\{\underline{\kappa}(\underline{x}) : \underline{x} \in C_\delta(\underline{x}^*)\}, \quad \max\{\bar{\kappa}(\underline{x}) : \underline{x} \in C_\delta(\underline{x}^*)\}],$$

by the continuity of $\underline{\kappa}(.,\ldots,.)$ and $\bar{\kappa}(.,\ldots,.)$ and the connectedness of $C_\delta(\underline{x}^*) \forall \delta \in (0,1]$ it follows that $\bigcup_{\underline{x} \in C_\delta(x^*)} [\underline{\kappa}(\underline{x}), \bar{\kappa}(\underline{x})]$ is connected and compact and therefore a closed interval.

Remark The concept of fuzzy confidence intervals applies also to one-dimensional transformed parameters $\tau(\theta) \in \mathbb{R}$.

4   Bayesian Confidence Regions

Bayesian confidence regions $\Theta_{1-\alpha}$ for a parameter θ in a stochastic model $X \sim f(.|\theta), \theta \in \Theta$ are based on the a posteriori distribution of the parameter.

$\Theta_{1-\alpha}$ is defined as a subset of Θ such that $\Pr(\Theta_{1-\alpha}|D) = 1 - \alpha$, where $0 < \alpha < 1$, and $1 - \alpha$ is called confidence level.

For continuous parameter space Θ usually so-called HPD-regions (Highest Posterior Density) are considered.

A HPD-region $\Theta_{HPD,1-\alpha}\subseteq\Theta$ is defined using the a posteriori density $\pi(.|D)$ on the parameter space Θ in the following way:

(1) $\int_{\Theta_{HPD,1-\alpha}}\pi(\theta|D)d\theta = 1 - \alpha$

(2) $\pi(\theta|D) \geq C_{max} \ \forall\theta \in \Theta_{HPD,1-\alpha}$

where C_{max} is the maximal constant such that condition (1) is fulfilled.

5 Fuzzy a Posteriori Densities

Bayes' theorem for classical probability densities reads

$$\pi(\theta|D) = \frac{\pi(\theta).l(\theta;D)}{\int_\Theta \pi(\theta).l(\theta;D)d\theta} \qquad \forall\theta \in \Theta.$$

Here $\pi(.)$ is the a priori density on the parameter space Θ, and $l(.;D)$ is the likelihood function which is in the simplest case of complete sample $D = (x_1,\ldots,x_n)$ given by its values

$$l(\theta;x_1,\ldots,x_n) = \prod_{i=1}^{n}f(x_i|\theta) \qquad \forall\theta \in \Theta.$$

In case of fuzzy data $D^* = (x_1^*,\ldots,x_n^*)$ Bayes' theorem has to be generalized. This is possible by using so-called *fuzzy densities* $f^*(.)$. These are functions defined on a measure space $(\Omega, \mathcal{A}, \mu)$ whose values $f^*(x)$ are fuzzy intervals with δ-cuts $\left[\underline{f}_\delta(x), \bar{f}_\delta(x)\right] \forall\delta \in (0, 1]$, such that the real valued functions $\underline{f}_\delta(.)$ and $\bar{f}_\delta(.)$ are integrable with finite integrals. Moreover there has to be a classical probability density $g(.)$ on $(\Omega, \mathcal{A}, \mu)$ obeying $\underline{f}_1(x) \leq g(x) \leq \bar{f}_1(x) \ \forall x \in \Omega$. The functions $\underline{f}_\delta(.)$ and $\bar{f}_\delta(.)$ are called *lower and upper δ-level functions* of $f^*(.)$.

For fuzzy a priori density $\pi^*(.)$ on Θ, and fuzzy sample $x_1^*,\ldots,x_n^*$ with combined fuzzy sample $\underline{x}^*$, Bayes' theorem can be generalized in the following way:

The likelihood function is generalized by application of the extension principle. From this fuzzy valued function $l^*(.;\underline{x}^*)$, the δ-level functions $\underline{l}_\delta(.;\underline{x}^*)$ and $\bar{l}_\delta(.;\underline{x}^*)$ are obtained. Then Bayes' theorem is generalized by using the δ-level functions $\underline{\pi}_\delta(.|\underline{x}^*)$ and $\bar{\pi}_\delta(.|\underline{x}^*)$ of the fuzzy a posteriori density $\pi^*(.|\underline{x}^*)$, whose values are defined in the following way:

$$\bar{\pi}_\delta(\theta|\underline{x}^*) := \frac{\bar{\pi}_\delta(\theta).\bar{l}_\delta(\theta;\underline{x}^*)}{\int_\Theta \frac{1}{2}[\underline{\pi}_\delta(\theta).\underline{l}_\delta(\theta;\underline{x}^*) + \bar{\pi}_\delta(\theta).\bar{l}_\delta(\theta;\underline{x}^*)]d\theta}$$

and

$$\pi_\delta(\theta|\underline{x}^*) := \frac{\pi_\delta(\theta).l_\delta(\theta;\underline{x}^*)}{\int_\Theta \frac{1}{2}[\underline{\pi}_\delta(\theta).\underline{l}_\delta(\theta;\underline{x}^*) + \bar{\pi}_\delta(\theta).\bar{l}_\delta(\theta;\underline{x}^*)]d\theta}$$

for all $\delta \in (0,1]$.

Based on the fuzzy a posteriori density, generalized Bayesian confidence regions as well as fuzzy HPD-regions can be constructed.

6   Fuzzy Bayesian Confidence Regions

In order to define fuzzy Bayesian confidence sets, the so-called *construction lemma* for fuzzy sets is important.

Construction Lemma *For universal set M and a family of classical subsets of M, i.e. $(A_\delta; \delta \in (0,1])$ which are nested, i.e. $\delta_1 < \delta_1 \Rightarrow A_{\delta_1} \supseteq A_{\delta_2}$, the membership function $\mu(.)$ of the generated fuzzy subset A^* of M is given by*

$$\mu(x) = \sup\{\delta.\mathbf{1}_{A_\delta}(x) : \delta \in [0,1]\} \quad \forall x \in M.$$

For the proof see [2].

Remark For the generated fuzzy subset A^* of M, the following equivalence is valid:

$$A_\delta = C_\delta[\mu(.)] \Leftrightarrow A_\delta \bigcap_{\beta < \delta} A_\beta$$

The proof of this equivalence is given in the forthcoming PhD thesis by L. Kovarova.

Now the generalized Bayesian confidence set can be defined:

Definition For a stochastic model $X \sim f(.|\theta), \theta \in \Theta$ with continuous parameter space $\Theta \subseteq \mathbb{R}^k$ and fuzzy a posteriori density $\pi^*(.|\underline{x}^*)$, let $\mathcal{D}_\delta$ denote the set of classical probability densities on Θ, for which the following holds:

$$\mathcal{D}_\delta = \{g : \Theta \to [0,\infty), g \text{ is a density on } \Theta \text{ obeying } \underline{\pi}_\delta(\theta|\underline{x}^*) \leq g(\theta)$$
$$\leq \bar{\pi}_\delta(\theta|\underline{x}^*) \,\forall \theta \in \Theta\}$$

Then for each $g \in \mathcal{D}_\delta$ and confidence level $1 - \alpha$ a standard Bayesian confidence set $B_{g,1-\alpha}$ is constructed. The union $A_\delta = \bigcup_{g \in \mathcal{D}_\delta} B_{g,1-\alpha}$ is taken as the generating family of the fuzzy confidence set $A^*_{1-\alpha}$. The membership function $\varphi(.)$ of the fuzzy confidence set is given by the construction lemma, i.e.

$$\varphi_{A^*_{1-\alpha}}(\theta) = \sup\{\delta.1_{A_\delta}(\theta) : \delta \in [0,1]\} \quad \forall \theta \in \Theta.$$

Remark For classical a posteriori densities this definition yields the indicator function of the standard Bayesian confidence region.

7 Fuzzy HPD-Regions

Based on the concept of HPD-regions from Sect. 4 the generalization of HPD-regions for fuzzy a posteriori distributions is possible, similar to the construction in Sect. 6.

For each $g \in \mathcal{D}_\delta$ an HPD-region $\Theta_{g,HPD,1-\alpha}$ can be constructed. Then the generating family of classical subsets $(B_\delta; \delta \in (0,1])$ is obtained by

$$B_\delta = \bigcup_{g\in\mathcal{D}_\delta} \Theta_{g,HPD,1-\alpha}.$$

The fuzzy HPD-region $\Theta^*_{HPD,1-\alpha}$ is the fuzzy subset of Θ whose membership function $\psi(.)$ is given by the construction lemma from Sect. 6 (see [3]).

8 Example

Let the stochastic quantity X have an exponential distribution $X \sim f(.|\theta), \theta > 0$ with density

$$f(x|\theta) = \frac{1}{\theta} e^{-\frac{x}{\theta}} 1_{(0,\infty)}(x).$$

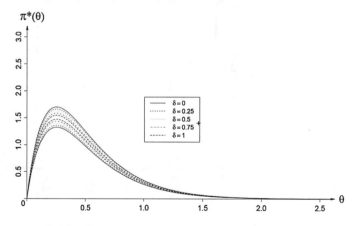

Fig. 3 Fuzzy a priori density

The fuzzy a priori density $\pi^*(.)$ is assumed to be a fuzzy Gamma density, which is depicted in Fig. 3.

A sample of eight fuzzy observations is given in Fig. 1. Applying the generalized Bayes' theorem from Sect. 5 the fuzzy a posteriori density is obtained. The result is displayed in Fig. 4.

The membership function of a fuzzy Bayesian confidence region with coverage probability 95%, based on the fuzzy sample from Fig. 1 is presented in Fig. 5.

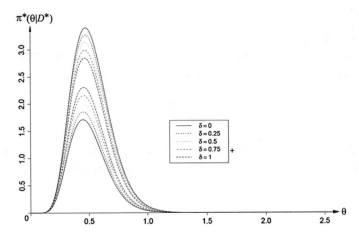

Fig. 4 Fuzzy a posteriori density

Fig. 5 Fuzzy HPD-interval for θ

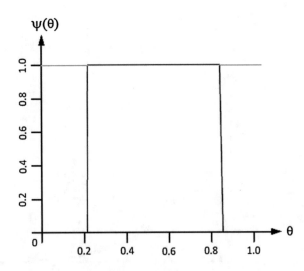

9 Conclusion

In this chapter, the concept of confidence regions has been generalized for fuzzy data. The construction has been explained for classical statistics as well as for Bayesian analysis. Bayesian confidence regions, fuzzy a posteriori densities, fuzzy Bayesian confidence regions, fuzzy Highest Posterior Density (HPD)-regions have been developed. For further studies, other extensions of fuzzy sets may be handled.

References

1. Viertl, R.: Statistical Methods for Fuzzy Data. Wiley, Chichester (2011)
2. Viertl, R., Hareter, D.: Beschreibung und Analyse unscharfer Information - Statistische Methoden für unscharfe Daten. Springer, Wien (2006)
3. Viertl, R., Mirzaei Yeganeh, S.: Fuzzy probability distributions in reliability analysis, fuzzy HPD-regions, and fuzzy predictive distributions. In: Borgelt, C., et al. (eds.) Towards Advanced Data Analysis. Springer, Berlin (2013)

Fuzzy Extensions of Confidence Intervals: Estimation for μ, σ², and p

Cengiz Kahraman, Irem Otay and Başar Öztayşi

Abstract Even though classical point and interval estimations (PIE) are one of the most studied fields in statistics, there are a few numbers of studies covering fuzzy point and interval estimations. In this pursuit, this study focuses on analyzing the works on fuzzy PIE for the years between 1980 and 2015. In the chapter, the literature is reviewed through Scopus database and the review results are given by graphical illustrations. We also use the extensions of fuzzy sets such as interval-valued intuitionistic fuzzy sets (IVIFS) and hesitant fuzzy sets (HFS) to develop the confidence intervals based on these sets. The chapter also includes numerical examples to increase the understandability of the proposed approaches.

Keywords Fuzzy · Point estimation · Interval estimation · Interval-valued intuitionistic fuzzy sets · Hesitant fuzzy sets

1 Introduction

A point estimator estimates the considered parameter as a specific numerical value. For instance, the best point estimate of the population mean (μ) is the sample mean ($\bar{x}$). An interval estimator of a parameter estimates the considered parameter as an interval or a range of values. This estimator may or may not contain the value of the parameter being estimated depending on the significance level (α). Confidence level can be expressed as ($1 - \alpha$). Confidence level of interval estimation is the probability that the parameter is within the confidence interval while a confidence interval is calculated based on the sample statistics and confidence level. An estimator

C. Kahraman (✉) · B. Öztayşi
Faculty of Management, Department of Industrial Engineering,
Istanbul Technical University, 34367 Macka, Istanbul, Turkey
e-mail: kahramanc@itu.edu.tr

I. Otay
Faculty of Management, Department of Management Engineering,
Istanbul Technical University, 34367 Macka, Istanbul, Turkey

© Springer International Publishing Switzerland 2016
C. Kahraman and Ö. Kabak (eds.), *Fuzzy Statistical Decision-Making*,
Studies in Fuzziness and Soft Computing 343,
DOI 10.1007/978-3-319-39014-7_9

should be unbiased, consistent and efficient. An unbiased estimator provides that the expected values of the sample statistics are equal to the corresponding population parameters. A consistent estimator provides that larger sample sizes give the better estimation of population parameters. A relatively efficient estimator has the smallest variance among all the estimators [12].

In the literature, confidence intervals are calculated for the population parameters: mean (μ), variance (σ^2), binomial proportion (p), difference in means ($\mu 1 - \mu 2$), ratio of two variances (σ_1^2/σ_2^2), and difference in binomial proportions (p1 − p2). In this chapter, we aim to estimate mean (μ), variance (σ^2), binomial proportion (p) using IVIFS and HFS. The classical estimation theory gives exact estimations for population parameters, which means the lower and upper confidence limits and point estimations are crisp numbers. However, in real life problems, there are uncertain, vague and incomplete data that can justify the usage of the fuzzy set theory. For instance, sale forecasting for a new product would be impossible with insufficient past data. In this case, we can express our forecasting by using linguistic expressions such as "around 150 units" or "between 100 and 200 units".

In the literature, there are a limited number of fuzzy works on point estimations and confidence intervals. These works will be explained in detail in Sect. 2. The aim of this chapter is to summarize the literature on fuzzy point and interval estimations (PIE) by giving numerical examples. The chapter will also propose some new approaches to fuzzy PIE.

The rest of this chapter is organized as follows: In Sect. 2, the publications on PIE are analyzed statistically. In Sect. 3, an extensive literature review on fuzzy confidence interval (FCI) is presented with numerical examples. In Sect. 4, a new approach for fuzzy estimation is proposed by using intuitionistic fuzzy sets while in Sect. 5 another approach for fuzzy estimation is presented by hesitant fuzzy sets. In the last section, conclusion remarks and future research directions are included.

2 Literature Review: Statistics of Publications on PIE

In this section, we first analyze the fuzzy papers on estimation theory by entering keywords "fuzzy confidence interval", "fuzzy point estimation", and "fuzzy interval estimation" into Scopus database for article title, abstract and keywords under search field type.

In Fig. 1, the publication frequencies of the papers published on fuzzy confidence intervals between 1980 and 2014 are illustrated. The years that the largest numbers of fuzzy PIE papers were published are 2005 and 2011. Figure 2 displays the journals publishing papers on fuzzy confidence intervals with respect to publishing frequencies. The journals that published the largest numbers of papers are Journal of Intelligent and Fuzzy Systems and Fuzzy Sets and Systems.

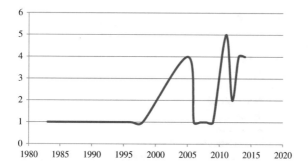

Fig. 1 Publication frequencies of the fuzzy PIE papers with respect to years

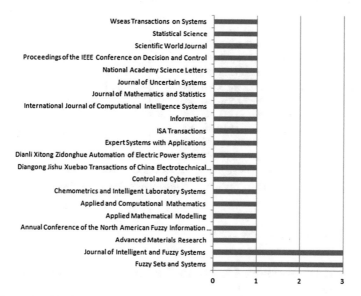

Fig. 2 Journals publishing fuzzy PIE papers

Figure 3 shows the distribution of papers on fuzzy PIE with respect to source countries and the number of papers published. The largest number of fuzzy PIE papers published is from Iran with nine papers and it is followed by China with four papers and United States with four papers.

Figure 4 demonstrates the subject areas of the fuzzy PIE papers by giving the percentages obtained from Scopus. The highest numbers of studies have been done especially in Engineering (63 %), Computer Science (59 %) and Mathematics (48 %) fields in the years between 1980 and 2014.

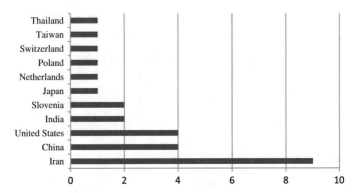

Fig. 3 Fuzzy PIE papers with respect to source countries

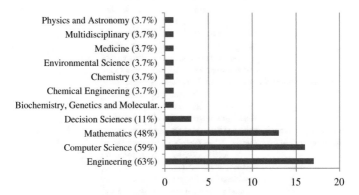

Fig. 4 Subject areas of fuzzy PIE papers

3 Literature Review: Ordinary Fuzzy Confidence Intervals

Earlier studies conducted on fuzzy estimation, particularly on confidence intervals are in 1980s. McCain [10] developed a fuzzy confidence interval. According to the author, the fuzzy concept is useful for optimization problems having fuzzy objectives and constraints. In the study, fuzzy confidence intervals are constructed from uncertain hypotheses. The author highlighted the importance of fuzzy confidence intervals on the applications of decision theory and economics. McCain [11] also studied a focal theory with regard to rational behavior by considering uncertainty. The author proposed a theory of choice, concentrated on fuzzy confidence intervals and highlighted empirical implication of the theory especially in kinky demand curves in an industrial economy. The theory was composed of the fuzzy confidence intervals and particular choice theory. In the study, demand elasticity is defined by using a fuzzy confidence interval and the concept of uncertainty is realized by the usage of the fuzzy set theory and particularly the fuzzy confidence interval.

However, many studies have been made starting from 2000s. For instance, Parchami and Mashinchi [13] made a study on process capability indices measuring the actual or the potential performance of processes. They compared the actual and desired performances considering the target and limits of specifications. Apart from the traditional methods analyzing precise estimations of process capabilities, the authors introduced a new algorithm based on Buckley's [3] estimation approach and used confidence intervals to estimate some process capability indices as C_p, C_{pk} and C_{pm} by employing triangular fuzzy numbers.

Let denote the upper and lower specification limits with U and L, respectively. For a random variable X having mean $\bar{x}$ and standard deviation of s, a $(1 - \beta)100\%$ confidence interval is calculated by Eq. (1) given below:

$$\left[\sigma_1^2(\beta), \sigma_2^2(\beta)\right] = \left[s^2(n-1)\Big/\chi_{n-1,1-\beta/2}^2, s^2(n-1)\Big/\chi_{n-1,\beta/2}^2\right] \qquad (1)$$

Fuzzy estimates for process capability indices (FEPCI) with α-cuts of $\widetilde{C}^p$ are:

$$\left(\widetilde{C}^p\right)_\alpha = \left[\frac{U-L}{6s}\sqrt{\frac{\chi_{n-1,\alpha/2}^2}{n-1}}, \frac{U-L}{6s}\sqrt{\frac{\chi_{n-1,1-\alpha/2}^2}{n-1}}\right] \qquad (2)$$

where $(0 \leq \alpha \leq 1)$

Example 1 We would like to estimate C_p. The upper and lower specification limits are 10 and 6, sequentially. Let say X is a random variable with $N(5, \sigma^2)$ and a standard deviation with ¾. A $\cdot (1 - \beta)100\%$ confidence interval for σ^2 can be calculated as follows:

$$\left[\sigma_1^2(\beta), \sigma_2^2(\beta)\right] = \left[9(n-1)\Big/16\chi_{n-1,1-\beta/2}^2, 9(n-1)\Big/16\chi_{n-1,\beta/2}^2\right]$$

$$\left(\widetilde{C}^p\right)_\alpha = \left[\frac{10-6}{6*(3/4)}\sqrt{\frac{\chi_{25-1,\alpha/2}^2}{25-1}}, \frac{10-6}{6*(3/4)}\sqrt{\frac{\chi_{25-1,1-\alpha/2}^2}{25-1}}\right]$$

where $(0 \leq \alpha \leq 1)$ (for n = 25).

Alex [1] concentrated on fuzzy point estimation for a supply chain management problem and made new definitions for arithmetic operations on fuzzy points. The paper proposed a new approach for dealing with uncertainties in the supply chain. According to the author, prediction of the market demand is very tough and important problem. Alex [1] presented a new fuzzy estimation approach and analyzed non-stationary supply chains.

Skrjanc [14] suggested a new approach for defining fuzzy confidence intervals by using applied statistics. The study aims to find the confidence interval employing the lower and upper fuzzy bounds by covering all the output measurements. The author emphasized that the method is useful for describing uncertain nonlinear functions or searching intervals for nonlinear process output. The developed model is applied for waste-water treatment plant modeling problem concerning uncertainties with regard to the composition of incoming waste water.

Wu [18] suggested fuzzy confidence intervals by considering unknown fuzzy parameters and fuzzy random variables and solved an optimization problem. In the study, $\tilde{X}_1, \tilde{X}_2, \ldots, \tilde{X}_n$ are independent, identically and normally distributed fuzzy random variables and $\tilde{\mu}$ is a fuzzy parameter. $\tilde{X}_{1h}^U, \ldots, \tilde{X}_{nh}^U$ and $\tilde{X}_{1h}^L, \ldots, \tilde{X}_{nh}^L$ are independent and identically distributed fuzzy random variables from $N\left(\tilde{\mu}_h^U, \sigma^2\right)$ and $N\left(\tilde{\mu}_h^L, \sigma^2\right)$, respectively. If σ is known and $\tilde{x}_{ih}^L$ and $\tilde{x}_{ih}^U$ are the observed values of $\tilde{X}_{ih}^L$ and $\tilde{X}_{ih}^U$, confidence intervals for $\tilde{\mu}_h^U$ and $\tilde{\mu}_h^L$ are $\left[L(\tilde{x}_h^U), U(\tilde{x}_h^U)\right]$ and $\left[L(\tilde{x}_h^L), U(\tilde{x}_h^L)\right]$ are given in Eqs. (3) and (4). In the study, triangular fuzzy number $\tilde{x}_i = \left(x_i^L, x_i, x_i^U\right)$ is employed.

$$L(\tilde{x}_h^L) = \frac{1}{n} \sum_{i=1}^n \tilde{x}_{ih}^L - z_{\alpha/2} \frac{\sigma}{\sqrt{n}}, \quad L(\tilde{x}_h^U) = \frac{1}{n} \sum_{i=1}^n \tilde{x}_{ih}^U - z_{\alpha/2} \frac{\sigma}{\sqrt{n}} \qquad (3)$$

$$U(\tilde{x}_h^L) = \frac{1}{n} \sum_{i=1}^n \tilde{x}_{ih}^L + z_{\alpha/2} \frac{\sigma}{\sqrt{n}}, \quad U(\tilde{x}_h^U) = \frac{1}{n} \sum_{i=1}^n \tilde{x}_{ih}^U + z_{\alpha/2} \frac{\sigma}{\sqrt{n}} \qquad (4)$$

Kaya and Kahraman [8] analyzed robust process capability indices (RPCIs) for a piston manufacturing company by incorporating the fuzzy set theory to increase PCIs' flexibility and sensitivity by defining fuzzy specification limits and fuzzy standard deviation. Then fuzzy RPCIs were obtained to express the process performance more realistically for the piston manufacturing stage. In their study, the authors estimated σ^2 by using the unbiased fuzzy confidence interval $(1 - \beta)100\%$ confidence level.

$$\tilde{\tilde{\sigma}}_c = \left[\frac{n\hat{\sigma}^2}{L(\lambda)}, \frac{n\hat{\sigma}^2}{R(\lambda)}\right], \quad (0 \leq \alpha \leq 1) \qquad (5)$$

$$(\hat{\sigma}_c)_\alpha = \left[\frac{n\hat{\sigma}^2}{(1-\alpha)\chi_{R,0.005}^2 + n\alpha}, \frac{n\hat{\sigma}^2}{(1-\alpha)\chi_{l,0.005}^2 + n\alpha}\right] \quad where \ 0 \leq \alpha \leq 1 \quad (6)$$

Skrjanc [15] proposed a method for identifying confidence interval for Takagi–Sugeno fuzzy models when the data have changeable variance. The authors implemented the model for the pH-titration curve. In the application part, pH processes have nonlinear behavior based on different titration curves. The interval fuzzy modeling is used for fault detection system. $y_j^* = \varphi_j^{*T} \theta_j + e_j^*$ and $\widehat{y}_j^* = \varphi_j^{*T} \widehat{\theta}_j$ where φ_j^{*T} indicates the regression matrix. The lower and the upper confidence intervals are as in Eqs. (7) and (8):

$$\underline{f_j}(z_i^*) = \varphi_{ij}^{*T}\theta_j - t_{\alpha,M-n}\hat{\sigma}_j(1 + \varphi_{ij}^{*T}\left(\varphi_j\varphi_j^T\right)^{-1}\varphi_{ij}^*)^{1/2} \qquad (7)$$

$$\overline{f_j}(z_i^*) = \varphi_{ij}^{*T}\theta_j + t_{\alpha,M-n}\hat{\sigma}_j(1 + \varphi_{ij}^{*T}\left(\varphi_j\varphi_j^T\right)^{-1}\varphi_{ij}^*)^{1/2} \qquad (8)$$

where $M - n$ represents the degree of freedom and $t_{\alpha,M-n}$ denotes for the percentile of t-distribution for $100(1 - 2\alpha)$ percentage confidence interval.

Chachi and Taheri [4] proposed the two-sided and one-sided fuzzy confidence intervals based on normal fuzzy random variables. In their study, h-level sets of fuzzy parameters were used to make fuzzy confidence intervals. In the study, by considering normal (Gaussian) fuzzy random variables with known variance σ^2 and for $X_1, X_2, \ldots, X_n \overset{i.i.d.}{\underset{\sim}{}} N\left(\tilde{\theta}, \sigma^2\right)$, a fuzzy confidence interval for the fuzzy parameter $(\widehat{\theta})$ was obtained. To make the calculations easier, triangular fuzzy numbers were employed. Two-sided fuzzy confidence intervals were calculated as follows:

Let say $X_{1h}^l, \ldots, X_{nh}^l \overset{i.i.d}{\underset{\sim}{}} N(\theta_h^l, \sigma^2)$ and $X_{1h}^u, \ldots, X_{nh}^u \overset{i.i.d}{\underset{\sim}{}} N(\theta_h^u, \sigma^2)$ for $h \in [0, 1]$.

Two-sided $100(1 - \alpha)$ % confidence intervals for θ_h^l and θ_h^u, respectively are as in Eqs. (9) and (10):

$$S_T(X_h^l) = \left\{\theta : \left|\frac{\sqrt{n}(X_h^l - \theta)}{\sigma}\right| \le z_{1-\frac{\alpha}{2}}\right\} = \left[\overline{X_h^l} - \frac{\sigma}{\sqrt{n}}z_{1-\frac{\alpha}{2}}, \overline{X_h^l} + \frac{\sigma}{\sqrt{n}}z_{1-\frac{\alpha}{2}}\right] \qquad (9)$$

$$S_T(X_h^u) = \left\{\theta : \left|\frac{\sqrt{n}(X_h^u - \theta)}{\sigma}\right| \le z_{1-\frac{\alpha}{2}}\right\} = \left[\overline{X_h^u} - \frac{\sigma}{\sqrt{n}}z_{1-\frac{\alpha}{2}}, \overline{X_h^u} + \frac{\sigma}{\sqrt{n}}z_{1-\frac{\alpha}{2}}\right] \qquad (10)$$

$$\text{where} \quad X_h^i = (X_{1h}^i, \ldots, X_{nh}^i)\, \bar{X}_h^l = \frac{\sum_{j=1}^n X_{jh}^l}{n} \text{ for } i = l, u. \qquad (11)$$

Example 2 The marketing department for a bulb producer company wants to estimate the average life of a bulb that the company recently developed. Only 24 new bulbs were tested. Because of some unexpected situations, we cannot measure the bulb life precisely, and we just obtain the bulb life around a number. The bulb life numbers are taken to be triangular fuzzy numbers as in Table 1. We assume that the data are observations from normally distributed fuzzy random variables with variance 149,400.

Using fuzzy arithmetics $\bar{X}$ is obtained as (8592, 8770, 8969) and $\bar{X}_h = [8592 + 178h, 8969 - 199h]$.

The two sided 0.95 confidence intervals are calculated using Eqs. (9) and (10) as follows;

Table 1 The data of 24 new bulbs as triangular fuzzy numbers

(8871,9061,9304)	(8558,8698,8868)	(8689,8814,9010)	(8458,8696,8932)
(8748,8911,9042)	(8528,8655,8809)	(8825,8923,9101)	(8519,8658,8857)
(8198,8433,8656)	(8579,8819,9058)	(8644,8834,9086)	(8774,8945,9156)
(8717,8860,9042)	(8087,8235,8385)	(8583,8693,8850)	(8316,8417,8596)
(8913,9076,9272)	(8856,9081,9300)	(8437,8689,8944)	(8252,8490,8730)
(8403,8611,8817)	(8516,8747,8919)	(8816,9025,9186)	(8924,9109,9327)

$$S_T\left(X_h^l\right) = \left[8592 - \frac{\sqrt{149400}}{\sqrt{24}} \times 1.96,\, 8592 + \frac{\sqrt{149400}}{\sqrt{24}} \times 1.96\right]$$
$$= [8482.21 + 178h,\, 8721.79 - 178h]$$

$$S_T\left(X_h^u\right) = \left[8969 - \frac{\sqrt{149400}}{\sqrt{24}} \times 1.96,\, 8969 + \frac{\sqrt{149400}}{\sqrt{24}} \times 1.96\right]$$
$$= [8839.21 + 199h,\, 9098.79 - 199h]$$

The intervals for different values of h are presented in Fig. 5.

Garg [6] presented a new methodology for evaluating repairable industrial systems by proposing availability-cost optimization model for improving the system efficiency with the proposed confidence interval based fuzzy Lambda-Tau (CIBFLT) methodology. The methodology was applied in a washing unit for the paper industry. The proposed methodology was composed of two stages. In the first stage, availability-cost optimization model was constructed and solved by using one of the meta-heuristic techniques called Particle swarm optimization (PSO) while in the second stage reliability parameters are calculated by using the proposed methodology and the results were compared with fuzzy Lambda-Tau methodology.

In their method, for $\tilde{A} = (a, b, c)$ as a triangular fuzzy number, the α-cut at $(1 - \eta)$ 100 % confidence level of $\tilde{A}$ is as in Eq. (12):

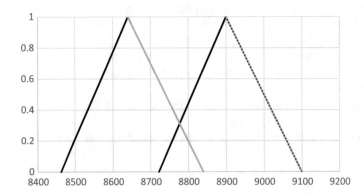

Fig. 5 Intervals for different values of h

$$A^{(\alpha)} = \left[a + \alpha(b - a) - \frac{\sigma}{\sqrt{k}} t_{k-1}\left(\frac{\eta}{2}\right), c - \alpha(c - b) + \frac{\sigma}{\sqrt{k}} t_{k-1}\left(\frac{\eta}{2}\right) \right] \qquad (12)$$

where σ is the standard deviation of b, and $k - 1$ is the degree of freedom.

Garg [7] also presented a confidence interval based lambda-tau methodology to handle with imprecise and vague information in the repairable industrial systems by using the fuzzy set theory and statistics. In the study, reliability parameters were used to analyze the repair system behavior of synthesis unit of a urea fertilizer plant in India, and compared the results with the results of fuzzy lambda-tau technique. The sensitivity analysis was conducted to check the effects of failure. In the paper, the two-sided confidence interval for the α-cut $(A^{(\alpha)})$ was stated as:

$$A^{(\alpha)} = \left[a + (b - a)\alpha - \frac{\sigma}{\sqrt{k}} t_{k-1}\left(\frac{\gamma}{2}\right), c - \alpha(c - b) + \frac{\sigma}{\sqrt{k}} t_{k-1}\left(\frac{\gamma}{2}\right) \right] \qquad (13)$$

where $\tilde{A} = (a, b, c)$ is a triangular fuzzy number, σ is the standard deviation of population b, $k - 1$ is the degree of freedom and T is a t distributed random variable.

The confidence interval for binomial proportion under fuzziness is another research area studied by various researchers in the literature. Buckley [3] proposed the following confidence interval for binomial proportion.

Let p be the probability of a success so that $q = 1 - p$ will be the probability of a failure. We want to estimate the value of p. We therefore gather a random sample which here is running the experiment in n independent times and counting the number of times we succeed. Let x be the number of times we observed a success in n independent repetitions of this experiment. Then our point estimate of p is $\hat{p} = x/n$. We know that $\left((\hat{p} - p)/\sqrt{p(1 - p)/n} \right)$ is approximately N(0, 1) if n is sufficiently large. Then,

$$P\left(z_{\frac{\beta}{2}} \leq \frac{\hat{p} - p}{\sqrt{p(1 - p)/n}} \leq z_{\frac{\beta}{2}}\right) \approx 1 - \beta \qquad (14)$$

Solving the inequality for p we have

$$\left[\hat{p} - z_{\frac{\beta}{2}}\sqrt{p(1 - p)/n}, \hat{p} + z_{\frac{\beta}{2}}\sqrt{p(1 - p)/n} \right] \qquad (15)$$

However, we have no value for p to use in this confidence interval. So, still assuming that n is sufficiently large, we substitute $\hat{p}$ for p in Eq. (15).

Using $\hat{q} = 1 - \hat{p}$, and we get the final $(1 - \beta)\ 100\ \%$ approximate confidence interval

Fig. 6 Fuzzy Estimator p̄,
$0.01 \leq \beta \leq 1$

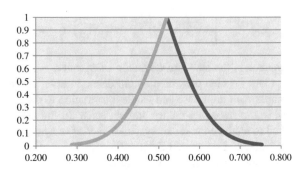

$$\left[\hat{p} - z_{\frac{\beta}{2}}\sqrt{\frac{\hat{p}\hat{q}}{n}}, \hat{p} + z_{\frac{\beta}{2}}\sqrt{\frac{\hat{p}\hat{q}}{n}}\right] \qquad (16)$$

Put these confidence intervals together, and we get $\bar{p}$, our triangular shape fuzzy number estimator of p.

Example 3 Assume that you have $\hat{p} = 0.52$ where n = 36. Figure 6 shows the fuzzy intervals depending on the α-cut levels.

4 Intuitionistic Fuzzy Confidence Intervals

4.1 Interval-Valued Intuitionistic Fuzzy Sets

In the fuzzy set theory, the membership of an element to a fuzzy set is a single value between zero and one. However, the degree of non-membership of an element in a fuzzy set may not be equal to 1 minus the membership degree since there may be some hesitation degree. Therefore, a generalization of fuzzy sets was proposed by Atanassov [2] as intuitionistic fuzzy sets (IFS) which incorporate the degree of hesitation, which is defined as 1 minus the sum of membership and non-membership degrees.

Definition 1 Let $X \neq \emptyset$ be a given set. An intuitionistic fuzzy set in X is an object A given by

$$\tilde{A} = \left\{ \langle x, \mu_{\tilde{A}}(x), \nu_{\tilde{A}}(x) \rangle ; x \epsilon X \right\}, \qquad (17)$$

where $\mu_{\tilde{A}} : X \rightarrow [0,1]$ and $\nu_{\tilde{A}} : X \rightarrow [0,1]$ satisfy the condition

$$0 \leq \mu_{\tilde{A}}(x) + \nu_{\tilde{A}}(x) \leq 1, \qquad (18)$$

for every $x \epsilon X$. Hesitancy is equal to "$1 - \left(\mu_{\tilde{A}}(x) + \nu_{\tilde{A}}(x)\right)$".

The definition of interval-valued intuitionistic fuzzy sets (IVIFS) is given as follows:

Definition 2 Let $D \subseteq [0, 1]$ be the set of all closed subintervals of the interval and X be a universe of discourse. An interval-valued intuitionistic fuzzy set in $\tilde{A}$ over X is an object having the form

$$\tilde{A} = \{\langle x, \tilde{\mu}_A(x), \tilde{v}_A(x) \rangle | x \in X\} \tag{19}$$

where $\tilde{\mu}_{\tilde{A}} \to D \subseteq [0, 1]$, $\tilde{v}_{\tilde{A}}(x) \to D \subseteq [0, 1]$ with the condition $0 \leq \sup \tilde{\mu}_{\tilde{A}}(x) + \sup \tilde{v}_{\tilde{A}}(x) \leq 1, \forall x \in X$.

The intervals $\tilde{\mu}_{\tilde{A}}(x)$ and $\tilde{v}_{\tilde{A}}(x)$ denote the membership function and the non-membership function of the element x to the set $\tilde{A}$, respectively. Thus, for each $x \in X$, $\tilde{\mu}_{\tilde{A}}(x)$ and $\tilde{v}_{\tilde{A}}(x)$ are closed intervals and their lower and upper end points are denoted by $\tilde{\mu}_{\widetilde{AL}}(x)$, $\tilde{\mu}_{\widetilde{AU}}(x)$, $\tilde{v}_{\widetilde{AL}}(x)$, $and \tilde{v}_{\widetilde{AL}}(x)$, respectively. Interval-valued intuitionistic fuzzy set $\tilde{A}$ is then defined by

$$\tilde{A} = \left\{ <x, \left[\tilde{\mu}_{\widetilde{AL}}(x), \tilde{\mu}_{\widetilde{AU}}(x)\right], [\tilde{v}_{\widetilde{AL}}(x), \tilde{v}_{\widetilde{AU}}(x)] > |x \in X \right\} \tag{20}$$

where $0 \leq \tilde{\mu}_{\widetilde{AU}}(x) + \tilde{v}_{\tilde{A}U}(x) \leq 1, \tilde{\mu}_{\widetilde{AL}}(x) \geq 0, \tilde{v}_{\tilde{A}L}(x) \geq 0$.

For each element x, we can compute the unknown degree (hesitancy degree) of an interval-valued intuitionistic fuzzy interval of $x \in X$ in $\tilde{A}$ defined as follows:

$$\pi_{\tilde{A}(x)} = 1 - \tilde{\mu}_{\tilde{A}}(x) - \tilde{v}_{\tilde{A}}(x) = \left([1 - \tilde{\mu}_{\widetilde{AU}}(x) - \tilde{v}_{\tilde{A}U}(x)], [1 - \tilde{\mu}_{\widetilde{AL}}(x) - \tilde{v}_{\widetilde{AL}}(x)] \right) \tag{21}$$

For convenience, let $\tilde{\mu}_{\tilde{A}}(x) = [\mu^-, \mu^+]$, $\tilde{v}_{\tilde{A}}(x) = [v^-, v^+]$, so $\tilde{A} = ([\mu^-, \mu^+], [v^-, v^+])$.

Some arithmetic operations with interval-valued intuitionistic fuzzy sets and $\lambda \geq 0$ are given in the following: Let $\tilde{I}_1 = \left([\mu_1^-, \mu_1^+], [v_1^-, v_1^+]\right)$ and $\tilde{I}_2 = \left([\mu_2^-, \mu_2^+], [v_2^-, v_2^+]\right)$ be two interval-valued intuitionistic fuzzy sets. Then,

$$\tilde{I}_1 \oplus \tilde{I}_2 = \left([\mu_1^- + \mu_2^- - \mu_1^-\mu_2^-, \mu_1^+ + \mu_2^+ - \mu_1^+\mu_2^+].[v_1^-v_2^-, v_1^+v_2^+]\right) \tag{22}$$

$$\tilde{I}_1 \otimes \tilde{I}_2 = \left([\mu_1^-\mu_2^-, \mu_1^+\mu_2^+], [v_1^- + v_2^- - v_1^-v_2^-, v_1^+ + v_2^+ - v_1^+v_2^+]\right) \tag{23}$$

$$\lambda\tilde{I}_1 = \left(\left[1 - (1 - \mu_1^-)^\lambda, 1 - (1 - \mu_1^+)^\lambda\right], \left[(v_1^-)^\lambda, (v_1^+)^\lambda\right]\right) \tag{24}$$

Using the extension principle, the arithmetic operations for interval-valued intuitionistic fuzzy numbers can be obtained by the general equation given in Eq. (25) [9].

$$\tilde{A} \circledast \tilde{B} = \left\{ \begin{array}{l} <z, \left[\max_{z=x*y} min\left\{ \mu_{\tilde{A}}^-(x), \mu_{\tilde{B}}^-(y) \right\}, \max_{z=x*y} min\left\{ \mu_{\tilde{A}}^+(x), \mu_{\tilde{B}}^+(y) \right\} \right], \\ \left[\min_{z=x*y} max\left\{ v_{\tilde{A}}^-(x), v_{\tilde{B}}^-(y) \right\}, \min_{z=x*y} max\left\{ v_{\tilde{A}}^+(x), v_{\tilde{B}}^+(y) \right\} \right] > | (x,y) \in X \times Y \end{array} \right\}$$

(25)

where the symbol "$*$" stands for one of the algebraic operations. For instance, the subtraction and summation operations for interval-valued intuitionistic fuzzy numbers are defined as in Eqs. (26) and (27):

$$\tilde{A} \ominus \tilde{B} = \left\{ \begin{array}{l} <z, \left[\max_{z=x-y} min\left\{ \mu_{\tilde{A}}^-(x), \mu_{\tilde{B}}^-(y) \right\}, \max_{z=x-y} min\left\{ \mu_{\tilde{A}}^+(x), \mu_{\tilde{B}}^+(y) \right\} \right], \\ \left[\min_{z=x-y} max\left\{ v_{\tilde{A}}^-(x), v_{\tilde{B}}^-(y) \right\}, \min_{z=x-y} max\left\{ v_{\tilde{A}}^+(x), v_{\tilde{B}}^+(y) \right\} \right] > | (x,y) \in X \times Y \end{array} \right\}$$

(26)

$$\tilde{A} \oplus \tilde{B} = \left\{ \begin{array}{l} <z, \left[\max_{z=x+y} min\left\{ \mu_{\tilde{A}}^-(x), \mu_{\tilde{B}}^-(y) \right\}, \max_{z=x+y} min\left\{ \mu_{\tilde{A}}^+(x), \mu_{\tilde{B}}^+(y) \right\} \right], \\ \left[\min_{z=x+y} max\left\{ v_{\tilde{A}}^-(x), v_{\tilde{B}}^-(y) \right\}, \min_{z=x+y} max\left\{ v_{\tilde{A}}^+(x), v_{\tilde{B}}^+(y) \right\} \right] > | (x,y) \in X \times Y \end{array} \right\}$$

(27)

In case of having multiple IVIFSs, the summation operation can be defined as in Eq. (28).

$$\sum_{j=1}^n \tilde{x}_{ij} = \left\{ \begin{array}{l} <z, \left[\begin{array}{l} \max_{z=\sum_{j=1}^n x_j} min\left\{ \mu_1^-(x_1), \mu_2^-(x_2), \ldots, \mu_n^-(x_n) \right\}, \\ \max_{z=\sum_{j=1}^n x_j} min\left\{ \mu_1^+(x_1), \mu_2^+(x_2), \ldots, \mu_n^+(x_n) \right\} \end{array} \right], \\ \left[\begin{array}{l} \min_{z=\sum_{j=1}^n x_j} max\left\{ v_1^-(x_1), v_2^-(x_2), \ldots, v_n^-(x_n) \right\}, \\ \min_{z=\sum_{j=1}^n x_j} max\left\{ v_1^+(x_1), v_2^+(x_2), \ldots, v_n^+(x_n) \right\} \end{array} \right] > \end{array} \right\}$$

(28)

Definition 3 Let $\tilde{\alpha}_j = \left([a_j, b_j], [c_j, d_j] \right)$ $(j = 1, 2, \ldots, n)$ be a collection of interval-valued intuitionistic fuzzy numbers and let IIFWA: $Q^n \to Q$, if

$$IIFWA_w(\tilde{\alpha}_1, \tilde{\alpha}_2, \ldots, \tilde{\alpha}_n) = w_1\tilde{\alpha}_1 \oplus w_2\tilde{\alpha}_2 \oplus \cdots \oplus w_n\tilde{\alpha}_n$$

(29)

Then IIFWA is called an interval-valued intuitionistic fuzzy weighted averaging (IIFWA) operator, where Q is the set of all IVIFNs, $w = (w_1, w_2, \ldots, w_n)$ is the weight vector of the IVIFNs $\tilde{\alpha}_j \, (j = 1, 2, \ldots, n)$, and $w_j > 0$, $\sum_{j=1}^{n} w_j = 1$. The IIFWA operator can be further transformed into the following form:

$$IIFWA_w(\tilde{\alpha}_1, \tilde{\alpha}_2, \ldots, \tilde{\alpha}_n) = \left(\left[1 - \left(\prod_{i=1}^{n} (1 - a_i) \right)^{w_i}, 1 - \left(\prod_{i=1}^{n} (1 - b_i) \right)^{w_i} \right], \left[\left(\prod_{i=1}^{n} c_i \right)^{w_i}, \left(\prod_{i=1}^{n} d_i \right)^{w_i} \right] \right)$$

$$(30)$$

Especially if $w = \left(\frac{1}{n}, \frac{1}{n}, \ldots, \frac{1}{n} \right)$, then the IIFWA operator reduces to an interval-valued intuitionistic fuzzy averaging (IIFA) operator, where

$$IIFA(\tilde{\alpha}_1, \tilde{\alpha}_2, \ldots, \tilde{\alpha}_n) = \frac{1}{n} (\tilde{\alpha}_1 \oplus \tilde{\alpha}_2 \oplus \cdots \oplus \tilde{\alpha}_n)$$

$$= \left(\left[1 - \left(\prod_{i=1}^{n} (1 - a_i) \right)^{1/n}, 1 - \left(\prod_{i=1}^{n} (1 - b_i) \right)^{1/n} \right], \left[\left(\prod_{i=1}^{n} c_i \right)^{1/n}, \left(\prod_{i=1}^{n} d_i \right)^{1/n} \right] \right)$$

$$(31)$$

4.2 A Proposed Method for Confidence Intervals with IVIFSs

In this section, we will estimate population mean, binomial proportion and population variance using interval-valued intuitionistic fuzzy sets and we will give numerical examples.

4.2.1 IVIF Confidence Interval for Population Mean

We will first obtain the interval-valued intuitionistic confidence interval for the population mean. Assume that we have an interval-valued intuitionistic fuzzy sample data set $\left(\tilde{x}_j, \left[\mu_j^-, \mu_j^+ \right], \left[v_j^-, v_j^+ \right] \right)$ as in Table 2.

Table 2 Interval-valued intuitionistic fuzzy sample data set

$\left(\tilde{x}_1, \left[\mu_1^-, \mu_1^+ \right], \left[v_1^-, v_1^+ \right] \right)$	$\left(\tilde{x}_2, \left[\mu_2^-, \mu_2^+ \right], \left[v_2^-, v_2^+ \right] \right)$	$\cdots$	$\left(\tilde{x}_n, \left[\mu_n^-, \mu_n^+ \right], \left[v_n^-, v_n^+ \right] \right)$

The sample mean from Table 2 is calculated as follows:

$$
\frac{\sum_{j=1}^{n} \tilde{x}_j}{n} = \frac{\left\{ <z, \begin{bmatrix} \max_{z=\sum_{j=1}^{n} x_j} \min\{\mu_1^-(x_1), \mu_2^-(x_2), \ldots, \mu_n^-(x_n)\}, \\ \max_{z=\sum_{j=1}^{n} x_j} \min\{\mu_1^+(x_1), \mu_2^+(x_2), \ldots, \mu_n^+(x_n)\} \end{bmatrix}, \begin{bmatrix} \min_{z=\sum_{j=1}^{n} x_j} \max\{v_1^-(x_1), v_2^-(x_2), \ldots, v_n^-(x_n)\}, \\ \min_{z=\sum_{j=1}^{n} x_j} \max\{v_1^+(x_1), v_2^+(x_2), \ldots, v_n^+(x_n)\} \end{bmatrix} > \right\}}{n}
$$

(32)

If σ is known and the significance level is α, then the interval estimation of fuzzy population mean is given by Eq. (33) based on Eq. (32).

$$
\left[\frac{\sum_{j=1}^{n} \tilde{x}_j}{n} \mp z_{\alpha/2} \times \frac{\sigma}{\sqrt{n}} \right]
$$

(33)

When there are multiple expert evaluations in terms of interval-valued intuitionistic fuzzy sets for a certain x_j value as seen in Table 3, the aggregation operation given in Eq. (31) can be applied. Thus, Table 3 is transformed to Table 2 composed of the aggregated values.

Example 4 Table 4 gives the interval-valued intuitionistic sample data composed of 49 random observations. Because of the space constraint, we only give a small part of the sample data. The standard deviation of the population is known to be 2.95 and the significance level is 5 %.

Table 3 Interval-valued intuitionistic fuzzy sample data in case of multi-experts

$(\tilde{x}_1, (([\mu_{11}^-, \mu_{11}^+], [v_{11}^-, v_{11}^+]), ([\mu_{12}^-, \mu_{12}^+], [v_{12}^-, v_{12}^+])))$

$(\tilde{x}_2, (([\mu_{21}^-, \mu_{21}^+], [v_{21}^-, v_{21}^+]), ([\mu_{22}^-, \mu_{22}^+], [v_{22}^-, v_{22}^+]), ([\mu_{23}^-, \mu_{23}^+], [v_{23}^-, v_{23}^+])))$

...

$(\tilde{x}_n, (([\mu_{n1}^-, \mu_{n1}^+], [v_{n1}^-, v_{n1}^+]), ([\mu_{n2}^-, \mu_{n2}^+], [v_{n2}^-, v_{n2}^+])))$

Table 4 Sample data for estimating population mean

$(72.6, [0.55, 0.75], [0, 0.15])$	$\cdots$	$(65.8, [0.75, 0.90], [0, 0.10])$
$\cdots$		$\cdots$
$(80.2, [0.35, 0.60], [0.25, 0.35])$	$\cdots$	$(63.4, [0.60, 0.85], [0.10, 0.15])$

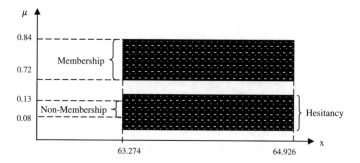

Fig. 7 Illustration of interval-valued intuitionistic confidence interval

From Table 4, we calculate the sample mean by using Eqs. (32) and (33), and find $(64.1, [0.72, 0.84], [0.08, 0.13])$. Then,

$$\left[(64.1, [0.72, 0.84], [0.08, 0.13]) \mp 1.96 \times \frac{2.95}{\sqrt{49}} \right]$$

From Eqs. (26) and (27), we obtain $(64.1, [0.72, 0.84], [0.08, 0.13]) \mp (0.826, [1.0, 1.0], [0, 0]))$. Thus, the confidence interval illustrated in Fig. 7 is expressed as

$$(63.274, [0.72, 0.84], [0.08, 0.13]) \leq \tilde{\mu} \leq (64.926, [0.72, 0.84], [0.08, 0.13]).$$

4.2.2 IVIF Confidence Interval for Binomial Proportion

Now we will develop interval-valued intuitionistic confidence interval for p (binomial proportion).

$$P \left(\begin{array}{c} \tilde{\tilde{p}}_{IVIFS} - z_{\alpha/2} \sqrt{\frac{\tilde{\tilde{p}}_{IVIFS}(1-\tilde{\tilde{p}}_{IVIFS})}{n}} \\ \leq \tilde{p}_{IVIFS} \leq \\ \tilde{\tilde{p}}_{IVIFS} + z_{\alpha/2} \sqrt{\frac{\tilde{\tilde{p}}_{IVIFS}(1-\tilde{\tilde{p}}_{IVIFS})}{n}} \end{array} \right) = 1 - \alpha \qquad (34)$$

$$P \left(\begin{array}{c} (\bar{p}, [\mu^-, \mu^+], [v^-, v^+]) - z_{\alpha/2} \sqrt{\frac{(\bar{p},[\mu^-,\mu^+],[v^-,v^+])(1-(\bar{p},[\mu^-,\mu^+],[v^-,v^+]))}{n}} \\ \leq \tilde{p}_{IVIFS} \leq \\ (\bar{p}, [\mu^-, \mu^+], [v^-, v^+]) + z_{\alpha/2} \sqrt{\frac{(\bar{p},[\mu^-,\mu^+],[v^-,v^+])(1-(\bar{p},[\mu^-,\mu^+],[v^-,v^+]))}{n}} \end{array} \right) = 1 - \alpha$$

$$(35)$$

where

$$
\tilde{P}_{IVIFS} = \left(
\begin{bmatrix}
\max\limits_{z=\frac{\sum_i x_i/n}{k}} \min\{\mu_1^-(x_1), \mu_2^-(x_2), \ldots, \mu_n^-(x_n)\}, \\
\max\limits_{z=\frac{\sum_i x_i/n}{k}} \min\{\mu_1^+(x_1), \mu_2^+(x_2), \ldots, \mu_n^+(x_n)\}
\end{bmatrix}^{<z,} ,
\begin{bmatrix}
\min\limits_{z=\frac{\sum_i x_i/n}{k}} \max\{v_1^-(x_1), v_2^-(x_2), \ldots, v_n^-(x_n)\}, \\
\min\limits_{z=\frac{\sum_i x_i/n}{k}} \max\{v_1^+(x_1), v_2^+(x_2), \ldots, v_n^+(x_n)\}
\end{bmatrix}^{>}
\right)
\tag{36}
$$

and x stands for the defectives in a sample and k denotes the number of possible defectives predicted by the experts.

Example 5 Consider the following data and compute the confidence interval for p. Let the sample size be n = 144 and the significance level be α = 5 %.

By using Eq. (31), IVIFSs in Table 5 are aggregated as follows:

$IIFA(n = 9)$
$$
= \left(\left[1 - ((1 - 0.3)(1 - 0.5))^{1/2}, 1 - ((1 - 0.5)(1 - 0.8))^{1/2}\right], \left[(0.4 \times 0)^{1/2}, (0.5 \times 0.2)^{1/2}\right] \right)
$$

$$
IIFA(n = 9) = [0.41, 0.68], [0, 0.32]
$$

Similarly,

$$
IIFA(n = 10) = \left(\left[1 - ((1 - 0.4)(1 - 0.5)(1 - 0.3))^{1/3}, 1 - ((1 - 0.7)(1 - 0.8)(1 - 0.7))^{1/3}\right], \right.
$$
$$
\left. \left[(0.1 \times 0 \times 0.2)^{1/3}, (0.3 \times 0.2 \times 0.3)^{1/3}\right] \right)
$$

$$
IIFA(n = 10) = [0.41, 0.73], [0, 0.04]
$$

$$
IIFA(n = 11) = \left(\left[1 - ((1 - 0.2)(1 - 0.3))^{1/2}, 1 - ((1 - 0.6)(1 - 0.5))^{1/2}\right], \right.
$$
$$
\left. \left[(0.1 \times 0.2)^{1/2}, (0.3 \times 0.4)^{1/2}\right] \right)
$$

$$
IIFA(n = 11) = [0.25, 0.55], [0.14, 0.35]
$$

Table 5 Interval-valued intuitionistic fuzzy sample data in case of multi-experts for the number of defectives

Number of defectives	IVIFSs
9	$(([0.3, 0.5], [0.4, 0.5]), ([0.5, 0.8], [0, 0.2]))$
10	$(([0.4, 0.7], [0.1, 0.3]), ([0.5, 0.8], [0, 0.2]), ([0.3, 0.7], [0.2, 0.3]))$
11	$(([0.2, 0.6], [0.1, 0.3]), ([0.3, 0.5], [0.2, 0.4]))$

Now, we apply Eq. (36) to obtain the mean value of defective numbers.

$$\tilde{\tilde{p}}_{IVIFS} = \left(\frac{\left(\frac{9}{144} + \frac{10}{144} + \frac{11}{144}\right)}{3}, [0.25, 0.55], [0.14, 0.35] \right)$$
$$= (0.07, [0.25, 0.55], [0.14, 0.35])$$

Then, using Eq. (35), the IVIF confidence interval for p is estimated as in the following:

$$(0.07, [0.25, 0.55], [0.14, 0.35])$$
$$\mp 1.96 \sqrt{\frac{(0.07, [0.25, 0.55], [0.14, 0.35])(1 - (0.07, [0.25, 0.55], [0.14, 0.35]))}{144}}$$

$$(0.03, [0.25, 0.55], [0.14, 0.35]) \le \tilde{p} \le (0.11, [0.25, 0.55], [0.14, 0.35])$$

This result indicates that the binomial proportion is between 0.03 and 0.11 with a membership of [0.25, 0.55] and a non-membership of [0.14, 0.35]).

4.2.3 IVIF Confidence Interval for Population Variance

Now we will develop interval-valued intuitionistic confidence interval for σ^2.

$$\frac{\tilde{s}^2_{IVIFS}(n-1)}{\chi^2_{n-1,1-\alpha/2}} \le \tilde{\sigma}^2_{IVIFS} \le \frac{\tilde{s}^2_{IVIFS}(n-1)}{\chi^2_{n-1,\alpha/2}} \tag{37}$$

$$P\left(\frac{(s^2, [\mu^-, \mu^+], [v^-, v^+])(n-1)}{\chi^2_{n-1,1-\alpha/2}} \le \tilde{\sigma}^2_{IVIFS} \le \frac{(s^2, [\mu^-, \mu^+], [v^-, v^+])(n-1)}{\chi^2_{n-1,\alpha/2}} \right)$$
$$= 1 - \alpha \tag{38}$$

where

$$\tilde{s}^2_{IVIFS} = \left(\begin{array}{c} \left[\begin{array}{c} \max\limits_{z=\frac{\sum_i s_i^2}{k}} \min\{\mu_1^-(x_1), \mu_2^-(x_2), \ldots, \mu_n^-(x_n)\}, \\ \max\limits_{z=\frac{\sum_i s_i^2}{k}} \min\{\mu_1^+(x_1), \mu_2^+(x_2), \ldots, \mu_n^+(x_n)\} \end{array} \right], \\ \left[\begin{array}{c} \min\limits_{z=\frac{\sum_i s_i^2}{k}} \max\{v_1^-(x_1), v_2^-(x_2), \ldots, v_n^-(x_n)\}, \\ \min\limits_{z=\frac{\sum_i s_i^2}{k}} \max\{v_1^+(x_1), v_2^+(x_2), \ldots, v_n^+(x_n)\} \end{array} \right] \end{array} \right) \tag{39}$$

and k represents the number of estimations for S^2.

Table 6 Interval-valued intuitionistic fuzzy sample data in case of multi-experts for S^2

S_i^2	IVIFSs
0.27	$([0.2, 0.4], [0.30, 0.50])$
0.29	$(([0.3, 0.6], [0.1, 0.4]), ([0.4, 0.6], [0.2, 0.4]), ([0.5, 0.7], [0.2, 0.3]))$
0.32	$(([0.4, 0.6], [0.3, 0.4]), ([0.2, 0.5], [0.4, 0.5]))$

Example 6 Consider the following data and compute the confidence interval for σ^2. Let the sample size be $n = 25$ and the significance level be $\alpha = 5\,\%$.

From the uncertain data, the various calculated values of S_i^2 are displayed in Table 6 with their IVIFSs.

By using Eq. (31), IVIFSs in Table 6 are aggregated as follows:

$$IIFA(S_2^2 = 0.29) = \left(\left[1 - ((1 - 0.3)(1 - 0.4)(1 - 0.5))^{1/3}, 1 - ((1 - 0.6)(1 - 0.6)(1 - 0.7))^{1/3} \right], \right.$$
$$\left. \left[(0.1 \times 0.2 \times 0.2)^{1/3}, (0.4 \times 0.4 \times 0.3)^{1/3} \right] \right)$$

$$IIFA(S_2^2 = 0.29) = [0.41, 0.64], [0, 16, 0.36]$$

$$IIFA(S_3^2 = 0.32) = \left(\left[1 - ((1 - 0.4)(1 - 0.2))^{1/2}, 1 - ((1 - 0.6)(1 - 0.5))^{1/2} \right], \right.$$
$$\left. \left[(0.3 \times 0.4)^{1/2}, (0.4 \times 0.5)^{1/2} \right] \right)$$

$$IIFA(S_3^2 = 0.32) = [0.31, 0.55], [0.35, 0.45]$$

The mean value of S^2 is calculated using Eq. (39) as follows:

$$\tilde{s}_{IVIFS}^2 = \left(\frac{0.27 + 0.29 + 0.32}{3}, [0.20, 0.40], [0.35, 0.50] \right)$$
$$= (0.293, [0.20, 0.40], [0.35, 0.50])$$

Then, using Eq. (38), the IVIF confidence interval for σ_{IVIFS}^2 is estimated as in the following:

$$P\left(\frac{(0.293, [0.20, 0.40], [0.35, 0.50])(24)}{39.364} \leq \tilde{\sigma}_{IVIFS}^2 \leq \frac{(0.293, [0.20, 0.40], [0.35, 0.50])(24)}{12.401} \right) = 1 - \alpha$$

$$(0.18, [0.20, 0.40], [0.35, 0.50]) \leq \tilde{\sigma}_{IVIFS}^2 \leq (0.57, [0.20, 0.40], [0.35, 0.50])$$

This result indicates that the population variance is between 0.18 and 0.57 with a membership of [0.20, 0.40] and a non-membership of [0.35, 0.50].

5 Hesitant Fuzzy Confidence Intervals

5.1 Hesitant Fuzzy Sets

Hesitant fuzzy sets (HFS) is a novel and recent extension of fuzzy sets that aim to model the uncertainty originated by the hesitation that might arise in the assignment of membership degrees of the elements to a fuzzy set. A HFS is defined in terms of a function that returns a set of membership values for each element in the domain.

Definition 4 Let X be a reference set, a HFS on X is a function $\mathfrak{h}$ that returns a subset of values in [0,1] [16]:

$$\mathfrak{h} : X \rightarrow \varphi([0,1]) \qquad (40)$$

Definition 5 Let $M = \{\mu_1, \mu_2, \ldots, \mu_n\}$ be a set of n membership functions. The HFS associated to M, $\mathfrak{h}_M$ is defined as [16]:

$$\mathfrak{h}_M : X \rightarrow \varphi([0,1])$$
$$\mathfrak{h}_M(x) = \cup_{\mu \in M} \{\mu(x)\} \qquad (41)$$

where $x \in X$.

Xia and Xu [19] completed the original definition of HFS by including the mathematical representation of a HFS as follows:

$$E = \{\langle x, h_E(x)\rangle : x \in X\} \qquad (42)$$

where $h_E(x)$ is a set of some values in [0,1], denoting the possible membership degrees of the element $x \in X$ to the set E. For convenience, Xia and Xu [19] noted $h = h_E(x)$ and called it Hesitant Fuzzy Element (HFE) of E and $H = \cup h_E(x)$, the set of all HFEs of E.

Chen et al. [5] presented the definition of Interval-Valued Hesitant Fuzzy Set (IVHFS), as a generalization of HFS in which the membership degrees of an element to a given set are defined by several possible interval values. An IVHFS is defined as follows:

Definiton 6 Let X be a reference set, and $I([0, 1])$ be a set of all closed subintervals of [0, 1]. An IVHFS on X is,

$$\tilde{A} = \{x_i, \tilde{h}_A(x_i) | x_i \in X, i = 1, 2, \ldots, n\} \qquad (43)$$

where $\tilde{h}_A(x_i)$ $X \rightarrow \varphi(I([0, 1]))$ denotes all possible interval-valued membership degrees of the element $x_i \in X$ to the set $\tilde{A}$. Aggregation of the interval-valued hesitant fuzzy sets is obtained using Definition 7.

Definition 7 Let $\tilde{h}_i(i = 1, 2, \ldots, n)$ be a collection of interval-valued hesitant fuzzy elements (IVHFE).Their aggregated value can be calculated by using hesitant interval-valued fuzzy weighted averaging (HIVFWA) operation given in Eq. (44) [17]:

$$HIVFWA\left(\tilde{h}_1, \tilde{h}_2, \ldots, \tilde{h}_n\right) = \left\{\left[1 - \prod_{j=1}^{n}\left(1 - \gamma_j^L\right)^{w_j}, 1 - \prod_{j=1}^{n}(1 - \gamma_j^R)^{w_j}\right]\right\} \quad (44)$$

where $w_j(j = 1, .., n)$ is the weighting vector of h_j and $w_j > 0, \sum_{j=1}^{n} w_j = 1$.

Using the extension principle, the arithmetic operations for interval-valued hesitant fuzzy numbers can be obtained by the general equation given in Eqs. (45) and (46).

$$\tilde{A} \ominus \tilde{B} = \left\{ <z, \left[\max_{z=x-y} min\left\{\mu_{\tilde{A}}^-(x), \mu_{\tilde{B}}^-(y)\right\}, \max_{z=x-y} min\left\{\mu_{\tilde{A}}^+(x), \mu_{\tilde{B}}^+(y)\right\}\right] > |(x, y) \in X \times Y \right\} \quad (45)$$

$$\tilde{A} \oplus \tilde{B} = \left\{ <z, \left[\max_{z=x+y} min\left\{\mu_{\tilde{A}}^-(x), \mu_{\tilde{B}}^-(y)\right\}, \max_{z=x+y} min\left\{\mu_{\tilde{A}}^+(x), \mu_{\tilde{B}}^+(y)\right\}\right] > |(x, y) \in X \times Y \right\} \quad (46)$$

5.2 A Proposed Method for Confidence Intervals with IVHFSs

In this section, we will estimate population mean, binomial proportion and population variance using interval-valued hesitant fuzzy sets and we will give numerical examples.

5.2.1 IVHF Confidence Interval for Population Mean

We will first obtain the interval-valued hesitant confidence interval for the population mean. Assume that we have an interval-valued hesitant fuzzy sample data set $(\tilde{x}_j, [\mu_j^-, \mu_j^+])$, $j = 1, \ldots, n$. as in Table 7, obtained from m experts.

First of all, the aggregation of hesitant fuzzy values is realized by using Eq. (44). Thus, Table 7 is transformed to Table 8. Then, Eq. (47) is used to calculate the sample mean.

Table 7 Interval-valued hesitant fuzzy sample data set

$$\left(\tilde{x}_1, \left(\left[\mu_{11}^-, \mu_{11}^+\right], \left[\mu_{12}^-, \mu_{12}^+\right]\right)\right)$$
$$\left(\tilde{x}_2, \left(\left(\left[\mu_{21}^-, \mu_{21}^+\right]\right), \left(\left[\mu_{22}^-, \mu_{22}^+\right]\right), \left(\left[\mu_{23}^-, \mu_{23}^+\right]\right)\right)\right)$$
…
$$\left(\tilde{x}_n, \left(\left[\mu_{n1}^-, \mu_{n1}^+\right], \left[\mu_{n2}^-, \mu_{n2}^+\right]\right)\right.$$

Table 8 Aggregated interval-valued hesitant fuzzy sample data set

$$\left(\tilde{x}_1, \left[\mu_{a1}^-, \mu_{a1}^+\right]\right)$$
$$\left(\tilde{x}_2, \left[\mu_{a2}^-, \mu_{a2}^+\right]\right)$$
…
$$\left(\tilde{x}_n, \left[\mu_{an}^-, \mu_{an}^+\right]\right)$$

$$\frac{\sum_{j=1}^{n} \tilde{x}_j}{n} = \frac{\left\{ <z, \begin{array}{c} \max\limits_{z=\sum_{j=1}^{n} x_j} min\{\mu_1^-(x_1), \mu_2^-(x_2), \dots, \mu_n^-(x_n)\}, \\ \max\limits_{z=\sum_{j=1}^{n} x_j} min\{\mu_1^+(x_1), \mu_2^+(x_2), \dots, \mu_n^+(x_n)\} \end{array} > \right\}}{n} \tag{47}$$

If σ is known and the significance level is α, then the interval estimation of fuzzy population mean is given by Eq. (48).

$$\frac{\sum_{j=1}^{n} \tilde{x}_j}{n} \mp z_{\alpha/2} \times \frac{\sigma}{\sqrt{n}} \tag{48}$$

Example 7 Table 9 gives the interval-valued hesitant sample data composed of 36 random observations. Because of the space constraint, we only give a small part of the sample data. The standard deviation of the population is known to be 1.75 and the significance level is 5 %.

From Table 9, we calculate the sample mean by using Eqs. (47) and (48), and find $(78.30, [0.65, 0.85])$. Then,

$$\left[(78.30, [0.65, 0.85]) \mp 1.96 \times \frac{1.75}{\sqrt{36}} \right]$$

Table 9 Sample data for numerical example

$(82.4, [0.55, 0.75], [0.60, 0.65], [0.45, 0.70])$	⋯	$(77.9, [0.75, 0.90], [0.65, 0.85])$
…		…
$(74.5, [0.35, 0.60], [0.40, 0.55])$	⋯	$(80.9, [0.60, 0.85], [0.65, 0.80], [0.60, 0.75])$

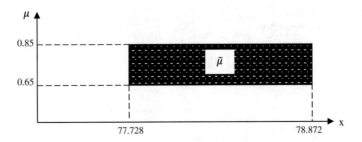

Fig. 8 Illustration of interval-valued hesitant confidence interval

From Eqs. (45) and (46), we obtain $(78.30, [0.65, 0.85]) \mp (0.572, [1.0, 1.0])$. Thus, the confidence interval illustrated in Fig. 8 is expressed as $(77.728, [0.65, 0.85]) \leq \tilde{\mu} \leq (78.872, [0.65, 0.85])$.

5.2.2 IVHF Confidence Interval for Binomial Proportion

Now we will develop interval-valued hesitant confidence interval for p (binomial proportion).

$$P\left(\begin{array}{c} \tilde{\tilde{p}}_{IVHFS} - z_{\alpha/2}\sqrt{\frac{\tilde{\tilde{p}}_{IVHFS}(1-\tilde{\tilde{p}}_{IVHFS})}{n}} \\ \leq \tilde{p}_{IVHFS} \leq \\ \tilde{\tilde{p}}_{IVHFS} + z_{\alpha/2}\sqrt{\frac{\tilde{\tilde{p}}_{IVHFS}(1-\tilde{\tilde{p}}_{IVHFS})}{n}} \end{array} \right) = 1 - \alpha \qquad (49)$$

$$P\left(\begin{array}{c} (\bar{p}, [\mu^-, \mu^+]) - z_{\alpha/2}\sqrt{\frac{(\bar{p}, [\mu^-, \mu^+])(1-(\bar{p}, [\mu^-, \mu^+]))}{n}} \\ \leq \tilde{p}_{IVHFS} \leq \\ (\bar{p}, [\mu^-, \mu^+]) + z_{\alpha/2}\sqrt{\frac{(\bar{p}, [\mu^-, \mu^+])(1-(\bar{p}, [\mu^-, \mu^+]))}{n}} \end{array} \right) = 1 - \alpha \qquad (50)$$

where

$$\tilde{\tilde{p}}_{IVHFS} = \left(\quad <z, \left[\begin{array}{c} \max\limits_{z=\frac{\sum_i x_i/n}{k}} \min\{\mu_1^-(x_1), \mu_2^-(x_2), \ldots, \mu_n^-(x_n)\}, \\ \max\limits_{z=\frac{\sum_i x_i/n}{k}} \min\{\mu_1^+(x_1), \mu_2^+(x_2), \ldots, \mu_n^+(x_n)\} \end{array} \right] > \right) \qquad (51)$$

Table 10 Interval-valued hesitant fuzzy sample data in case of multi-experts	Number of defectives	IVHFSs
	2	$\{[0.2, 0.4], [0.3, 0.5]\}$
	3	$\{[0.4, 0.8], [0.6, 0.8], [0.7, 0.9]\}$
	4	$\{[0.2, 0.3], [0.3, 0.4], [0.2, 0.4]\}$

and x stands for the defectives in a sample and k denotes the number of possible defectives predicted by the experts.

Example 8 Consider the following data and compute the confidence interval for p. Let the sample size be n = 169 and the significance level be α = 5 %.

By using Eq. (44), IVHFSs in Table 10 are aggregated as follows:

$$HIVFWA(n = 2) = \left(\left[1 - ((1 - 0.2)(1 - 0.3))^{1/2} \right], \left[1 - ((1 - 0.4)(1 - 0.5))^{1/2} \right] \right)$$

$$HIVFWA(n = 2) = [0.25, 0.45]$$

$$HIVFWA(n = 3)$$
$$= \left(\left[1 - ((1 - 0.4)(1 - 0.6)(1 - 0.7))^{1/3}, 1 - ((1 - 0.8)(1 - 0.8)(1 - 0.9))^{1/3} \right] \right)$$

$$HIVFWA(n = 3) = [0.58, 0.86]$$

$$HIVFWA(n = 4)$$
$$= \left(\left[1 - ((1 - 0.2)(1 - 0.3)(1 - 0.2))^{1/3}, 1 - ((1 - 0.3)(1 - 0.4)(1 - 0.4))^{1/3} \right] \right)$$

$$HIVFWA(n = 4) = [0.23, 0.40]$$

Now, we apply Eq. (51) to obtain the mean value of defective numbers.

$$\tilde{\tilde{p}}_{IVHFS} = \left(\frac{\left(\frac{2}{169} + \frac{3}{169} + \frac{4}{169} \right)}{3}, [0.23, 0.40] \right) = (0.05, [0.23, 0.40])$$

Then, using Eq. (50), the IVHF confidence interval for p is estimated as in the following:

$$(0.05, [0.23, 0.40]) \mp 1.96 \sqrt{\frac{(0.05, [0.23, 0.40])(1 - (0.05, [0.23, 0.40]))}{169}}$$

$$(0.02, [0.23, 0.40]) \leq \tilde{p}_{IVHFS} \leq (0.08, [0.23, 0.40])$$

This result indicates that the binomial proportion is between 0.02 and 0.08 with a membership of $[0.23, 0.40]$

5.2.3 IVHF Confidence Interval for Population Variance

Now we will develop interval-valued hesitant confidence interval for σ^2.

$$\frac{\tilde{s}_{IVHFS}^2(n-1)}{\chi_{n-1,1-\alpha/2}^2} \leq \tilde{\sigma}_{IVHFS}^2 \leq \frac{\tilde{s}_{IVHFS}^2(n-1)}{\chi_{n-1,\alpha/2}^2} \tag{52}$$

$$P\left(\frac{(s^2, [\mu^-, \mu^+])(n-1)}{\chi_{n-1,1-\alpha/2}^2} \leq \tilde{\sigma}_{IVHFS}^2 \leq \frac{(s^2, [\mu^-, \mu^+])(n-1)}{\chi_{n-1,\alpha/2}^2}\right) = 1 - \alpha \tag{53}$$

where

$$\tilde{s}_{IVHFS}^2 = \left(\frac{\sum_i S_i^2}{k}, \quad <z, \quad \begin{bmatrix} \max\limits_{z=\frac{\sum_i S_i^2}{k}} \min\{\mu_1^-(x_1), \mu_2^-(x_2), \ldots, \mu_n^-(x_n)\}, \\ \max\limits_{z=\frac{\sum_i S_i^2}{k}} \min\{\mu_1^+(x_1), \mu_2^+(x_2), \ldots, \mu_n^+(x_n)\} \\ > \end{bmatrix}\right), \tag{54}$$

$s^2 = \frac{\sum_i S_i^2}{k}$ and k represents the number of estimations for S^2.

Example 9 Consider the following data and compute the confidence interval for σ^2. Let the sample size be $n = 28$ and the significance level be $\alpha = 5\%$.

From the uncertain data, the various calculated values of S_i^2 are displayed in Table 11 with their IVHFSs.

By using Eq. (44), IVHFSs in Table 11 are aggregated as follows:

$$HIVFWA(S_1^2 = 1.45)$$
$$= \left(\left[1 - ((1-0.2)(1-0.4))^{1/2}\right], \left[1 - ((1-0.4)(1-0.5))^{1/2}\right]\right)$$

$$HIVFWA(S_1^2 = 1.45) = [0.31, 0.45]$$

Table 11 Interval-valued hesitant fuzzy sample data in case of multi-experts for S^2

S_i^2	IVHFS
1.45	$[0.2, 0.4], [0.4, 0.5]$
1.53	$[0.5, 0.7], [0.45, 0.65], [0.55, 0.75]$
1.59	$[0.35, 0.55], [0.25, 0.45]$

$$HIVFWA(S_2^2 = 1.53) = \left(\left[1 - ((1 - 0.5)(1 - 0.45)(1 - 0.55))^{1/3} \right], \right.$$
$$\left. \left[1 - ((1 - 0.7)(1 - 0.65)(1 - 0.75))^{1/3} \right] \right)$$

$$HIVFWA(S_2^2 = 1.53) = [0.50, 0.70]$$

$$HIVFWA(S_3^2 = 1.59) = \left(\left[1 - ((1 - 0.35)(1 - 0.25))^{1/2} \right], \left[1 - ((1 - 0.55)(1 - 0.45))^{1/2} \right] \right)$$

$$HIVFWA(S_3^2 = 1.59) = [0.30, 0.50]$$

The mean value of S^2 is calculated using Eq. (53) as follows:

$$\tilde{s}_{IVHFS}^2 = \left(\frac{1.45 + 1.53 + 1.59}{3}, [0.30, 0.45] \right) = (1.52, [0.30, 0.45])$$

Then, using Eq. (54), the IVHF confidence interval for $\tilde{\sigma}_{IVHFS}^2$ is estimated as in the following:

$$P\left(\frac{(1.52, [0.30, 0.45])(27)}{43.195} \leq \tilde{\sigma}_{IVIFS}^2 \leq \frac{(1.52, [0.30, 0.45])(27)}{14.573} \right) = 1 - \alpha$$

$$(0.95, [0.30, 0.45]) \leq \tilde{\sigma}_{IVIFS}^2 \leq (2.82, [0.30, 0.45])$$

This result indicates that the population variance is between 0.95 and 2.82 with a membership of $[0.30, 0.45]$.

6 Conclusion Remarks and Future Research Suggestions

This chapter deals with the interval estimation of a single parameter such as μ, σ, and p. Various approaches to ordinary fuzzy confidence intervals have been proposed by several researchers by today. The chapter summarizes these approaches by giving numerical examples. It also includes new approaches for confidence intervals using the extensions of fuzzy sets, namely hesitant fuzzy sets and intuitionistic fuzzy sets. Each proposed approach has been illustrated by a numerical example. The future studies may be based on the other extensions of fuzzy sets such as fuzzy multisets, and type-II fuzzy sets, etc. Other types of hesitant and intuitionistic fuzzy sets may be also used for the development of fuzzy confidence intervals such as triangular hesitant fuzzy sets, and triangular intuitionistic fuzzy sets, etc.

References

1. Alex, R.: Fuzzy point estimation and its application on fuzzy supply chain analysis. Fuzzy Sets Syst. **158**(14), 1571–1587 (2007)
2. Atanassov, K.T.: Intuitionistic fuzzy sets. Fuzzy Sets Syst. (Elsevier) **20**, 87–96 (1986)
3. Buckley, J.J.: Fuzzy Statistics. Springer, Berlin (2004)
4. Chachi, J., Taheri, S.M.: Fuzzy confidence intervals for mean of Gaussian fuzzy random variables. Expert Syst. Appl. **38**, 5240–5244 (2011)
5. Chen, N., Xu, Z.S., Xia, M.M.: Interval-valued hesitant preference relations and their applications to group decision making. Knowl.-Based Syst. **37**(1), 528–540 (2013)
6. Garg, H.: Performance analysis of complex repairable industrial systems using PSO and fuzzy confidence interval based methodology. ISA Trans. **52**, 171–183 (2013)
7. Garg, H.: Analyzing the behavior of an industrial system using fuzzy confidence interval based methodology. Natl. Acad. Sci. Lett. **37**(4), 359–370 (2014)
8. Kaya, İ., Kahraman, C.: A new perspective on fuzzy process capability indices: robustness. Expert Syst. Appl. **37**(6), 4593–4600 (2010)
9. Li, D.F.: Extension principles for interval-valued intuitionistic fuzzy sets and algebraic operations. Fuzzy Optim. Decis. Making **10**(1), 45–58 (2011)
10. McCain, R.A.: Fuzzy confidence intervals. Fuzzy Sets Syst. **10**(1–3), 281–290 (1983)
11. McCain, R.A.: Fuzzy confidence intervals in a theory of economic rationality. Fuzzy Sets Syst. **23**(2), 205–218 (1987)
12. Montgomery, D.C., Runger, G.C.: Applied Statistics and Probability for Engineers. John Wiley & Sons, New York (1994)
13. Parchami, A., Mashinchi, M.: Fuzzy estimation for process capability indices. Inf. Sci. **177**(6), 1452–1462 (2007)
14. Skrjanc, I.: Confidence interval of fuzzy models: an example using a waste-water treatment plant. Chemom. Intell. Lab. Syst. **96**(2), 15, 182–187 (2009)
15. Skrjanc, I.: Fuzzy confidence interval for pH titration curve. Appl. Math. Model. **35**, 4083–4090 (2011)
16. Torra, V.: Hesitant fuzzy sets. Int. J. Intell. Syst. **25**(6), 529–539 (2010)
17. Wei, G., Zhao, X., Lin, R.: Some hesitant interval-valued fuzzy aggregation operators and their applications to multiple attribute decision making. Knowl.-Based Syst. **46**, 43–53 (2013)
18. Wu, H.C.: Statistical confidence intervals for fuzzy data. Expert Syst. Appl. **36**(2), 2670–2676 (2009)
19. Xia, M.M., Xu, Z.S.: Hesitant fuzzy information aggregation in decision making. Int. J. Approximate Reasoning **52**, 395–407 (2011)

Testing Fuzzy Hypotheses: A New *p*-value-based Approach

Abbas Parchami, S. Mohmoud Taheri, Bahram Sadeghpour Gildeh
and Mashaallah Mashinchi

Abstract In this paper, on the basis of Zadeh's probability measure of fuzzy
events, the *p*-value concept is generalized for testing fuzzy hypotheses. We prove
that the introduced *p*-value has uniform distribution over (0, 1) when the null fuzzy
hypothesis is true. Then, based on such a *p*-value, a procedure is illustrated to test
various types of fuzzy hypotheses. Several applied examples are given to show the
performance of the method.

Keywords Fuzzy statistics · Testing hypothesis · Fuzzy hypothesis · *p*-value

1 Introduction

In testing statistical hypotheses, we may face with situations in which hypotheses
are imprecise (fuzzy) rather than crisp. For instance, suppose that the interested
parameter θ is the proportion of a population which has a certain disease. We take a
random sample from population and study the sample for having some idea about θ.
In ordinary hypotheses testing, one uses the hypotheses of the form:
"$H_0 : \theta = 0.15$", versus "$H_1 : \theta \neq 0.15$"; or of the form: "$H_0 : \theta \geq 0.15$", versus
"$H_1 : \theta < 0.15$"; and so on. However, we would sometimes like to test more realistic hypotheses, such as small, very small, large, approximately 0.15, and so on.
In this situation, fuzzy expressions are more suitable to real life problems than

A. Parchami (✉) · M. Mashinchi
Department of Statistics, Faculty of Mathematics and Computer,
Shahid Bahonar University of Kerman, Kerman, Iran
e-mail: parchami@uk.ac.ir

S.M. Taheri
Faculty of Engineering Science, College of Engineering,
University of Tehran, Tehran, Iran

B. Sadeghpour Gildeh
Department of Statistics, Faculty of Mathematical Science,
Ferdowsi University of Mashhad, Mashhad, Iran

© Springer International Publishing Switzerland 2016
C. Kahraman and Ö. Kabak (eds.), *Fuzzy Statistical Decision-Making*,
Studies in Fuzziness and Soft Computing 343,
DOI 10.1007/978-3-319-39014-7_10

classical crisp hypotheses. Therefore, more suitable formulation of the hypotheses might be, "$\tilde{H}_0 : \theta$ is small", versus "$\tilde{H}_1 : \theta$ is not small". We call such vague expressions as fuzzy hypotheses. In this paper we are going to present a new p-value-based approach for testing fuzzy hypotheses.

Let us review some works on the topic of testing hypotheses in fuzzy environment. Tanaka et al. [26] investigated hypotheses testing problem with fuzzy data in the decision problem framework. In decision theory framework, Casals et al. [7] proposed Bayes and minimax fuzzy tests for testing statistical hypotheses based on vague data. By using Zadeh's probabilistic definition, Casals et al. [6] extended the Neyman-Pearson Lemma and Bayesian approach to testing hypotheses, where the observations are fuzzy. Arnold [3, 4] worked on testing fuzzy hypotheses with crisp data. He gave new definitions for probability of type I and type II errors and presented a best test for the one-parameter exponential family. Grzegorzewski and Hryniewicz [13] reviewed some methods in testing statistical hypotheses in fuzzy environment, pointing out their advantages or disadvantages and practical problems. Taheri and Behboodian [24] formulated the problem of testing fuzzy hypotheses when the hypotheses are fuzzy and the observations are crisp. In order to establish optimality criteria, they gave new definitions for probability of type I and type II errors. Then, on the basis of these new errors, they stated and proved Neyman-Pearson Lemma for testing fuzzy hypotheses. Also they studied on the problem of hypotheses testing, from a Bayesian point of view, when the observations are ordinary and the hypotheses are fuzzy, see Taheri and Behboodian [25]. Arnold and Gerke [5] studied testing fuzzy linear hypotheses in linear regression models. Holeňa [14] presented a principally different approach for testing fuzzy hypotheses, which was motivated by observational logic and its success in automated knowledge discovery. Using a generalized metric for fuzzy numbers, Montenegro et al. [17] proposed a method to test the fuzzy mean of a fuzzy random variable. Torabi and Behboodian [27] introduced the likelihood ratio test for testing fuzzy hypotheses problem. González-Rodríguez el al. [12] introduced a bootstrap approach to the one-sample test of mean for imprecisely valued sample data. Filzmoser and Viertl [9] worked on the problem of testing hypotheses, and introduced a fuzzy p-value when the observations are fuzzy and hypotheses are crisp. In a similar framework, and using the extension principle, Parchami et al. [20] worked on testing fuzzy hypotheses problem with crisp data and introduced a concept of fuzzy p-value for such situations. For combining the ideas of two recent works, see Parchami et al. [22]. González-Rodríguez el al. [11] developed a one-way ANOVA test for fuzzy observations in which the fuzzy observations are treated as functional data of a functional Hilbert space.

In this paper, a new p-value-based method is introduced for testing fuzzy hypotheses. The proposed method is on the basis of the concept of probability measure of fuzzy events by Zadeh [30]. Also, it must be mentioned that all results of this study coincide to the results of testing classical hypotheses, when the hypotheses reduce to two crisp sets of the parameter space.

This paper is organized as follows. In Sect. 2, we review p-value approach to testing classical hypotheses. In Sects. 3 and 4, we recall and redefine some preliminaries and concepts about fuzzy hypotheses and measuring the probability under a fuzzy hypothesis. In Sect. 5, we present a new p-value-based approach for testing fuzzy hypotheses problem. The distribution of the introduced p-value under null fuzzy hypothesis is discussed in Sect. 6. Some applied examples are presented in Sect. 7. A conclusion is given in the final section.

2 Testing Statistical Hypotheses

Let $\mathbf{X} = (X_1, \ldots, X_n)$ be a random sample with the observed value $\mathbf{x} = (x_1, \ldots, x_n)$, where X_i has the probability density function (p.d.f.) or the probability mass function (p.m.f.) $f(x_i; \theta)$, $i = 1, \ldots, n$ with the unknown parameter $\theta \in \Theta \subseteq R$. It will be assumed that the functional form of $f(x; \theta)$ is known. The problem of testing statistical hypotheses is to decide whether to accept (or reject) the null hypothesis $H_0 : \theta \in \Theta_0 \subset \Theta$ against $H_1 : \theta \in \Theta_0^c = \Theta - \Theta_0$, based on the random sample $\mathbf{X}$. Usually, statistical hypotheses are one of the following forms:

  (i) $H_0 : \theta = \theta_0$ versus $H_1 : \theta = \theta_1$ $(\theta_0 > \theta_1)$
  (ii) $H_0 : \theta = \theta_0$ versus $H_1 : \theta = \theta_1$ $(\theta_0 < \theta_1)$
  (iii) $H_0 : \theta \geq \theta_0$ versus $H_1 : \theta < \theta_0$
  (iv) $H_0 : \theta \leq \theta_0$ versus $H_1 : \theta > \theta_0$
  (v) $H_0 : \theta = \theta_0$ versus $H_1 : \theta \neq \theta_0$

in which θ_0 and θ_1 are two known real numbers and we named them the boundary of the null and alternative hypotheses, respectively. A test φ is said to be a test of (significance) level $\alpha \in [0, 1]$ if $\alpha_\varphi \leq \alpha$, where $\alpha_\varphi = \sup_{\theta \in \Theta_0} P_\theta$ (Rejection of H_0). Commonly, the statistical tests are based on a so called test statistic $T(\mathbf{X})$. In a nonrandomized test, the space of possible values of the test statistic T is decomposed into a rejection region and its complement, the acceptance region. Under some certain conditions, the rejection region usually takes one of the following forms:

$$\textbf{(a) } T \leq t_l \qquad \textbf{(b) } T \geq t_r \qquad \textbf{(c) } T \notin (t_1, t_2) \tag{1}$$

where, t_l, t_r, or t_1 and t_2 are certain quantiles of the distribution of T, so that $\alpha_\varphi = \alpha$. In case (c), we may obtain t_1 and t_2 by the equal tails method, so that $P_\theta(T \leq t_1) = P_\theta(T \geq t_2) = \alpha/2$. The hypothesis H_0 is rejected if the value $t = t(\mathbf{x})$ falls into the rejection region. In usual tests, the critical regions of testing hypotheses (i) and (iii) are of form (a), the critical region of testing hypotheses (ii) and (iv) are of form (b), and the critical region of testing hypotheses (v) is of form (c). For more details see [9] and Page 381 of [15].

There exist different statistical approaches to the problem of testing hypotheses. One approach is to use p-value, which is defined as the smallest significance level leading to rejection of the null hypothesis (see Page 381 of [15]). The p-value for cases (1.a), (1.b) and (1.c) could be obtained as follows

(a) $P_{\theta_0}(T \leq t)$

(b) $P_{\theta_0}(T \geq t)$

(c) $2 \min[P_{\theta_0}(T \leq t), P_{\theta_0}(T \geq t)] = \begin{cases} 2P_{\theta_0}(T \geq t) & \text{if } t \geq m \\ 2P_{\theta_0}(T \leq t) & \text{if } t \leq m \end{cases}$

where θ_0 is the boundary of the null hypothesis, m is the median of the distribution of T and t is the observed value of test statistic (T); see Filzmoser and Viertl [9] and Parchami et al. [20]. If the p-value is less than α, then null hypothesis is rejected at the significance level α, otherwise null hypothesis is not rejected.

3   Fuzzy Hypotheses: Motivations and Basic Concepts

Traditionally, the hypotheses for which we wish to provide a test should be formulized in precise assertions. This limitation, sometimes, forces a statistician to make decision procedure in an unrealistic manner. But, in realistic problems, we may come across non-precise (fuzzy) hypotheses.

Definition 1 [24] Any hypothesis of the form "$\tilde{H} : \theta$ is $H(\theta)$" is called a fuzzy hypothesis, where "θ is $H(\theta)$" implies that θ is in a fuzzy set of Θ, the parameter space, with membership function $H(\theta)$.

Note that the ordinary hypothesis "$H : \theta \in \Theta_0$" is a fuzzy hypothesis with the membership function $H(\theta) = 1$ at $\theta \in \Theta_0$, and zero otherwise, i.e., the indicator function of the crisp set Θ_0.

Example 1 Let θ be the parameter of a binomial distribution. Consider the following function

$$H_0(\theta) = \begin{cases} 2\theta & \text{if } 0 < \theta < \frac{1}{2}, \\ 2 - 2\theta & \text{if } \frac{1}{2} \leq \theta < 1. \end{cases}$$

The hypothesis "$\tilde{H}_0 : \theta$ is $H(\theta)$" is a fuzzy hypothesis and it means that $\theta \simeq \frac{1}{2}$, i.e. "$\theta$ is approximately $\frac{1}{2}$" (see Fig. 1).

Example 2 Suppose that we are going to have an investigation on the amount of Cadmium (Cd) absorption in a plant from a polluted soil. The unknown parameter is the amount of Cd uptake in a plant (in term of mg kg^{-1} DM) from soil which we denoted it with μ. The optimum range Cd absorbed in a plant have been proposed by Pais and Benton [19] as [0.05, 0.2]; and also its maximum have been specified by 3 mg kg^{-1} DM. The experimenter wants to investigate on the following question: Whether the mean of Cd absorption coincides to the proposed suitable

Fig. 1 The membership functions of the fuzzy hypothesis in Example 1

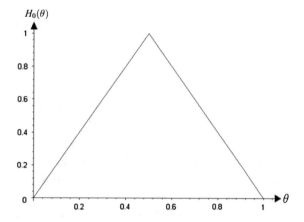

amounts by Pais and Benton or not? If one decides to formulate this problem by either of the following two testing classical hypotheses, then he/she will face to a contradiction in the results of

Test 1 $H_0 : \mu \geq 0.2$, against, $H_1 : \mu < 02$,
          and
Test 2 $H_0 : \mu \geq 3$, against, $H_1 : \mu < 3$,

for more details see Tables 3–5 of [21]. The presented contradiction in the result of Test 1 and Test 2 comes from the difference between the null hypotheses in two tests; in other words it comes from very vague proposed information in [19].

In this applied example, one cannot represent the whole above presented information by Pais and Benton with a classical (precise) set. But, using fuzzy set theory, one can show the optimum range and the maximum amount of Cd uptakes in a plant by the following fuzzy set, in which the membership is considered to be 1 on interval $[0, 0.05]$, since the lower Cd absorption is better for any plant (see Fig. 2).

Fig. 2 The membership functions of the fuzzy hypotheses in Example 2

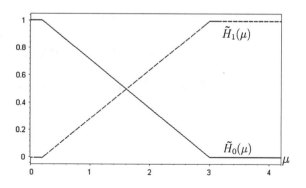

$$H_0(\mu) = \begin{cases} 1 & \text{if } 0 \le \mu < 0.2, \\ \frac{3-\mu}{2.8} & \text{if } 0.2 \le \mu < 3, \\ 0 & \text{if } \mu \ge 3. \end{cases} \qquad (9)$$

Now, the experimenter can test fuzzy hypotheses "$\tilde{H}_0 : \mu$ is $H_0(\mu)$", against "$\tilde{H}_1 : \mu$ is $H_1(\mu)$" without facing any contradiction in the result, where the membership functions of $H_0(\mu)$ and $H_1(\mu) = 1 - H_0(\mu)$ are depicted in Fig. 2.

Definition 2 (See also Arefi and Taheri [1, 2]) **(a)** Let the fuzzy hypothesis "$\tilde{H} : \theta$ is $H(\theta)$" be such that (*i*) H is an increasing (decreasing) function of θ, and (*ii*) there exists $\theta_1 \in \Theta$ so that $H(\theta) = 1$ for $\theta \ge \theta_1$ ($\theta \le \theta_1$). Then, $\tilde{H}$ is called a *fuzzy one-sided hypothesis*.

(b) Let the fuzzy hypothesis "$\tilde{H} : \theta$ is $H(\theta)$" be such that (*i*) there exists an interval $[\theta_1, \theta_2] \subset \Theta$ so that $H(\theta) = 1$ for $\theta \in [\theta_1, \theta_2]$, (*ii*) H is an increasing function of θ for $\theta \le \theta_1$ and is a decreasing function for $\theta \ge \theta_2$. Then, $\tilde{H}$ is called a *fuzzy two-sided hypothesis*.

Now, we recall the concept of the boundary of fuzzy hypothesis from Parchami et al. [20], which is needed in next sections.

Definition 3 The *boundary of the fuzzy hypothesis* $\tilde{H}$ is a fuzzy subset of Θ with membership function H_b defined as follows,

(i) $H_b(\theta) = \begin{cases} H(\theta) & \text{for } \theta \le \theta_1 \\ 0 & \text{for } \theta > \theta_1 \end{cases}$, if $\tilde{H}$ is one-sided and H is non-decreasing,

(ii) $H_b(\theta) = \begin{cases} H(\theta) & \text{for } \theta \le \theta_1 \\ 0 & \text{for } \theta > \theta_1 \end{cases}$, if $\tilde{H}$ is one-sided and H is non-increasing,

(iii) $H_b(\theta) = H(\theta)$, if $\tilde{H}$ is two-sided.

Remark 1 In Definition 3, we call H_b the boundary of the fuzzy hypothesis $\tilde{H}$, since as $\tilde{H}$ reduces to a crisp hypothesis, H_b becomes just a single point which is the boundary of the crisp hypothesis $\tilde{H}$.

Example 3 Let θ be the parameter of a binomial distribution. Suppose that

$$H(\theta) = \begin{cases} 1 & \text{if } 0 \le \theta \le 0.2 \\ \frac{6-10\theta}{4} & \text{if } 0.2 \le \theta \le 0.6 \\ 0 & \text{if } 0.6 \le \theta \le 1. \end{cases}$$

Then, the hypothesis "$\tilde{H} : \theta$ is $H(\theta)$" is a fuzzy one-sided hypothesis. So, by Definition 3,

Fig. 3 The membership
functions of the fuzzy
hypothesis and its boundary
in Example 3

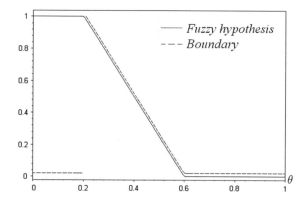

$$H_b(\theta) = \begin{cases} \frac{6-10\theta}{4} & if \ 0.2 \le \theta \le 0.6 \\ 0 & elsewhere, \end{cases}$$

is the membership function of the boundary of the fuzzy hypothesis $\tilde{H}$ (see Fig. 3).

4   Probability Measure Under a Fuzzy Hypothesis

Definition 4 [27] Let the random variable X have p.d.f. or p.m.f. $f(x; \theta)$ and
"$\tilde{H} : \theta$ is $H(\theta)$" be a fuzzy hypothesis for which $\int_\theta H(\theta) \, d\theta < \infty$. The weighted
probability density function of X, under fuzzy hypothesis $\tilde{H}$, is defined by

$$f(x; \tilde{H}) = \int_\theta H^*(\theta) f(x; \theta) \, d\theta,$$

where $H^*(\theta) = \frac{H(\theta)}{\int_\theta H(\theta) \, d\theta}$ is the *normalized* membership function of $H(\theta)$. Replace
integration by summation in discrete case.

Note that the normalized membership function is not necessarily a membership
function, i.e., it may be greater than 1 for some values of θ.

The advantage of Definition 4 is that the weighted p.d.f. can integrate all possible p.d.f.s with different weights. The value of $H^*(\theta)$ can be understood as the
weight of $f(x; \theta)$, and the weighted p.d.f. can let different possible $f(x; \theta)$'s play
different roles in this integration.

Remark 2 [27] Note that $f(x; \tilde{H})$ in Definition 4 is a p.d.f., since $f(x; \tilde{H})$ is non-
negative and

$$\int_x f(x; \tilde{H}) \, dx = \int_x \int_\theta H^*(\theta) f(x; \theta) \, d\theta \, dx$$

$$= \int_\theta H^*(\theta) \left(\int_x f(x; \theta) \, dx \right) d\theta$$

$$= \int_\theta H^*(\theta) \, d\theta = 1.$$

Remark 3 If H is the crisp hypothesis "$H : \theta = \theta_0$", then $f(x; \tilde{H}) = f(x; \theta_0)$.

Example 4 Let $X \sim N(\mu, 0.7)$, μ , and be unknown. The weighted probability density function of X under fuzzy hypothesis $\mu \simeq 4$ is as follows

$$f(x; \mu \simeq 4) = \int_\mu H^*(\mu) f(x; \mu) \, d\mu$$

$$= \frac{2}{3} \int_\mu H(\mu) \frac{1}{\sqrt{1.4\pi}} e^{-\frac{(x-\mu)^2}{1.4}} \, d\mu, \quad x \in R$$

where $\mu \simeq 4$ is defined by the following membership function

$$H(\mu) = \begin{cases} \frac{\mu-2}{2} & if \ 2 \leq \mu \leq 4 \\ 2(4.5 - \mu) & if \ 4 < \mu \leq 4.5 \\ 0 & elsewhere. \end{cases}$$

The weighted p.d.f. of X under fuzzy hypothesis $\mu \simeq 4$ is drown in Fig. 4.

Example 5 Let $X \sim \text{bin}(n = 8, \theta)$; i.e.

$$f(x; \theta) = \binom{8}{x} \theta^x (1 - \theta)^{8-x}, \ x = 0, 1, \ldots, 8, \ 0 < \theta < 1.$$

Fig. 4 The weighted p.d.f. of X under fuzzy hypothesis $\mu \simeq 4$ and the p.d.f of X under hypothesis $\mu = 4$ in Example 4

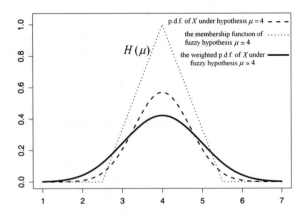

p.d.f. of X under hypothesis $\mu = 4$ – – –

the membership function of · · · · · · ·
fuzzy hypothesis $\mu \simeq 4$

the weighted p.d.f. of X under ———
fuzzy hypothesis $\mu \simeq 4$

$H(\mu)$.

Then, under fuzzy hypothesis "$\tilde{H} : \theta$ is approximately $\frac{1}{2}$" where $H_0(\theta)$ is defined in Example 1, the weighted probability density function of X is as follows

$$f(x; \tilde{H}_0) = f\left(x; \theta \simeq \frac{1}{2}\right)$$

$$= \int_\theta H_0^*(\theta) f(x; \theta) \, d\theta$$

$$= 2 \int_\theta H_0(\theta) \left[\binom{8}{x} \theta^x (1 - \theta)^{8-x}\right] d\theta, \quad x = 0, 1, \ldots, 8.$$

Now, for example, under fuzzy hypothesis $\tilde{H}_0$ the value of $P(X \leq 2)$ is obtained as follows

$$P_{\tilde{H}_0}(X \leq 2) = P_{\theta \simeq \frac{1}{2}}(X \leq 2)$$

$$= \int_\theta H_0^*(\theta) P_\theta(X \leq 2) \, d\theta$$

$$= 2 \int_\theta H_0(\theta) \left[P_\theta(X = 0) + P_\theta(X = 1) + P_\theta(X = 2)\right] d\theta$$

$$= 2 \int_0^{1/2} 2\theta \left[(21\theta^2 + 6\theta + 1)(1 - \theta)^6\right] d\theta$$

$$+ 2 \int_{1/2}^1 (2 - 2\theta) \left[(21\theta^2 + 6\theta + 1)(1 - \theta)^6\right] d\theta$$

$$= 0.238 + 0.021 = 0.259.$$

5 Testing Fuzzy Hypotheses: A *p*-value Approach

The main problem. The main problem studied in this work is to test fuzzy hypotheses

$$\begin{cases} \tilde{H}_0 : \theta \text{ is } H_0(\theta) \\ \tilde{H}_1 : \theta \text{ is } H_1(\theta) \end{cases}$$

based on a random sample from a p.d.f. or p.m.f. $f(x; \theta)$, $\theta \in \Theta$. This problem is called the problem of testing fuzzy hypotheses. In the sequence, by inspiration of [30], we propose a new *p*-value-based approach to such a problem.

First, note that, similar to the kinds of hypotheses given in Sect. 2, the fuzzy hypotheses can be modeled by one of the following forms in practice:

(i) $\begin{cases} \tilde{H}_0 : \theta \text{ is approximately } \theta_0, \\ \tilde{H}_1 : \theta \text{ is approximately } \theta_1, \end{cases}$ $(Def(H_0) > Def(H_1))$

(ii) $\begin{cases} \tilde{H}_0 : \theta \text{ is approximately } \theta_0, \\ \tilde{H}_1 : \theta \text{ is approximately } \theta_1, \end{cases}$ $(Def(H_0) < Def(H_1))$

(iii) $\begin{cases} \tilde{H}_0 : \theta \text{ is approximately bigger than } \theta_0, \\ \tilde{H}_1 : \theta \text{ is approximately smaller than } \theta_0, \end{cases}$

(iv) $\begin{cases} \tilde{H}_0 : \theta \text{ is approximately smaller than } \theta_0, \\ \tilde{H}_1 : \theta \text{ is approximately bigger than } \theta_0, \end{cases}$

(v) $\begin{cases} \tilde{H}_0 : \theta \text{ is near to } \theta_0, \\ \tilde{H}_1 : \theta \text{ is away from } \theta_0, \end{cases}$

where θ_0 and θ_1 are two known numbers and $Def(.)$ is a defuzzifier function. It is obvious that the critical regions of testing fuzzy hypotheses are similar to the critical regions of testing precise hypotheses which are formulated in (1). In other words, the critical regions of testing fuzzy hypotheses (*i*) and (*iii*) is of form (1.a), the critical regions of testing fuzzy hypotheses (*ii*) and (*iv*) is of form (1.b), and the critical region of testing fuzzy hypotheses (*v*) is of form (1.c). It must be mentioned that the critical regions of testing fuzzy hypotheses (*i*) and (*ii*) are determined after defuzzification of fuzzy hypotheses and they depend on the defuzzifier function.

Definition 4 In testing fuzzy hypotheses problem, for any critical region of forms (1.a), (1.b) and (1.c), the *p*-value is respectively defined as

$$\textbf{(a)} \quad p\text{-value} = P_{H_{0b}}(T \le t)$$
$$= \int_\theta H_{0b}^*(\theta) \, P_\theta(T \le t) \, d\theta, \tag{2}$$

$$\textbf{(b)} \quad p\text{-value} = P_{H_{0b}}(T \ge t)$$
$$= \int_\theta H_{0b}^*(\theta) \, P_\theta(T \ge t) \, d\theta, \tag{3}$$

and

$$\textbf{(c)} \quad p\text{-value} = \begin{cases} 2P_{H_{0b}}(T \ge t) & \text{if } t \ge m_r \\ 2P_{H_{0b}}(T \le t) & \text{if } t \le m_l \end{cases}$$
$$= \begin{cases} 2\int_\theta H_{0b}^*(\theta) \, P_\theta(T \ge t) \, d\theta & \text{if } t \ge m_r \\ 2\int_\theta H_{0b}^*(\theta) \, P_\theta(T \le t) \, d\theta & \text{if } t \le m_l \end{cases} \tag{4}$$

where H_{0b}^* is the normalized membership function of the boundary of the fuzzy null hypothesis. Moreover,

$$m_l = \inf\{m : m \in Supp(\mathbf{m})\}, \tag{5}$$

$$m_r = \sup\{m : m \in Supp(\mathbf{m})\}, \tag{6}$$

where the fuzzy set $\mathbf{m}$, with membership function $\mathbf{m}(m) = H_{0b}(\theta)$, is called the median of the distribution of the test statistic under the fuzzy null boundary H_{0b} and m is the median of the distribution $T(\mathbf{X})$ under θ. Replace integration by summation in discrete case.

Remark 4 When the hypotheses are crisp rather than fuzzy, the membership function of the fuzzy boundary is reduced to the indicator function of a single point, i.e. the indicator function of the boundary θ_0. In these cases, considering formulas (2–4), the introduced p-values in Definition 4 are reduced to the classical p-values in Sect. 2.

Remark 5 As we mentioned in Sect. 1, some works have been done by researchers on the p-value-based methods in fuzzy environments. Denœux et al. [8] and Filzmoser and Viertl [9] introduced the fuzzy p-value for testing hypotheses based on fuzzy data; Parchami et al. [20] introduced a fuzzy p-value for testing fuzzy hypotheses; and Parchami et al. [22] introduced the fuzzy p-value for testing fuzzy hypotheses based on fuzzy data. In contrast with the above works, it should be mentioned that, the p-value introduced in this study is a real number on unit interval which is formulated on the basis of the probability measure under fuzzy hypothesis.

Remark 6 Considering Formula (4), in a two-sided test we have a three-decision testing problem as follows

  (i)   accept $\tilde{H}_1$ and reject $\tilde{H}_0$,
  (ii) reject $\tilde{H}_1$ and accept $\tilde{H}_0$,
 (iii) neither accept nor reject both $\tilde{H}_0$ and $\tilde{H}_1$.

In the case (*iii*), the uncertainty of the decision is expressed by fuzziness of $\mathbf{m}$, i.e. when $t \in Supp(\mathbf{m})$ we cannot come to a clear decision, because there is not a significant difference between the observed value of the statistic and the median of the T under $\tilde{H}_0$. In such cases, one may take more samples and follow the procedure until acceptance or rejection of $\tilde{H}_0$. Note that, Filzmoser and Viertl [9] proposed to consider a no-decision region in their fuzzy p-value-based approach to test crisp hypotheses with fuzzy data, where they say "The need for formulating a three-decision testing problem was already indicated by Neyman and Pearson [18]". It should be mentioned that, with decreasing fuzziness of the null hypothesis in case (c), the no-decision region decreases. On the other hand, this region is deleted in testing usual (crisp) hypotheses.

6 Distribution of the Introduced p-value Under Fuzzy Null Hypothesis

Theorem 1 In testing fuzzy hypotheses "$\tilde{H}_0 : \theta$ is $H_0(\theta)$" versus "$\tilde{H}_1 : \theta$ is $H_1(\theta)$", if the test statistic has a continuous distribution, then p-value has uniform distribution over $(0, 1)$ under the boundary of fuzzy null hypothesis.

Proof We prove Theorem 1 for critical region of form (1.a). The proof is similar for cases (1.b) and (1.c). Regarding to Remark 2, we denote the weighted p.d.f. (or p.m. f.) of X under fuzzy hypothesis $\tilde{H}$ by $f(x; \tilde{H})$, and we denote its weighted cumulative density function by $F_{X;\tilde{H}}(x) = P_{\tilde{H}}(X \leq x)$ under fuzzy hypothesis $\tilde{H}$. By Definition 4, P-value $= P_{H_{0b}}(T \leq t) = F_{T;H_{0b}}(t)$ in case (a), and hence before observing data we can denote the random p-value by

$$P\text{-value} = F_{T;H_{0b}}(T). \tag{7}$$

Therefore, we have

$$
\begin{aligned}
F_{P-\text{value};H_{0b}}(u) &= P_{H_{0b}}[P\text{-value} \leq u] \\
&= P_{H_{0b}}\left[F_{T;H_{0b}}(T) \leq u\right], \quad \text{by (7)} \\
&= P_{H_{0b}}\left[T \leq F_{T;H_{0b}}^{-1}(u)\right], \quad \text{since } F_{T;H_{0b}} \text{ is a } 1-1 \text{ function} \\
&= F_{T;H_{0b}}\left(F_{T;H_{0b}}^{-1}(u)\right) = u, \quad \forall\, 0 < u < 1,
\end{aligned}
$$

and the proof is finished.□

Remark 7 Considering Remark 4, the result of Theorem 1 for precise hypotheses is reduced to Lemma 3.3.1 in Page 64 of (Lehmann and Romano [16]).

Corollary 1 *The fact that p-value has uniform distribution over (0, 1) under H_{0b} can be useful in practice. For example, suppose that we have p-values $p_1, \ldots, p_k$ for k independent tests of the same null fuzzy hypothesis. By assuming that the p-values have uniform distribution over (0, 1), one can combine the p-values using the test statistic*

$$T = -2\sum_{i=1}^{k} \ln(p_i) \tag{8}$$

and conclude that T has chi-square distribution with 2k degrees of freedom under the boundary of null fuzzy hypothesis.

7 Illustrative Examples

In this section, we illustrate different situations of testing fuzzy hypotheses by four examples. In Examples 6 and 8 we consider one-sided tests, and in Examples 7 and 9 we consider two-sided tests.

Example 6 The lifetime X of certain lamps (in term of hour) produced by a factory is distributed normally with unknown mean μ and standard deviation $\sigma = 120$. In a random sample of size $n = 36$ lamps, we get $\bar{x} = 1327$. Suppose that we wish to test the following fuzzy hypotheses

$$\begin{cases} \tilde{H}_0 : \mu \text{ is approximately } 1300, \\ \tilde{H}_1 : \mu \text{ is approximately bigger than } 1300, \end{cases}$$

where the membership functions of $\tilde{H}_0$ and $\tilde{H}_1$ are considered by an expert as follows (see Fig. 5)

$$H_0(\mu) = \begin{cases} \frac{\mu - 1275}{25} & \text{if } 1275 \leq \mu < 1300, \\ \frac{1325 - \mu}{25} & \text{if } 1300 \leq \mu < 1325, \\ 0 & \text{elsewhere}, \end{cases}$$

$$H_1(\mu) = \begin{cases} 0 & \text{if } \mu < 1275, \\ \frac{\mu - 1275}{50} & \text{if } 1275 \leq \mu < 1325, \\ 1 & \text{if } \mu \geq 1325. \end{cases}$$

In this example, the rejection region is of the form (1.b). Note that in this case $H_{0b}(\mu) = H_0(\mu)$ is the boundary of $\tilde{H}_0$, and considering (3), one can compute *p*-value as follows

Fig. 5 The membership functions of the fuzzy hypotheses in Example 6

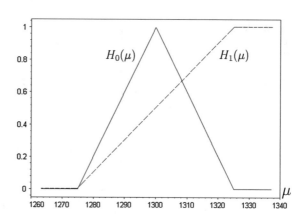

$$p\text{-value} = P_{H_{0b}}(\bar{X} \geq \bar{x})$$

$$= \int_{\mu} H_{0b}^*(\mu)\, P_{\mu}(\bar{X} \geq 1327)\, d\mu$$

$$= \frac{1}{\int_{\mu} H_{0b}(\mu)\, d\mu} \int_{\mu} H_{0b}(\mu)\, P_{\mu}(\bar{X} \geq 1327)\, d\mu$$

$$= \frac{1}{25} \int_{1275}^{1300} \frac{\mu - 1275}{25} P_{\mu}(\bar{X} \geq 1327)\, d\mu$$

$$+ \frac{1}{25} \int_{1300}^{1325} \frac{1325 - \mu}{25} P_{\mu}(\bar{X} \geq 1327)\, d\mu$$

$$= \frac{1}{25^2} \int_{1275}^{1300} (\mu - 1275) \int_{1327}^{\infty} \frac{1}{20\sqrt{2\pi}} e^{-\frac{(\bar{x}-\mu)^2}{40}}\, d\bar{x}\, d\mu$$

$$+ \frac{1}{25^2} \int_{1300}^{1325} (1325 - \mu) \int_{1327}^{\infty} \frac{1}{20\sqrt{2\pi}} e^{-\frac{(\bar{x}-\mu)^2}{40}}\, d\bar{x}\, d\mu$$

$$= 0.0223 + 0.0926$$

$$= 0.1149.$$

Therefore, for instance, one can accept fuzzy hypothesis $\tilde{H}_0$ at the significance level 0.10.

Example 7 Suppose that, in Example 6, we wish to test

$$\begin{cases} \tilde{H}_0 : \mu \text{ is near to } 1300, \\ \tilde{H}_1 : \mu \text{ is away from } 1300, \end{cases}$$

where the membership function $H_0(\mu)$ introduced in Example 6 and $H_1(\mu) = 1 - H_0(\mu)$ (see Fig. 6). In this example, we only change the form of fuzzy hypotheses of Example 6 into one of five proposed forms for fuzzy hypotheses in Sect. 5. In this case, the rejection region is of the form (1.c). Note that

Fig. 6 The membership functions of the fuzzy hypotheses in Example 7

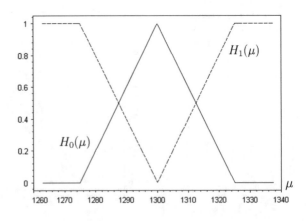

$\bar{X} \sim N(\mu, \frac{120^2}{36})$, and since for the normal distribution the median coincides with the mean, therefore under $\tilde{H}_0$ the membership function of **m** is

$$\mathbf{m}(m) = \begin{cases} \frac{m-1275}{25} & \text{if } 1275 \leq m < 1300, \\ \frac{1325-m}{25} & \text{if } 1300 \leq m < 1325, \\ 0 & \text{elsewhere.} \end{cases}$$

Here $t = \bar{x} = 1327 \geq m_r = 1325$, and hence,

$$\begin{aligned} p\text{-value} &= 2 \int_\mu H_{0b}^*(\mu) P_\mu(\bar{X} \geq 1327) \, d\mu \\ &= \frac{2}{25^2} \int_{1275}^{1300} (\mu - 1275) \int_{1327}^\infty \frac{1}{20\sqrt{2\pi}} e^{-\frac{(\bar{x}-\mu)^2}{40}} d\bar{x} \, d\mu \\ &\quad + \frac{2}{25^2} \int_{1300}^{1325} (1325 - \mu) \int_{1327}^\infty \frac{1}{20\sqrt{2\pi}} e^{-\frac{(\bar{x}-\mu)^2}{40}} d\bar{x} \, d\mu \\ &= 0.23 \end{aligned}$$

So, based on the observed data, the fuzzy null hypothesis is accepted at any significance level $\alpha \leq 0.23$.

Example 8 The manager of a factory has reinstalled a new system to upgrade the security of his personnel. We can suppose that the number of monthly accidents has the poisson distribution with mean λ. A study shows that there occurred 27 accidents during the past year. After installation of the new system the manager wants to test if the average of the monthly accidents is approximately bigger than 3. That is to test

$$\begin{cases} \tilde{H}_0 : \lambda \text{ is approximately bigger than 3}, \\ \tilde{H}_1 : \lambda \text{ is approximately smaller than 3}, \end{cases}$$

where $\tilde{H}_0$ and $\tilde{H}_1$ have the following membership functions (see Fig. 7)

$$H_0(\lambda) = \begin{cases} 0 & \text{if } \lambda < 2.75, \\ 2(\lambda - 2.75) & \text{if } 2.75 \leq \lambda < 3.25, \\ 1 & \text{if } \lambda \geq 3.25, \end{cases}$$

$$H_1(\lambda) = \begin{cases} 1 & \text{if } \lambda < 2.75, \\ 2(3.25 - \lambda) & \text{if } 2.75 \leq \lambda < 3.25, \\ 0 & \text{if } \lambda \geq 3.25. \end{cases}$$

The rejection region is of the form (1.a), so the membership function of the fuzzy boundary is

Fig. 7 The membership
functions of the fuzzy
hypotheses in Example 8

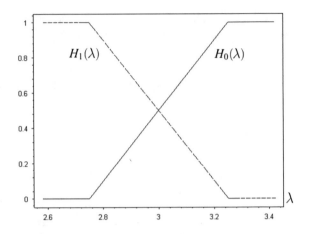

$$H_{0b}(\lambda) = \begin{cases} 2(\lambda - 2.75) & \text{if } 2.75 \le \lambda < 3.25, \\ 0 & \text{elsewhere.} \end{cases}$$

Therefore, considering $T = \sum_{i=1}^{12} X_i \sim P(12\lambda)$,

$$
\begin{aligned}
p\text{-value} &= P_{H_{0b}}(T \le t) \\
&= \int_{\lambda} H_{0b}^*(\lambda)\, P_{\lambda}(T \le 27)\, d\lambda \\
&= \frac{1}{\int_{\lambda} H_{0b}(\lambda)\, d\lambda} \int_{\lambda} H_{0b}(\lambda)\, P_{\lambda}(T \le 27)\, d\lambda \\
&= 4 \int_{2.75}^{3.25} 2(\lambda - 2.75) \left[\sum_{t=0}^{27} \frac{e^{-12\lambda}(12\lambda)^t}{t!} \right] d\lambda \\
&= 0.0588.
\end{aligned}
$$

Hence, the null fuzzy hypothesis is accepted at each level $\alpha \le 0.0588$ and is rejected at each level $\alpha > 0.0588$.

Example 9 The amount of an adverse substance extracted from a sample of size 14 cigarettes of a special brand is given in milligrams as 15.3, 13.5, 13.4, 12.7, 14.4, 14.4, 13.9, 14.3, 13.8, 16.2, 15.4, 15.6, 12.8 and 13.1. From the previous experiments it is known that the random variable of interest is distributed normally with standard deviation $\sigma = 1.2$. Suppose that, we wish to test the following fuzzy hypotheses about unknown mean μ

$$\begin{cases} \tilde{H}_0 : \mu \text{ is near to } 15, \\ \tilde{H}_1 : \mu \text{ is away from } 15, \end{cases}$$

where $\tilde{H}_0$ is given by the membership function

$$H_0(\mu) = \begin{cases} \frac{\mu - 14.3}{0.7} & \text{if } 14.3 \leq \mu < 15, \\ \frac{15.7 - \mu}{0.7} & \text{if } 15 \leq \mu < 15.7, \\ 0 & \text{elsewhere}, \end{cases}$$

and $H_1(\mu) = 1 - H_0(\mu)$ (see Fig. 7). In this case, the rejection region is of the form (1.c). Since $T = \bar{X} \sim N(\mu, \frac{\sigma^2}{n})$ and for a normal distribution the median coincides with the mean, therefore, under $\tilde{H}_0$ the membership function of **m** is

$$\mathbf{m}(m) = \begin{cases} \frac{\mu - 14.3}{0.7} & \text{if } 14.3 \leq \mu < 15, \\ \frac{15.7 - \mu}{0.7} & \text{if } 15 \leq \mu < 15.7, \\ 0 & \text{elsewhere}. \end{cases}$$

In this example we see that $t \leq m_l$, since $t = \bar{x} = 14.2$ and $m_l = 14.3$. Also, note that $H_{0b}(\mu) = H_0(\mu)$, and so

$$\begin{aligned} p\text{-value} &= 2 \int_{\mu} H_{0b}^*(\mu) P_{\mu}(\bar{X} \leq 14.2) \, d\mu \\ &= \frac{2}{\int_{\mu} H_{0b}(\mu) \, d\mu} \int_{\mu} H_{0b}(\mu) P_{\mu}(Z \leq \frac{14.2 - \mu}{1.2}) \, d\mu \\ &= \frac{20}{7} \int_{14.3}^{15} \frac{\mu - 14.3}{0.7} \int_{-\infty}^{\frac{14.2 - \mu}{1.2}} (2\pi)^{-\frac{1}{2}} \exp(-\frac{z^2}{2}) \, dz \, d\mu \\ &\quad + \frac{20}{7} \int_{15}^{15.7} \frac{15.7 - \mu}{0.7} \int_{-\infty}^{\frac{14.2 - \mu}{1.2}} (2\pi)^{-\frac{1}{2}} \exp(-\frac{z^2}{2}) \, dz \, d\mu \\ &= 0.01920 + 0.00299 \\ &= 0.0222. \end{aligned}$$

Fig. 8 The membership functions of the fuzzy hypotheses in Example 9

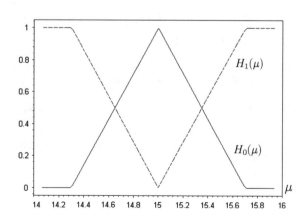

Therefore, for instance, the fuzzy hypothesis $\tilde{H}_0$ is rejected at significance level 0.05 and is accepted at significance level 0.01 (see Fig. 8).

8 Conclusion

In this paper, a new p-value-based approach was presented for testing statistical hypotheses when the hypotheses are fuzzy rather than crisp. The proposed approach, which is based on the concept of the probability measure of fuzzy events, is an extension of the classical p-value approach. It was shown that the introduced p-value has uniform distribution over $(0,1)$ under the null fuzzy hypothesis. Numerical examples were provided to illustrate the performance of the method.

The study of the applicability of the proposed approach to test about the parameters of fuzzy regression models is a possible topic for further research. In addition, the investigation of the p-value approach to test of fuzzy hypotheses from a Bayesian perspective is another potential subject for more study.

References

1. Arefi, M., Taheri, S.M.: Testing fuzzy hypotheses using fuzzy data based on fuzzy test statistic. J. Uncertain Syst. **5**, 45–61 (2011)
2. Arefi, M., Taheri, S.M.: A new approach for testing fuzzy hypotheses based on fuzzy data. Int. J. Comput. Intell. Syst. **6**, 318–327 (2013)
3. Arnold, B.F.: An approach to fuzzy hypothesis testing. Metrika **44**, 119–126 (1996)
4. Arnold, B.F.: Testing fuzzy hypothesis with crisp data. Fuzzy Sets Syst. **94**, 323–333 (1998)
5. Arnold, B.F., Gerke, O.: Testing fuzzy hypotheses in linear regression models. Metrika **57**, 81–95 (2003)
6. Casals, M.R., Gil, M.A., Gil, P.: On the use of Zadeh's probabilistic definition for testing statistical hypotheses from fuzzy information. Fuzzy Sets Syst. **20**, 175–190 (1986)
7. Casals, M.R., Gil, M.A., Gil, P.: The fuzzy decision problem: an approach to the problem of testing statistical hypotheses with fuzzy information. Eur. J. Oper. Res. **27**, 371–382 (1986)
8. Denœux, T., Masson, M.H., Hébert, P.A.: Nonparametric rank-based statistics and significance tests for fuzzy data. Fuzzy Sets and Systems 153, 1–28 (2005)
9. Filzmoser, P., Viertl, R.: Testing hypotheses with fuzzy data: the fuzzy p-value. Metrika **59**, 21–29 (2004)
10. González-Rodríguez, G., Colubi, A., Gil, M.A.: A fuzzy representation of random variables: an operational tool in exploratory analysis and hypothesis testing. Comput. Stat. Data Anal. **51**, 163–176 (2006)
11. González-Rodríguez, G., Colubi, A., Gil, M.A.: Fuzzy data treated as functional data: a one-way ANOVA test approach. Comput. Stat. Data Anal. **56**, 943–955 (2012)
12. González-Rodríguez, G., Montenegro, M., Colubi, A., Gil, M.A.: Bootstrap techniques and fuzzy random variables: synergy in hypothesis testing with fuzzy data. Fuzzy Sets Syst. **157**, 2608–2613 (2006)
13. Grzegorzewski, P., Hryniewicz, O.: Testing hypotheses in fuzzy environment. Math. Soft Comput. **4**, 203–217 (1997)

14. Holeňa, M.: Fuzzy hypotheses testing in the framework of fuzzy logic. Fuzzy Sets Syst. **145**, 229–252 (2004)
15. Knight, K.: Mathematical Statistics. Chapman and Hall, Boca Raton (2000)
16. Lehmann, E.L., Romano, J.P.: Testing Statistical Hypotheses, 3rd edn. Springer, New York (2005)
17. Montenegro, M., Colubi, A., Casals, M.R., Gil, M.A.: Asymptotic and bootstrap techniques for testing the expected value of a fuzzy random variable. Metrika **59**, 31–49 (2004)
18. Neyman, J., Pearson, E.S.: The theory of statistical hypotheses in relation to probabilities a priori. Proc. Camb. Phil. Soc. **29**, 492–510 (1933)
19. Pais, I., Benton, J.J.: The Handbook of Trace Elements. St. Lucie Press, Boca Raton, Florida (1997)
20. Parchami, A., Taheri, S.M., Mashinchi, M.: Fuzzy *p*-value in testing fuzzy hypotheses with crisp data. Stat. Pap. **51**, 209–226 (2010)
21. Parchami, A., Ivani, R., Mashinchi, M.: An application of testing fuzzy hypotheses: a soil study on bioavailability of Cd. Scientia Iranica **18**, 470–478 (2011)
22. Parchami, A., Taheri, S.M., Mashinchi, M.: Testing fuzzy hypotheses based on vague observations: a *p*-value approach. Stat. Pap. **53**, 469–484 (2012)
23. Rohatgi, V.K., Ehsanes Saleh, A.K.: An Introduction to Probability and Statistics, 2nd edn. Wiley, New York (2001)
24. Taheri, S.M., Behboodian, J.: Neyman-Pearson Lemma for fuzzy hypotheses testing. Metrika **49**, 3–17 (1999)
25. Taheri, S.M., Behboodian, J.: A Bayesian approach to fuzzy hypotheses testing. Fuzzy Sets Syst. **123**, 39–48 (2001)
26. Tanaka, H., Okuda, T., Asai, K.: Fuzzy information and decision in a statistical model. In: Gupta, M.M., et al. (eds.) Advances in Fuzzy Set Theory and Applications, pp. 303–320. North-Holland, Amsterdam (1979)
27. Torabi, H., Behboodian, J.: Sequential probability ratio test for fuzzy hypotheses testing with vague data. Austrian J. Stat. **34**, 25–38 (2005)
28. Torabi, H., Behboodian, J.: Likelihood ratio tests for fuzzy hypotheses testing. Stat. Pap. **48**, 509–522 (2007)
29. Viertl, R.: Statistical Methods for Fuzzy Data. Wiley, New York (2011)
30. Zadeh, L.A.: Probability measures of fuzzy events. J. Math. Anal. Appl. **23**, 421–427 (1968)

Fuzzy Regression Analysis: An Actuarial Perspective

Jorge de Andrés-Sánchez

Abstract The first objective of this paper is to describe from a critical point of view the main types of fuzzy regression methods: those based on minimum fuzziness principle, those that are built up by minimising the squared distance between observations and estimates and models that mix both methodologies. Finally, we revise the actuarial applications of fuzzy regression proposed in the literature and develop in detail two of them: estimating the yield curve and calculating claim reserves.

Keywords Fuzzy regression analysis · Actuarial applications · Minimum fuzziness principle · Fuzzy least squares · Yield curve model

1 Introduction

Fuzzy Data Analysis (FDA) literature has been very productive in last two decades. It includes theoretical papers as well as empirical applications. Following Dubois and Prade [1] FDA includes fuzzy analysis of data and also analysis of fuzzy data. Concretely, this chapter is devoted to fuzzy regression (FR) which is one of the most common FDA used in empirical applications in several fields as medicine [2], management [3, 4, 5] or economics [6, 7]. Concretely, in chapter we describe the most relevant applications of FR on insurance issues.

In our opinion, there are several advantages of FR methods over conventional regression methods that can explain their success in empirical applications:

(a) The estimates that are obtained after adjusting the coefficients are not random variables, which are difficult to manipulate in arithmetical operations, but fuzzy numbers, which are easier to handle arithmetically. So, when starting

J. de Andrés-Sánchez (✉)
Social and Business Research Laboratory, Rovira i Virgili University, Tarragona, Spain
e-mail: jorge.deandres@urv.cat

© Springer International Publishing Switzerland 2016   175
C. Kahraman and Ö. Kabak (eds.), *Fuzzy Statistical Decision-Making*,
Studies in Fuzziness and Soft Computing 343,
DOI 10.1007/978-3-319-39014-7_11

from magnitudes estimated by random variables (e.g. as a result of its prediction from a least squares regression) these random variables are often
reduced to their mathematical expectation (which may or may not be corrected
by their variance) so that they are easier to handle. If fuzzy numbers are used,
this loss of information is not needed.

(b) If the phenomena investigated are economic or social, the observations are a
consequence of the interaction between the economic agents' beliefs and
expectations, which are highly subjective and vague. Fuzzy Set Theory,
therefore, is a good way of treating this information.

(c) Observations are often not crisp numbers; they are confidence intervals or not
well defined quantities. If econometric methods are to be used, the observations for the explained variable and/or the explanatory variable must be represented by a single value which involves losing a great deal of information.
Fuzzy regression, however, does not necessarily reduce the values of each
variable to a crisp number, i.e., all the observed values can be used in the
regression analysis.

The literature has proposed several FR methods throughout time, i.e. we cannot
properly speak about "Fuzzy Regression" but about "Fuzzy Regression Methods".
Following Chang and Ayyub [8] we can classify FR methods in three categories,
depending on the criteria used to fit coefficients:

(a) Methods that use Minimum Fuzziness Principle [9, 10].
(b) Methods that generalise Ordinary Least Squares to the presence of fuzziness in
the sample [11, 14].
(c) Methods that use sequentially above criteria [12, 13].

The chapter is structured as follows. In the next section we will describe the
basic concepts on fuzzy numbers to develop FR methods. In the third section we
expose the most usual FR method in actuarial applications as well as we reflect on
their advantages and drawbacks. Fourth and fifth sections develop two usual
applications of FR methods in actuarial literature: fitting yield curve for actual
financial pricing and estimating claiming behaviour in a non-life insurance context.
Finally, we remark the most relevant conclusions of this survey.

2 Basics on Fuzzy Numbers

A Fuzzy Number (FN) is a fuzzy subset $\tilde{a}$ defined over real numbers. It is the main
instrument used in Fuzzy Set Theory (FST) for quantifying uncertain quantities.
Two properties are required for a FN. The first one is that it must be a normal fuzzy
set (i.e. $\sup_{\forall x \in \Re} \mu_{\tilde{a}}(x) = 1$). The second is that it must be convex (i.e. its α-cuts
must be convex sets).

For practical purposes, Triangular Fuzzy Numbers (TFNs) are widely used since
they are easy to handle arithmetically and they can be interpreted intuitively. We

shall symbolise a TFN $\tilde{a}$ as $\tilde{a} = (a, l_a, r_a)$ where a is the centre and l_a and r_a are the left and right spreads, respectively. For example, a subjective judgement by an economist "I expect that for the next two years the inflation rate will be around 2 % and deviations no greater than 1 %" may be quantified in a very natural way as (0.02, 0.01, 0.01). Analytically, a TFN is characterised by its membership function $\mu_{\tilde{a}}(x)$ or, alternatively, by its α-cuts, a_α, as:

$$\mu_{\tilde{a}}(x) = \begin{cases} \frac{x-a+l_a}{l_a} & a - l_a < x \le a \\ \frac{a+r_a-x}{r_a} & a < x \le a + r_a \\ 0 & \text{otherwise} \end{cases} \tag{1a}$$

$$a_\alpha = \lfloor \underline{a}(\alpha), \overline{a}(\alpha) \rfloor = [a - l_a(1 - \alpha), a + r_a(1 - \alpha)] \tag{1b}$$

In a fuzzy regression context is very usual to use symmetrical TFNs (STFNs), i.e. $l_a = r_a$ In this case the spread will be symbolised as s_a and a STFN will be denoted as $\tilde{a} = (a, s_a)$. Analytically:

$$\mu_{\tilde{a}}(x) = \begin{cases} \frac{|x-a|}{s_a} & a - s_a \le x \le a + s_a \\ 0 & \text{otherwise} \end{cases} \tag{2a}$$

$$a_\alpha = \lfloor \underline{a}(\alpha), \overline{a}(\alpha) \rfloor = [a - s_a(1 - \alpha), a + s_a(1 - \alpha)] \tag{2b}$$

In many quantitative analyses, it is often necessary to evaluate functions which we shall name $y = f(x_1, x_2, \ldots, x_n)$. Then, if $x_1, x_2, \ldots, x_n$ are not crisp numbers but the FNs $\tilde{a}_1, \tilde{a}_2, \ldots, \tilde{a}_n, f(\cdot)$ induces the FN $\tilde{b} = f(\tilde{a}_1, \tilde{a}_2, \ldots, \tilde{a}_n)$. To obtain its α-cuts, b_α, from $a_{1_\alpha}, a_{2_\alpha}, \ldots, a_{n_\alpha}$, it is necessary to evaluate:

$$a_\alpha = f(\tilde{a}_1, \tilde{a}_2, \ldots, \tilde{a}_n)_\alpha = f(a_{1_\alpha}, a_{2_\alpha}, \ldots, a_{n_\alpha}) \tag{3a}$$

Many functional relationships are continuously increasing or decreasing with respect to all the variables involved in such a way that it is easy to evaluate the α-cuts of $\tilde{b}$. Buckley and Qu [14] demonstrate that if the function $f(\cdot)$ that induces $\tilde{b}$ is increasing with respect to the first m variables, where $m \le n$, and decreasing with respect to the last n-m variables, then b_α (3a) turns into:

$$b_\alpha = \left[\underline{b}(\alpha), \overline{b}(\alpha) \right] = \left[f\left(\underline{a_1}(\alpha), \ldots, \underline{a_m}(\alpha), \overline{a_{m+1}}(\alpha), \ldots, \overline{a_n}(\alpha) \right), f\left(\overline{a_1}(\alpha), \ldots, \overline{a_m}(\alpha), \underline{a_{m+1}}(\alpha), \ldots, \underline{a_n}(\alpha) \right) \right]$$
$$\tag{3b}$$

If a FN $\tilde{b}$ is obtained from a linear combination of the TFNs $\tilde{a}_i = (a_i, l_{a_i}, r_{a_i})$, $i = 1, \ldots, n$, i.e. $\tilde{b} = \sum_{i=1}^n k_i \tilde{a}_i, k_i \in \Re, \tilde{b}$ will be the TFN, $\tilde{b} = (b, l_b, r_b)$, where:

$$b = \sum_{i=1}^{n} a_i k_i, \ l_b = \sum_{i=1, k_i \geq 0}^{n} l_{a_i} |k_i| + \sum_{i=1, k_i < 0}^{n} r_{a_i} |k_i|, \ r_b = \sum_{i=1, k_i \geq 0}^{n} r_{a_i} |k_i| + \sum_{i=1, k_i < 0}^{n} l_{a_i} |k_i|$$

$$(4a)$$

If $\tilde{b}$ is obtained from a linear combination of the STFNs $\tilde{a}_i = (a_i, s_{a_i})$, $i = 1, ...,$ n, we obtain the STFN, $\tilde{b} = (b, l_b)$, where:

$$b = \sum_{i=1}^{n} a_i k_i, \ s_b = \sum_{i=1}^{n} s_{a_i} |k_i|$$

$$(4b)$$

A key concept in several regression models is the level of inclusion of a FN $\tilde{b}$ within another FN $\tilde{a}$, $\mu(\tilde{b} \subseteq \tilde{a})$. By using α-cuts $a_\alpha = \lfloor \underline{a}(\alpha), \overline{a}(\alpha) \rfloor$ and $b_\alpha = \lfloor \underline{b}(\alpha), \overline{b}(\alpha) \rfloor$, $\mu(\tilde{b} \subseteq \tilde{a}) \geq \alpha$ if $b_\alpha \subseteq a_\alpha$, i.e.:

$$\underline{a}(\alpha) \leq \underline{b}(\alpha) \quad \text{and} \quad \overline{a}(\alpha) \geq \overline{b}(\alpha)$$

$$(5)$$

Other relevant concept in some fuzzy regression methods is the squared distance between FNs. There is no unique definition for it. Whereas Diamond [15] proposes the following measure for TFNs:

$$d_D^2(\tilde{a}, \tilde{b}) = [a - b - (l_a - l_b)]^2 + (a - b)^2 + [a - b + (r_a - r_b)]^2$$

$$(6a)$$

On the other hand, Chang [11], by using weighted fuzzy arithmetic, defines squared distance between $\tilde{a}$ and $\tilde{b}$ as:

$$d_C^2(\tilde{a}, \tilde{b}) = \int_0^1 [\underline{b}(\alpha) - \underline{a}(\alpha)]^2 \alpha \, d\alpha + \int_0^1 [\overline{b}(\alpha) - \overline{a}(\alpha)]^2 \alpha \, d\alpha$$

$$(6b)$$

where (6b) is:

$$d_C^2(\tilde{a}, \tilde{b}) = (a - b)^2 + \frac{1}{3}[(r_a - r_b) - (l_a - l_b)] + \frac{1}{12}\left[(r_a - r_b)^2 + (l_a - l_b)^2\right]$$

$$(6c)$$

when $\tilde{a}$ and $\tilde{b}$ are TFN.

3   Fuzzy Regression Models

Like any regression technique, the aim of Fuzzy Regression (FR) is to determine a functional relationship between a dependent variable and a set of independent ones. FR allows not only obtaining functional relationships between some variables when they are crisp but also when the observations are quantified with FNs.

As in econometric linear regression, we shall suppose that the explained variable is a linear combination of the explanatory variables. This relationship should be obtained from a sample of n observations $\{(Y_1,X_1), (Y_2,X_2), \ldots,(Y_j,X_j),\ldots, (Y_n,X_n)\}$ where X_j is the j-th observation of the explanatory variable, $X_j = (X_{0j}, X_{1j}, X_{2j}, \ldots, X_{ij}, \ldots, X_{mj})$. Moreover, $X_{0,j} = 1\ \forall j$, and X_{ij} is the observed value for the i-th variable in the j-th case of the sample. Y_j is the j-th observation of the explained variable, $j = 1, 2, \ldots, n$. The j-th observation may either be a crisp value or a TFN. In either case, it can be represented through the TFN representation $\tilde{Y}_j = (Y_j, l_{Y_j}, r_{Y_j})$ where, for a crisp observations, $l_{Y_j} = r_{Y_j} = 0$.

Finally, we must estimate the following fuzzy linear function:

$$\tilde{Y}_j^* = \tilde{a}_0 X_{0j} + \tilde{a}_1 X_{1j} + \cdots + \tilde{a}_m X_{mj} \tag{7}$$

where $\tilde{Y}_j^*$ is the estimate of the true observation $\tilde{Y}_j$ after fitting the coefficients $\tilde{a}_0, \tilde{a}_1, \ldots, \tilde{a}_m$. Of course, the final objective is obtaining the estimates of dependent variables as closed as possible to their corresponding observed values. Literature has proposed several models to fit those coefficients. The significant difference between them is the criteria considered to measure neighbourhood between $\tilde{Y}_j^*$ and $\tilde{Y}_j$. So, we distinguish:

(a) Models that use Minimum Fuzziness Criteria (MFC) to fit (7). We will describe the model developed in Tanaka [9] and Tanaka and Ishibuchi [10].
(b) Models that minimise squared distances, in such a way that classical least squares criteria is generalised to FR. This paper will use the results in Chang [11].
(c) Models that combine criteria (a) and (b). We will describe developments by Savic and Pedrycz [13] and Ishibuchi and Nii [12].

3.1   Fuzzy Regression with Minimum Fuzziness Principle

The fuzzy regression model based on the minimum fuzziness principle (MFP) is developed in Tanaka [9] and Tanaka and Ishibuchi [10]. It has been widely used models in economic applications. Concretely, in actuarial applications we can outline Andrés and Terceño [16] and Berry-Sölzle et al. [17]. Andrés and Terceño [16] present a model to fit TSIR that combine cubic spline model by McCulloch

[18] and FR under MFP. Subsequently they reflect on how to apply fuzzy estimates of discount rates in actuarial pricing. Berry Stölzle et al. [17] use this model of FR to detect fuzziness in the relation between solvency of property-liability insurance companies and some accounting indicators.

In this case, the parameters $\tilde{a}_i$, $i = 0, 1, 2, \ldots, m$ must be STFNs. These parameters can therefore be written as $\tilde{a}_i = (a_i, s_{a_i})$, $i = 0, 1, \ldots, m$. When we have obtained $\tilde{a}_i$, the estimates of, $\tilde{Y}_j$, $\tilde{Y}_j^*$ are also STFN $\tilde{Y}_j^* = \left(Y_j^*, s_{Y_j^*} \right)$, will be, from (4b):

$$\tilde{Y}_j^* = \left(Y_j^*, s_{Y_j^*} \right) = \sum_{i=0}^{m} (a_i, s_{a_i}) X_{ij} = \left(\sum_{i=0}^{m} a_i X_{ij}, \sum_{i=0}^{m} s_{a_i} |X_{ij}| \right) \qquad (8a)$$

whose α-cuts for a level α' are:

$$
\begin{aligned}
Y_{j_{\alpha'}}^* &= \left[Y_j^* - s_{Y_j^*}(1 - \alpha'), Y_j^* + s_{Y_j^*}(1 - \alpha') \right] \\
&= \left[\sum_{i=0}^{m} a_i X_{ij} - (1 - \alpha') \sum_{i=0}^{m} s_{a_i} |X_{ij}|, \sum_{i=0}^{m} a_i X_{ij} + (1 - \alpha') \sum_{i=0}^{m} s_{a_i} |X_{ij}| \right]
\end{aligned} \qquad (8b)
$$

The parameters a_i and s_{a_i}, must minimise the spreads of $\tilde{Y}_j^*$, and simultaneously maximise (8b) the congruence of $\tilde{Y}_j^*$ with $\tilde{Y}_j$, which is measured as $\mu\left(\tilde{Y}_j \subseteq \tilde{Y}_j^* \right)$ (see Eq. 5). Specifically, we must solve the following multiple objective programme:

$$\underset{a_i, s_{a_i}, i=0,1,\ldots,m}{\text{Minimise}} \ z = \sum_{j=1}^{n} s_{Y_j^*} = \sum_{j=1}^{n} \sum_{i=0}^{m} s_{a_i} |X_{ij}|, \quad \underset{a_i, s_{a_i}, i=0,1,\ldots,m}{\text{Maximise}} \ \alpha \qquad (9a)$$

subject to:

$$\mu\left(\tilde{Y}_j \subseteq \tilde{Y}_j^* \right) \geq \alpha \quad j = 1, 2, \ldots, n \quad s_{a_i} \geq 0 \quad i = 0, 1, \ldots, m, \quad \alpha \in [0, 1] \qquad (9b)$$

If for the second objective we require a minimum accomplishment level α', the above programme is transformed into the following linear one:

$$\underset{a_i, s_{a_i}, i=0,1,\ldots,m}{\text{Minimise}} \ z = \sum_{j=1}^{n} \sum_{i=0}^{m} s_{a_i} |X_{ij}| \qquad (10a)$$

subject to:

$$\sum_{i=0}^{m} a_i X_{ij} - (1 - \alpha') \sum_{i=0}^{m} s_{a_i} |x_{ij}| \leq Y_j - l_{Y_j}(1 - \alpha'), j = 1, 2, \ldots, n \qquad (10b)$$

$$\sum_{i=0}^{m} a_i X_{ij} + (1 - \alpha') \sum_{i=0}^{m} s_{a_i} |x_{ij}| \geq Y_j + r_{Y_j}(1 - \alpha'), j = 1, 2, \ldots, n \qquad (10c)$$

$$s_{a_i} \geq 0 \quad i = 0, 1, \ldots, m \qquad (10d)$$

Let us point out several drawbacks of this FR model:

(a) This model usually has a lack of connection with ordinary least squares (OLS) as it is outlined in Shapiro [19]. The modal value of observations on explained variable and their estimates may be not very close. However, as we will expose bellow, Savic and Pedrycz [13] and Ishibuchi and Nii [12] propose extensions to this regression method that partially use least squares criteria and solve this problem.

(b) This fuzzy regression method is very sensitive to outliers and so, the spreads of coefficients can take an excessive great value. So, the predictions from the model may be too uncertain to be useful. Following Hung and Yang [20], *"Once an outlier has been detected, it should be put under scrutiny. One should not mechanically reject outliers and proceed with the analysis. If the outliers are bona fide observations, they may indicate the inadequacy of the model under some specific conditions. They often provide valuable clues to the analyst for constructing a better model. It is important for a data analyst to be able to identity outliers and assess their effect on various aspects of the analysis."* So, those reasons explain why the literature has proposed a great number of methods to detect and solve those problem. For a wide survey, see Chen [21] or Hung and Yang [20].

(c) Chang and Ayyub [8] and Hojati et al. [22] point out that the linear programming problem to solve increases its complexity very much since for every new observation the two new constraints must be added an so, the minimising problem must be reformulated.

(d) The criteria to choose the level α' to solve (10a)–(10d) is arbitrary. So, Savic and Pedrycz [13] indicates that it is usual taking $\alpha' = 0.5$ but it is not necessarily the best choice and it has not a deep foundation. However, Moskowitz and Kim [23], propose a method that assesses practitioners choosing α' rationally.

3.2 Fuzzy Least Squares

Fuzzy Least Squares (FLS) uses as a criteria to fit coefficients $\tilde{a}_i$, $i = 0, 1, \ldots, m$ to minimise the squared distance between $\tilde{Y}_j$ and $\tilde{Y}_j^*$. In actuarial issues, Koissi and Shapiro [24] fit a fuzzy temporal structure of interest rates whereas Apaydin and Baser [25] adjust claiming behaviour in a non-life context with geometric separation method by Taylor [26].

FLS does not require the coefficients to be STFN, i.e. $\tilde{a}_i = (a_i, l_{a_i}, r_{a_i}), i = 0, 1, \ldots,$ m. On the other hand, we will suppose that $X_{i,j} \geq 0, i = 0, 1, \ldots, m$ and $j = 1, 2, \ldots,$ n. Notice that it does not suppose any loose of generality. If any observation of the ith explanatory variable is negative, $X_{i,j} < 0$, we can rescale the observations of the ith explanatory variable as $X_{i,j}^* = X_{i,j} - \text{Min}_{j=1,2,\ldots,n}\{X_{i,j}\}$. So, $\tilde{Y}_j^* = \left(Y_j^*, l_{Y_j^*}, r_{Y_j^*}\right)$ is, from (4a):

$$\tilde{Y}_j^* = \left(Y_j^*, l_{Y_j^*}, r_{Y_j^*}\right) = \sum_{i=0}^{m}(a_i, l_{a_i}, r_{a_i})X_{ij} = \left(\sum_{i=0}^{m}a_iX_{i,j}, \sum_{i=0}^{m}l_{a_i}X_{i,j}, \sum_{i=0}^{m}r_{a_i}X_{i,j}\right)$$

(11a)

whose α-cuts are:

$$Y_{j_\alpha}^* = \left[Y_j^* - l_{Y_j^*}(1-\alpha), Y_j^* + r_{Y_j^*}(1-\alpha)\right]$$
$$= \left[\sum_{i=0}^{m}a_iX_{ij} - (1-\alpha)\sum_{i=0}^{m}l_{a_i}X_{i,j}, \sum_{i=0}^{m}a_iX_{ij} + (1-\alpha)\sum_{i=0}^{m}r_{a_i}X_{i,j}\right]$$

(11b)

The parameters a_i, l_{a_i} and r_{a_i} must minimise the distance between $\tilde{Y}_j^*$ and the real observation of response variable $\tilde{Y}_j$. If we use the squared distance defined by Diamond [15], as it is done by Koissi and Shapiro [24], the following quadratic programming problem must be solved:

$$\underset{a_i,l_{a_i},r_{a_i}i=0,1,\ldots,m}{\text{Minimise}} \sum_{j=1}^{n}d_D^2\left(\tilde{Y}_j, \tilde{Y}_j^*\right) = \sum_{j=1}^{n}d_D^2\left(\tilde{Y}_j, \sum_{i=0}^{m}\tilde{a}_iX_{i,j}\right)$$

(12a)

where, from (6c) we find:

$$d_D^2\left(\tilde{Y}_j, \sum_{i=0}^{m}\tilde{a}_iX_{i,j}\right) = \left[Y_j - \sum_{i=0}^{m}a_iX_{i,j} - \left(l_{Y_j} - \sum_{i=0}^{m}l_{a_i}X_{i,j}\right)\right]^2$$
$$+ \left(Y_j - \sum_{i=0}^{m}a_iX_{i,j}\right)^2 + \left[Y_j - \sum_{i=0}^{m}a_iX_{i,j} + \left(r_{Y_j} - \sum_{i=0}^{m}r_{a_i}X_{i,j}\right)\right]^2$$

(12b)

Apaydin and Baser [25] use the definition of distance by Chang [11]. In this case the quadratic programming problem to solve is:

$$\underset{a_i,l_{a_i},r_{a_i}i=0,1,\ldots,m}{\text{Minimise}} \sum_{j=1}^{n}d_C^2\left(\tilde{Y}_j, \tilde{Y}_j^*\right) = \sum_{j=1}^{n}d_C^2\left(\tilde{Y}_j, \sum_{i=0}^{m}\tilde{a}_iX_{i,j}\right)$$

(13a)

where, from (6a) we find:

$$
d_C^2\left(\tilde{Y}_j, \sum_{i=0}^m \tilde{a}_i X_{i,j}\right) = \left(Y_j - \sum_{i=0}^m a_i X_{i,j}\right)^2 + \frac{1}{3}\left[\left(r_{Y_j} - \sum_{i=0}^m r_{a_i} X_{i,j}\right) - \left(l_{Y_j} - \sum_{i=0}^m l_{a_i} X_{i,j}\right)\right]
$$
$$
+ \frac{1}{12}\left[\left(r_{Y_j} - \sum_{i=0}^m r_{a_i} X_{i,j}\right)^2 - \left(l_{Y_j} - \sum_{i=0}^m l_{a_i} X_{i,j}\right)^2\right]
$$

(13b)

In both cases FLS is reduced to three independent OLS estimates. The centres are obtained by using OLS and taking as observed responses Y_j, $j = 1, 2, \ldots,$ n. Likewise, the left (right) widths l_{a_i} (r_{a_i}), $i = 0, 1, \ldots, m$ are also obtained with OLS by considering for the observations of dependent variable l_{Y_j} (r_{Y_j}), $j = 1, 2, \ldots,$ n. So, following Chang [11], for the centres we must solve:

$$
\sum_{k=0}^m \left(\sum_{j=1}^n X_{i,j} X_{k,j}\right) a_k = \sum_{j=1}^n X_{i,j} Y_j, \quad i = 0, 1, 2, \ldots, m
$$

(14a)

Analogously, for the left spreads:

$$
\sum_{k=0}^m \left(\sum_{j=1}^n X_{i,j} X_{k,j}\right) l_{a_k} = \sum_{j=1}^n X_{i,j} l_{Y_j}, \quad i = 0, 1, 2, \ldots, m
$$

(14b)

And for the right spreads, the system to solve is:

$$
\sum_{k=0}^m \left(\sum_{j=1}^n X_{i,j} X_{k,j}\right) r_{a_k} = \sum_{j=1}^n X_{i,j} r_{Y_j}, \quad i = 0, 1, 2, \ldots, m
$$

(14c)

In our opinion, the principal criticism to this model is that it does not reflect all the uncertainty of the observed system as we can check in the examples of Sect. 3.4. It does not quantify the uncertainty in the relation between input and output variables when observations are crisp. So, if the observed responses are crisp, the difference between $\tilde{Y}_j^*$ and $\tilde{Y}_j$ is explained neither from fuzziness nor from randomness. Actually, as we can check in the second example of subsection 3.4., when observations of responses are fuzzy, the total fuzziness of the observations about $\tilde{Y}_j$ measured as $\sum_{j=1}^n \left(l_{Y_j} + r_{Y_j}\right)$ is not completely explained by the width sum of estimates $\sum_{j=1}^n \left(l_{Y_j^*} + r_{Y_j^*}\right)$ since $\sum_{j=1}^n \left(l_{Y_j} + r_{Y_j}\right) > \sum_{j=1}^n \left(l_{Y_j^*} + r_{Y_j^*}\right)$.

3.3   Combining Least Squares and Minimum Fuzziness Principle

A reasonable way to avoid some problems of MFP and FLS is to combine both criteria in order to take advantage of their strengths. Savic and Pedrycz [13] suppose, as pure MFC model, that the coefficients $\tilde{a}_i$ are STFN, in such a way that $\tilde{a}_i = (a_i, s_{a_i})$, $i = 0, 1, \ldots, m$. On the other hand, centres are adjusted by using least squares principle. Subsequently, spreads are obtained by using MFP. In life insurance field, Koissi and Shapiro [27] follows this methodology to forecast mortality combining Lee-Carter schema with fuzzy regression.

So, to adjust coefficients, Savic and Pedrycz [13] propose the following two steps.

Step 1   Fit the centres $a_{i,}$ $i = 0, 1, \ldots, m$ by using OLS. As observed values of explained variable take their centres, $Y_j, j = 1, 2, \ldots, n$. So, the values fitted for $a_0, a_1, \ldots a_m$ the centres are equal to those obtained with FLS

Step 2   Fit the widths s_{a_i}, $i = 0, 1, \ldots, m$ with MFP. We must solve the linear programme (10a)–(10d) but taking into account that decision variables are only the spreads. Of course, α' must be fixed beforehand:

$$\underset{s_{a_i}, i=0,1,\ldots,m}{\text{Minimise } z} = \sum_{j=1}^{n} \sum_{i=0}^{m} s_{a_i} |X_{ij}| \tag{15a}$$

subject to:

$$\sum_{i=0}^{m} a_i X_{ij} - (1 - \alpha') \sum_{i=0}^{m} s_{a_i} |x_{ij}| \leq Y_j - l_{Y_j}(1 - \alpha'), \quad j = 1, 2, \ldots, n \tag{15b}$$

$$\sum_{i=0}^{m} a_i X_{ij} + (1 - \alpha') \sum_{i=0}^{m} s_{a_i} |x_{ij}| \geq Y_j + r_{Y_j}(1 - \alpha'), \quad j = 1, 2, \ldots, n \tag{15c}$$

$$s_{a_i} \geq 0 \quad i = 0, 1, \ldots, m \tag{15d}$$

Ishibuchi and Nii [12] extend Savic and Pedrycz results to the case of non-symmetrical TFN, i.e., $\tilde{a}_i = (a_i, l_{a_i}, r_{a_i})$, $i = 0, 1, \ldots, m$. In actuarial applications, Andrés-Sánchez 28 propose combining exponential splines and this FR model to fit temporal structure of interest rates. In a non-life context Andrés-Sánchez [29] predicts claiming behaviour by using a trending fuzzified function and Andrés-Sánchez [30] extends ANOVA schema by Kremer [31] for claim reserving to the presence of fuzziness.

To adjust the centres, we proceed exactly as Savic and Pedrycz [13] or FLS method. Likewise, to obtain widths we also use MFP, but now we must solve a slightly different linear programme to (15a)–(15d). Concretely:

$$\underset{l_{a_i},r_{a_i},i=0,1,\ldots,m}{\text{Minimise}}\ z = \sum_{j=1}^{n}\sum_{i=0}^{m} l_{a_i}|x_{ij}| + \sum_{j=1}^{n}\sum_{i=0}^{m} r_{a_i}|x_{ij}| \qquad (16a)$$

subject to:

$$\sum_{i=0}^{m} a_i X_{ij} - \left(\sum_{i=0,x_{ij}\geq 0}^{m} l_{a_0}|X_{ij}| + \sum_{i=0,x_{ij}<0}^{m} r_{a_i}|X_{ij}| \right)(1-\alpha') \leq Y_j - l_{Y_j}(1-\alpha'), \quad (16b)$$
$$j = 1,2,\ldots,n$$

$$\sum_{i=0}^{m} a_i X_{ij} + \left(\sum_{i=0,x_{ij}\geq 0}^{m} r_{a_i}|x_{ij}| + \sum_{i=0,x_{ij}<0}^{m} l_{a_i}|x_{ij}| \right)(1-\alpha') \geq Y_j + r_{Y_j}(1-\alpha'), \quad (16c)$$
$$j = 1,2,\ldots,n$$

$$l_{a_i}, r_{a_i} \geq 0 \quad i = 0,1,\ldots,m \qquad (16d)$$

3.4 Numerical Examples

In this subsection we will develop two simple numerical examples to show in detail how to implement the regression models exposed above. In both examples the model to be fitted is: $\tilde{Y}_j^* = \tilde{a}_0 X_{0j} + \tilde{a}_1 X_{1j}$, where $X_{0,j} = 1$, $j = 1, 2, 3, 4$. In Example 1 the output observations are crisp numbers and the data is taken from Table 1. So, the fuzziness in the linear relation reflects only the deviations between observations Y_j, and the centres of their estimates, Y_j^*. With pure MFC (Tanaka [9] and Ishibuchi and Tanaka [10]) we must fit a model governed by STFNs:

$$\tilde{Y}_j^* = \left(Y_j^*, s_{Y_j^*} \right) = (a_0, s_{a_0}) + (a_1, s_{a_1})X_{1,j} = \left(a_0 + a_1 X_{1j}, s_{a_0} + s_{a_1}|X_{1j}| \right)$$

Table 1 Data for Example 1 and estimates from each regression model

j	$\tilde{Y}_j^*$ (TK)	$\tilde{Y}_j^*$ (FLS)	$\tilde{Y}_j^*$ (S&P)	$\tilde{Y}_j^*$ (I&N)	$\tilde{Y}_j$	$X_{0,j}$	$X_{1,j}$
1	(1000, 50)	(992.5, 0, 0)	(992.5, 68.33)	(992.50, 68.33, 65)	(1000, 0, 0)	1	1
2	(1500, 100)	(1510, 0, 0)	(1510, 136.67)	(1510, 136.67, 80)	(1550, 0, 0)	1	2
3	(2000, 150)	(2027.5, 0, 0)	(2027.5, 205)	(2027.5, 205, 95)	(1925, 0, 0)	1	3
4	(2500, 200)	(2545, 0, 0)	(2545, 273.33)	(2545, 273.33, 110)	(2600, 0, 0)	1	4

TK Tanaka's model, FLS Fuzzy Least Squares, S&P Savic and Pedrycz's (1992) model and I&N Ishibuchi and Nii's [12] model

To fit the parameters a_i and s_{a_i} we will require a minimum accomplishment level $\alpha' = 0.5$ and so, the linear programme to solve for this model is:

$$\underset{a_0,a_1,s_{a_0},s_{a_1}}{\text{Minimise }} z = 4s_{a_0} + 10s_{a_1}$$

subject to:

$$a_0 + a_1 \cdot 1 - (1 - 0.5)(s_{a_0} + s_{a_1} \cdot 1) \le 1000 \le a_0 + a_1 \cdot 1 + (1 - 0.5)(s_{a_0} + s_{a_1} \cdot 1)$$
$$a_0 + a_1 \cdot 2 - (1 - 0.5)(s_{a_0} + s_{a_1} \cdot 2) \le 1550 \le a_0 + a_1 \cdot 2 + (1 - 0.5)(s_{a_0} + s_{a_1} \cdot 2)$$
$$a_0 + a_1 \cdot 3 - (1 - 0.5)(s_{a_0} + s_{a_1} \cdot 3) \le 1925 \le a_0 + a_1 \cdot 3 + (1 - 0.5)(s_{a_0} + s_{a_1} \cdot 3)$$
$$a_0 + a_1 \cdot 4 - (1 - 0.5)(s_{a_0} + s_{a_1} \cdot 4) \le 2600 \le a_0 + a_1 \cdot 4 + (1 - 0.5)(s_{a_0} + s_{a_1} \cdot 4)$$
$$s_{a_0}, s_{a_1} \ge 0$$

and solving this linear programme we find $\tilde{Y}_j^* = \left(Y_j^*, s_{Y_j^*} \right) = (500, 0) + (500, 50) X_{1,j}$.

With FLS we must fit the following model:

$$\tilde{Y}_j^* = \left(Y_j^*, 0, 0 \right) = (a_0, 0, 0) + (a_1, 0, 0)X_{1,j} = \left(a_0 + a_1 X_{1j}, 0, 0 \right)$$

Notice that there is no fuzziness in estimates since the observations are crisp and so (14b)–(14c) must not be applied. And so, by using OLS we finally fit $\tilde{Y}_j^* = Y_j^* = 475 + 517.5X_{1,j}$

The model by Savic and Pedrycz [13], suppose symmetrical coefficients but estimates their centres by using OLS, i.e. $a_0 = 475$ and $a_1 = 517.5$ as FLS. Subsequently we must fit s_{a_i}, $i = 0, 1$, by solving:

$$\underset{s_{a_0},s_{a_1}}{\text{Minimise }} z = 4s_{a_0} + 10s_{a_1}$$

subject to:

$$475 + 517.5 \cdot 1 - (1 - 0.5)(s_{a_0} + s_{a_1} \cdot 1) \le 1000 \le 475 + 517.5 \cdot 1 + (1 - 0.5)(s_{a_0} + s_{a_1} \cdot 1)$$
$$475 + 517.5 \cdot 2 - (1 - 0.5)(s_{a_0} + s_{a_1} \cdot 2) \le 1550 \le 475 + 517.5 \cdot 2 + (1 - 0.5)(s_{a_0} + s_{a_1} \cdot 2)$$
$$475 + 517.5 \cdot 3 - (1 - 0.5)(s_{a_0} + s_{a_1} \cdot 3) \le 1925 \le 475 + 517.5 \cdot 3 + (1 - 0.5)(s_{a_0} + s_{a_1} \cdot 3)$$
$$475 + 517.5 \cdot 4 - (1 - 0.5)(s_{a_0} + s_{a_1} \cdot 4) \le 2600 \le 475 + 517.5 \cdot 4 + (1 - 0.5)(s_{a_0} + s_{a_1} \cdot 4)$$
$$s_{a_0}, s_{a_1} \ge 0$$

and solving this linear programme we find $\tilde{Y}_j^* = \left(Y_j^*, s_{Y_j^*} \right) = (475, 0) + (517.5, 68.33)X_{1,j}$

Ishibuchi and Nii [12] suppose that coefficients are non-symmetrical. Likewise given that the observations of the explanatory variable are positive, we can write:

$$\tilde{Y}_j^* = \left(Y_j^*, l_{Y_j^*}, r_{Y_j^*}\right) = (a_0, l_{a_0}, r_{a_0}) + (a_1, l_{a_1}, r_{a_1})X_{ij} =$$
$$= \left(a_0 + a_1 X_{1j}, l_{a_0} + l_{a_1} X_{1j}, r_{a_0} + r_{a_1} X_{1j}\right)$$

The centres of the coefficients are fitted with OLS, i.e. $a_0 = 475$ and $a_1 = 517.5$. Subsequently we must fit l_{a_i}, r_{a_i} $i = 0,1$, by solving:

$$\underset{l_{a_0}, l_{a_1}, r_{a_0}, r_{a_1}}{\text{Minimise}\, z} = 4l_{a_0} + 10l_{a_1} + 4r_{a_0} + 10r_{a_1}$$

subject to:

$$475 + 517.5 \cdot 1 - (1 - 0.5)(l_{a_0} + l_{a_1} \cdot 1) \leq 1000 \leq 475 + 517.5 \cdot 1 + (1 - 0.5)(r_{a_0} + r_{a_1} \cdot 1)$$
$$475 + 517.5 \cdot 2 - (1 - 0.5)(l_{a_0} + l_{a_1} \cdot 2) \leq 1550 \leq 475 + 517.5 \cdot 2 + (1 - 0.5)(r_{a_0} + r_{a_1} \cdot 2)$$
$$475 + 517.5 \cdot 3 - (1 - 0.5)(l_{a_0} + l_{a_1} \cdot 3) \leq 1925 \leq 475 + 517.5 \cdot 3 + (1 - 0.5)(r_{a_0} + r_{a_1} \cdot 3)$$
$$475 + 517.5 \cdot 4 - (1 - 0.5)(l_{a_0} + l_{a_1} \cdot 4) \leq 2600 \leq 475 + 517.5 \cdot 4 + (1 - 0.5)(r_{a_0} + r_{a_1} \cdot 4)$$
$$l_{a_0}, l_{a_1}, r_{a_0}, r_{a_1} \geq 0$$

and solving this minimising programme we find $\tilde{Y}_j^* = \left(Y_j^*, l_{Y_j^*}, r_{Y_j^*}\right) = (475, 0, 50) + (517.5, 68.33, 15)X_{1,j}$

Results in Table 1 suggest the following questions:

(a) The mean of the absolute difference between Y_j^* and Y_j is greater with pure MFP than with LS estimates. With MFP we find $(0 + 50 + 75 + 100)/4 = 56.25$. Otherwise, LS estimates produce the following mean absolute error: $(7.5 + 40 + 102.5 + 55)/4 = 51.25$.

(b) When output observations are crisp, FLS gives a crisp estimate. So, the difference between the crisp observation and the linear system is not explained.

(c) So, the models by Savic and Pedrycz [13] and Ishibuchi and Nii [12] allow avoiding the greatest discordance between the centres of the observed and estimated output that arises when using strictly MFP but also explain the difference between $\tilde{Y}_j^*$ and $\tilde{Y}_j$ from a fuzzy perspective. Likewise, when coefficients $\tilde{a}_0$ and $\tilde{a}_1$ are not constrained to be symmetrical the estimates, $\tilde{Y}_j^*$ incorporate less uncertainty and, simultaneously, allow reflecting asymmetrical structure of observations.

In Example 2 the output observations are fuzzy numbers (see Table 2). So, the fuzziness in the linear relation must take into account not only the deviations between observations Y_j, and the centres of their estimates, Y_j^*, but also the fuzziness of observations. In all the models that use MFC we stablish a minimum

Table 2 Data for Example 2 and estimates from each regression model

j	$\tilde{Y}_j^*$ (TK)	$\tilde{Y}_j^*$(FLS)	$\tilde{Y}_j^*$(S&P)	$\tilde{Y}_j^*$(I&N)	$\tilde{Y}_j$	$X_{0,j}$	$X_{1,j}$
1	(992.36, 90.28)	(992.5, 45, 47.5)	(992.50, 101.67)	(992.50, 101.67, 90.00)	(1000, 50, 75)	1	1
2	(1501.39, 180.56)	(1510, 77.5, 95)	(1510.00, 203.33)	(1510.00, 203.33, 163.33)	(1550, 75, 50)	1	2
3	(2010.42, 270.83)	(2027.5, 110, 142.5)	(2027.50, 305.00)	(2027.50, 305.00, 236.67)	(1925, 100, 150)	1	3
4	(2519.44, 361.11)	(2545, 142.5, 190)	(2545.00, 406.67)	(2545.00, 406.67, 310.00)	(2600, 150, 200)	1	4

TK Tanaka's model, *FLS* Fuzzy Least Squares, *S&P* Savic and Pedrycz's [12] model and *I&N* Ishibuchi and Nii's [12] model

accomplishment level $\alpha' = 0.5$ for the inclusion constraints. To fit the parameters a_i and s_{a_i} with pure MFC, we must solve the following linear programming problem:

$$\underset{a_0, a_1, s_{a_0}, s_{a_1}}{\text{Minimise}} \; z = 4s_{a_0} + 10s_{a_1}$$

subject to:

$$a_0 + a_1 \cdot 1 - (1 - 0.5)(s_{a_0} + s_{a_1} \cdot 1) \leq 1000 - 50(1 - 0.5)$$

$$\cdots$$

$$a_0 + a_1 \cdot 4 - (1 - 0.5)(s_{a_0} + s_{a_1} \cdot 4) \leq 2600 - 150(1 - 0.5)$$

$$a_0 + a_1 \cdot 1 + (1 - 0.5)(s_{a_0} + s_{a_1} \cdot 1) \geq 1000 + 75(1 - 0.5)$$

$$\cdots$$

$$a_0 + a_1 \cdot 4 + (1 - 0.5)(s_{a_0} + s_{a_1} \cdot 4) \geq 2600 + 200(1 - 0.5)$$

$$s_{a_0}, s_{a_1} \geq 0$$

and solving this linear programme, we find $\tilde{Y}_j^* = \left(Y_j^*, s_{Y_j^*} \right) = (483.33, \; 0) + (509.03, 90.28)X_{1,j}$

With FLS, we must adjust:

$$\tilde{Y}_j^* = \left(Y_j^*, l_{Y_j^*}, r_{Y_j^*} \right) = (a_0, l_{a_0}, r_{a_0}) + (a_1, l_{a_1}, r_{a_1})X_{ij}$$
$$= \left(a_0 + a_1 X_{1,j}, l_{a_0} + l_{a_1} X_{1j}, r_{a_0} + r_{a_1} X_{1j} \right)$$

To adjust the parameters we must fit three OLS models. For the centres we find with (14a):

$$Y_j^* = 475 + 517.5X_{1,j}.$$

For the left and right spreads, by using (14b)–(14c), we find $l_{Y_j^*} = 12.5 + 32.5X_{1,j}$ and $r_{Y_j^*} = 42.5X_{1,j}$.

To fit the model by Savic and Pedrycz [13], we suppose that the centres of the parameters are estimated with OLS, i.e. $a_0 = 475$ and $a_1 = 517.5$. Subsequently we fit s_{a_i}, $i = 0, 1$, by solving:

$$\underset{s_{a_0}, s_{a_1}}{\text{Minimise}} \ z = 4s_{a_0} + 10s_{a_1}$$

subject to:

$$475 + 517.5 \cdot 1 - (1 - 0.5)(s_{a_0} + s_{a_1} \cdot 1) \le 1000 - 50(1 - 0.5)$$

$$\cdots$$

$$475 + 517.5 \cdot 4 - (1 - 0.5)(s_{a_0} + s_{a_1} \cdot 4) \le 2600 - 150(1 - 0.5)$$
$$475 + 517.5 \cdot 1 + (1 - 0.5)(s_{a_0} + s_{a_1} \cdot 1) \ge 1000 + 75(1 - 0.5)$$

$$\cdots$$

$$475 + 517.5 \cdot 4 + (1 - 0.5)(s_{a_0} + s_{a_1} \cdot 4) \ge 2600 + 200(1 - 0.5)$$
$$s_{a_0}, s_{a_1} \ge 0$$

being the final solution for this model $\tilde{Y}_j^* = \left(Y_j^*, s_{Y_j^*} \right) = (475, 0) + (517.5, 101.67) X_{1,j}$

Given that coefficients are non-symmetrical and taking into account that the centres are the same as in the case of FLS, i.e. $a_0 = 475$ and $a_1 = 517.5$, to fit the widths with Ishibuchi and Nii [12] we must solve:

$$\underset{l_{a_0}, l_{a_1}, r_{a_0}, r_{a_1}}{\text{Minimise}} \ z = 4l_{a_0} + 10l_{a_1} + 4r_{a_0} + 10r_{a_1}$$

subject to:

$$475 + 517,5 \cdot 1 - (1 - 0.5)(l_{a_0} + l_{a_1} \cdot 1) \le 1000 - 50(1 - 0.5)$$

$$\cdots$$

$$475 + 517.5 \cdot 4 - (1 - 0.5)(l_{a_0} + l_{a_1} \cdot 4) \le 2600 - 150(1 - 0.5)$$
$$475 + 517.5 \cdot 1 + (1 - 0.5)(r_{a_0} + r_{a_1} \cdot 1) \ge 1000 + 75(1 - 0.5)$$

$$\cdots$$

$$475 + 517.5 \cdot 4 + (1 - 0.5)(s_{a_0} + s_{a_1} \cdot 4) \ge 2600 + 200(1 - 0.5)$$
$$l_{a_0}, l_{a_1}, r_{a_0}, r_{a_1} \ge 0$$

Solving this linear programme we find $\tilde{Y}_j^* = \left(Y_j^*, l_{Y_j^*}, r_{Y_j^*} \right) = (475, 0, 101.67) + (517.5, 16.67, 73.33)X_{1,j}$.

The results in Table 2 suggest the following questions:

(a) The mean of the absolute deviation between Y_j^* and Y_j is still greater with pure MFP than with OLS estimates. With MFP we find $(7.64 + 48.61 + 85.42 + 80.56)/4 = 55.5$. Otherwise, the mean deviation of OLS estimates is 51.25.

(b) We must also remark that despite the centres of observations over response variable do not change respect example 1, the centres of the parameters (and so, mode of the predictions) are different in pure MFP model respect to the width of observations. That is to say, the estimates for the most feasible values are sensible in pure MFC regression. On the other hand, FLS and models that mix LS with MFP are not sensible in this sense.

(c) FLS neither reflects in a fuzzy way the uncertainty about the relation of response and input variables nor the total fuzziness of the observations about independent variable. Whereas the global uncertainty of the observations, measured as the sum of the spreads, is $50 + 75 + 75 + \cdots + 200 = 425$, the uncertainty reflected in FLS is $45 + 47.5 + 77.5 + \cdots + 190 = 375$. On the other hand, the constraints of inclusion lead to the models that use MFP to capture all the uncertainty of observations. So, in all the cases the sum of the widths from response observations is smaller than the sum of spreads of their estimates.

4 Estimating a Fuzzy Yield Curve for Fuzzy Financial Pricing

Several papers in the financial literature as Kaufmann [32], Buckley [33] or Li Calzi [34] propose using fuzzy numbers to model interest rate uncertainty. In an insurance context Ostaszewski [35] and Cummins and Derrig [36] develop life and non-life insurance financial pricing with fuzzy parameters. In this way Andrés and Terceño [16] and Koissi and Shapiro [24] propose estimating a fuzzy Temporal Structure of Interest Rates (TSIR) with FR as a basis for fuzzy financial pricing since fuzzy TSIR enable to quantify the anticipated rates in the fixed income markets for the future with fuzzy numbers. In all the cases the authors modelise the discount factor described by spot rates with a spline curve with parameters quantified via TFNs. However, in each paper fit different FR models. Andrés and Terceño [16] use pure MFP model and Andrés and Terceño (2004) estimate Ishibuchi and Nii's [12] FR model. On the other hand, Koissi and Shapiro [24] fits TSIR with FLS.

In this section the financial basis of TSIR model to fuzzify is slightly different to abovementioned papers. Concretely we will adjust the yield curve model by Echols and Elliot [37]. The yield curve of a bond market relates the Internal Rate of Return (IRR) of these bonds with their maturity and it is a good approximation of the real

TSIR that implicitly governs the market. Concretely, Echols and Elliott's formulation of yield curve is:

$$IRRj = a_0 + a_1 (t_j)^{-1} + a_2 t_j + a_3 c_j \qquad (17)$$

where:

$IRRj$ =IRR of the jth bond.

tj= maturity of the jth bond.

c_j = Coupon of the jth bond. The objective of this variable is to measure the coupon bias of the yield curve respect to the TSIR.

Likewise, a_0, a_1, a_2 and a_3 stand for the parameters of the model.

In our opinion, FR have a number of advantages over traditional regression techniques in this problem. For our purposes Andrés and Terceño [19] point out the following:

(a) The interest rates that we will fit after adjusting regression model is easier to use in subsequent calculations if they are fuzzy numbers rather that they are adjusted with random variables.

(b) The asset prices that are determined in the markets depend on the agents' expectations of future inflation, the issuers' credibility, etc., i.e., information that is highly subjective and, sometimes, also vague.

(c) Observations are often not crisp numbers; they are confidence intervals. For instance, the price (or alternatively, the IRR) of a financial asset throughout one session often oscillates within an interval, rarely does it remain the same.

In our case, we will assume that the observations about the IRR of the bonds are TFN. So, for the ith bond, we have that $\tilde{IRR}_j = (IRR_j, l_{IRR_j}, r_{IRR_j})$. For the centre, IRR_j we will take the IRR when the price of the bond is negotiated at its mean price weighted by the liquidity. The left spread is the difference between IRR_j and the minimum IRR negotiated for this bond during the session. Analogously, r_{IRR_j} comes from the difference between the maximum IRR of the bond within the day and IRR_j.

In our application we will adjust the yield curve to Spanish public debt market the 01/25/2013 (see data in Table 3). If we take IRR_j, $j = 1, 2, …, 34$ for the observations of dependent variable, with for the conventional OLS (14a) we fit:

$$IRRj = 1.155 - 0.085 (t_j)^{-1} + 0.217 t_j + 0.257 c_j$$

To use Tanaka's Model, we formulate Echols and Elliot equation in the following manner:

$$\left(IRR_j^*, s_{IRR_j^*}\right) = (a_0, s_{a_0}) + (a_1, s_{a_1})(t_j)^{-1} + (a_2, s_{a_2}) t_j + (a_3, s_{a_3}) c_j \qquad (18)$$

Table 3 Fixed income instruments negotiated in the Spanish bond market on 01/25/2013

Asset	Coupon	Maturity	Mean Price	Maximun Price	Minimum Price	IRR_j (%)	l_{IRR_j} (%)	r_{IRR_j} (%)
T-Bill	0	02/15/2013	99.989	99.989	99.989	0.192	0.000	0.000
T-Bill	0	04/19/2013	99.898	99.898	99.898	0.445	0.000	0.000
T-Bill	0	07/19/2013	99.657	99.657	99.657	0.720	0.000	0.000
T-Bill	0	08/23/2013	99.476	99.477	99.476	0.918	0.002	0.000
T-Bill	0	12/13/2013	98.908	98.935	98.891	1.253	0.031	0.020
T-Bill	0	01/24/2014	98.717	98.769	98.717	1.304	0.054	0.000
T-Bill	0	06/20/2014	97.592	97.592	97.592	1.758	0.000	0.000
Bond	2.3	04/30/2013	100.441	100.445	100.42	0.593	0.015	0.080
Bond	4.2	07/30/2013	101.65	101.65	101.65	0.928	0.000	0.000
Bond	3.4	04/30/2014	102.04	102.04	102.04	1.744	0.000	0.000
Bond	4.75	07/30/2014	104.154	104.16	104.1	1.922	0.004	0.035
Bond	3.3	10/31/2014	102.03	102.042	102.029	2.111	0.007	0.001
Bond	4.4	01/31/2015	104.266	104.285	104.18	2.212	0.009	0.043
Bond	2.75	03/31/2015	100.68	100.68	100.68	2.423	0.000	0.000
Bond	3	04/30/2015	101.173	101.2	101.13	2.457	0.012	0.020
Bond	4	07/30/2015	103.32	103.32	103.32	2.610	0.000	0.000
Bond	3.75	10/31/2015	102.544	102.62	102.42	2.777	0.028	0.046
Bond	3.15	01/31/2016	100.672	100.76	100.65	2.914	0.031	0.008
Bond	3.25	04/30/2016	100.7	100.7	100.699	3.018	0.000	0.000
Bond	4.25	10/31/2016	103.451	103.47	103.436	3.258	0.005	0.004
Bond	3.8	01/31/2017	101.543	101.55	101.53	3.382	0.002	0.003
Bond	5.5	07/30/2017	108.165	108.18	108.15	3.505	0.003	0.003
Bond	4.5	01/31/2018	103.23	103.35	103.2	3.781	0.026	0.007
Bond	4.1	07/30/2018	101.173	101.3	101.03	3.856	0.026	0.029
Bond	4.6	07/30/2019	102.8	102.8	102.8	4.097	0.000	0.000
Bond	4.3	10/31/2019	99	99	99	4.472	0.000	0.000
Bond	4	04/30/2020	97.8	97.8	97.8	4.357	0.000	0.000
Bond	4.85	10/31/2020	101.713	101.81	101.628	4.580	0.015	0.013
Bond	5.5	04/30/2021	105.066	105.3	105.008	4.741	0.034	0.008
Bond	5.85	01/31/2022	106.569	106.802	106.09	4.930	0.031	0.065
Bond	5.43	01/31/2023	101.971	102.23	101.855	5.173	0.033	0.015
Bond	4.8	01/31/2024	96.287	96.321	96.272	5.252	0.004	0.002
Bond	5.9	07/30/2026	103.82	103.82	103.82	5.488	0.000	0.000
Bond	6	01/31/2029	105.35	105.35	105.35	5.489	0.000	0.000
Bond	5.75	07/30/2032	101.84	102	101.68	5.589	0.013	0.014
Bond	4.9	07/30/2040	88.6	88.6	88.6	5.731	0.000	0.000

where the estimate of $I\tilde{R}Rj$ after adjusting (8), $I\tilde{R}R_j^*$, is a STFN. In our sample, we find:

$$\left(IRR_j^*, s_{IRR_j^*}\right) = (1.679, 0.000) + (-0.311, 0.224)(t_j)^{-1} + (0.286, 0.098)t_j$$
$$+ (0.048, 0.000)c_j$$

FLS formulation for (17) is:

$$\left(IRR_j^*, l_{IRR_j^*}, r_{IRR_j^*}\right) = (a_0, l_{a_0}, r_{a_0}) + (a_1, l_{a_1}, r_{a_1})t_j^{-1} \tag{19}$$
$$+ (a_2, l_{a_2}, r_{a_2})t_j + (a_3, l_{a_3}, r_{a_3})c_j$$

and regressing with OLS centres and radius of $I\tilde{R}R_j$ respect t_j, $(t_j)^{-1}$ and c_j we obtain:

$$\left(IRR_j^*, l_{IRR_j^*}, r_{IRR_j^*}\right) = (1.155, 0, 0) + (-0.085, 0.001, 0.075)(t_i)^{-1}$$
$$+ (0.217, 0, 0.007)t_i + (0.257, 0, 0.012)c_i$$

To use Savic and Pedrycz's extension of Tanaka's model we take for the centres the OLS estimates of the coefficients, i.e. $a_0 = 1.155$; $a_1 = -0.085$; $a_2 = 0.217$; $a_3 = 0.257$, and so:

$$\left(IRR_j^*, s_{IRR_j^*}\right) = (1.155, 0.033) + (-0.085, 0.033)(t_j)^{-1} + (0.217, 0.391)t_j$$
$$+ (0.257, 0.103)c_j$$

Likewise, with Ishibuchi and Nii [12], the model to be estimated is also (19) where, as in the case of Savic [13], $a_0 = 1.155$; $a_1 = -0.085$; $a_2 = 0.217$; $a_3 = 0.257$ but the coefficients are non STFN.

$$\left(IRR_j^*, l_{IRR_j^*}, r_{IRR_j^*}\right) = (a_0, l_{a_0}, r_{a_0}) + (a_1, l_{a_1}, r_{a_1})t_j^{-1}$$
$$+ (a_2, l_{a_2}, r_{a_2})t_j + (a_3, l_{a_3}, r_{a_3})c_j$$

Concretely:

$$\left(IRR_i^*, l_{IRR_i^*}, r_{IRR_i^*}\right) = (1.155, 0.030, 0.243) + (-0.085, 0.030, 0.075)(t_j)^{-1}$$
$$+ (0.217, 0.153, 0.014)t_j + (0.257, 0.116, 0.000)c_j$$

The TSIR is the functional relation between spot rates (that for a given maturity t years we will symbolise as r_t) and their maturity (t years). That function is also known as zero coupon rate since r_t is the IRR of a bond with that maturity that does

Table 4 Estimates of spot rates, for maturities from 6 months to 30 years

t	TK	FLS	S&P	I&N
0.5	(1.200, 0.497)	(1.094, 0.000, 0.014)	(0.986, 0.832)	(1.094, 0.350, 0.309)
1	(1.654, 0.322)	(1.288, 0.001, 0.007)	(1.288, 0.457)	(1.288, 0.212, 0.333)
2	(2.095, 0.309)	(1.548, 0.002, 0.003)	(1.548, 0.295)	(1.548, 0.166, 0.401)
3	(2.433, 0.370)	(1.779, 0.003, 0.002)	(1.779, 0.262)	(1.779, 0.170, 0.474)
4	(2.744, 0.450)	(2.004, 0.003, 0.002)	(2.004, 0.263)	(2.004, 0.188, 0.548)
5	(3.045, 0.537)	(2.225, 0.004, 0.001)	(2.225, 0.276)	(2.225, 0.210, 0.623)
6	(3.341, 0.628)	(2.446, 0.005, 0.001)	(2.446, 0.296)	(2.446, 0.235, 0.698)
7	(3.634, 0.721)	(2.665, 0.006, 0.001)	(2.665, 0.320)	(2.665, 0.261, 0.773)
8	(3.925, 0.815)	(2.884, 0.007, 0.001)	(2.884, 0.346)	(2.884, 0.288, 0.848)
9	(4.215, 0.911)	(3.103, 0.008, 0.001)	(3.103, 0.373)	(3.103, 0.316, 0.923)
10	(4.504, 1.006)	(3.321, 0.008, 0.001)	(3.321, 0.402)	(3.321, 0.344, 0.999)
20	(7.376, 1.979)	(5.500, 0.017, 0.000)	(5.500, 0.712)	(5.500, 0.636, 1.752)
30	(10.237, 2.960)	(7.677, 0.025, 0.000)	(7.677, 1.035)	(7.677, 0.933, 2.506)

TK Tanaka's model, *FLS* Fuzzy Least Squares, *S&P* Savic and Pedrycz's [13] model and *I&N* Ishibuchi and Nii's [12] model

not pay any coupon. So, with Echols and Elliot [37] formulation of yield curve, TSIR is approximated without considering the coupon. In a crisp environment, we will approximate in our numerical application, $r_t \approx 1.155 - 0.085t^{-1} + 0.217t$. Likewise, Table 4 shows several TSIR fitted with FR models.

5   Fuzzy Regression to Trend Behaviour of Claiming

Claim provisions are crucial for the financial stability of insurance companies. This is why actuarial literature has proposed numerous claim reserving methods, which are usually based on statistical concepts. However, Straub [38] points out that the mutant and uncertain behaviour of insurance environments does not make advisable to use a wide data-base when calculating claim reserves. For example, if claims are related to bodily injuries, the future losses for the company will depend on the growth of the wage index (which will be used to determine the amount of indemnification due), changes in court practices and public awareness of liability matters. On the one hand, this involves a considerable loss in reliability of statistical methods but, on the other, it makes the use of Fuzzy Set Theory very attractive. In this section we expose the FR method by Andrés-Sánchez [29] to trend claiming growth rate throughout development periods.

The data-base about the evolution of the claims, is usually presented in a *run-off* triangle similar to Table 1, where $Z_{i,j}$ is the accumulated claim cost of the insurance contracts underwritten in the *i*th period ($i = 0, 1, ..., n$) at the end of the *j*th claiming period ($j = 0, 1, ..., n$). We would like to point out that it is common in actuarial literature to use a "squared" claim triangle like ours: i.e. a claim triangle in which the number of origin and development periods is equal ($n + 1$ in Table 1). So the accumulated value of claims we know for the ith origin period ends at a jth period, where $i + j = n$ or, equivalently, $j = n - i$.

The differences between the methods proposed in actuarial literature for claim reserving are not in the way in which the data are presented but how the cumulated claims are predicted in development periods where they are unknown (i.e., to determine $Z_{i,j}$, $i = 0, 1, \ldots, n$, $j > n - i$ in Table 5 and $i = 0, 1, \ldots, 4$, $j > 4 - i$ in Table 6).

Many claim reserving methods require that the link ratio triangle be determined firstly. The link ratio of the ith underwriting period at claiming period j, $r_{i,j}$, is the growth ratio of cumulated claims at the ith origin period between development periods j and $j + 1$:

$$r_{i,j} = \frac{Z_{i,j+1}}{Z_{i,j}} \tag{20}$$

So, by applying (20) in Table 1, we deduce the general form of a link ratio triangle, i.e. Table 7. Table 8 shows the link ratio triangle in our numerical example.

The key to predict future claiming behaviour consists in fitting a representative link-ratio for any underwriting period for the pairs of claiming periods j and $j + 1$, r_j, $j = 0, 1, \ldots, n - 1$. However, if independent estimates are obtained for r_j, $j = 0, 1, \ldots, n - 1$, an overparameterisation problem may be produced. So, the capability

Table 5 Run off triangle

		Claiming/development period						
		0	1	...	j	...	n-1	n
Underwriting/origin period	0	$Z_{0,0}$	$Z_{0,1}$	...	$Z_{0,j}$	...	$Z_{0,n-1}$	$Z_{0,n}$
	1	$Z_{1,0}$	$Z_{1,1}$	...	$Z_{1,j}$	...	$Z_{1,n-1}$	
	$\vdots$	$\vdots$	$\vdots$	$\vdots$	$\vdots$	$\vdots$		
	i	$Z_{i,0}$	$Z_{i,1}$	...	$Z_{i,j}$	...		
	$\vdots$	$\vdots$	$\vdots$	$\vdots$				
	n-1	$Z_{n-1,0}$	$Z_{n-1,1}$	...				
	n	$Z_{n,0}$		...				

Table 6 Run off triangle in our numerical applications

Underwriting/origin period	Claiming/development period					
	year	i/j	0	1	2	3	4
	2000	0	1120	2090	2610	2920	3130
	2001	1	1030	1920	2370	2710	
	2002	2	1090	2140	2610		
	2003	3	1300	2650			
	2004	4	1420				

Table 7 Link ratio triangle

		Claiming/development period						
		0	1	...	j	...	$n-1$	n
Underwriting/origin period	0	$r_{0,0} = Z_{0,1}/Z_{0,0}$	$r_{0,1} = Z_{0,2}/Z_{0,1}$	...	$r_{0,j} = Z_{0,j+1}/Z_{0,j}$	...	$r_{0,n-1} = Z_{0,n}/Z_{0,n-1}$	
	1	$r_{1,0} = Z_{1,1}/Z_{1,0}$	$r_{1,1} = Z_{1,2}/Z_{0,1}$	...	$r_{1,j} = Z_{1,j+1}/Z_{1,j}$	...	–	–
	⋮	⋮	⋮	⋮	⋮	⋮		
	i	$r_{i,0} = Z_{i,1}/Z_{i,0}$	$r_{i,1} = Z_{i,2}/Z_{i,1}$	...	$r_{i,j} = Z_{i,j+1}/Z_{i,j}$	...	–	–
	⋮	⋮	⋮	⋮	⋮			
	$n-1$	$r_{n-1,0} = Z_{n-1,1}/Z_{n-1,0}$						
	n	–	–	...	–		–	–

Table 8 Link ratio in our numerical example

			Claiming/development year				
		i/j	0	1	2	3	4
Underwriting/origin year	2000	0	1.866	1.249	1.119	1.072	–
	2001	1	1.864	1.234	1.143	–	–
	2002	2	1.963	1.220	–	–	–
	2003	3	2.038	–	–	–	–
	2004	4	–	–	–	–	–

of generalising the historical data of this usual procedure may be not very good. To obtain a better generalisation capability, the number of parameters to be fitted should be reduced as much as possible. To solve this drawback, Sherman [39] proposes to smooth link ratios r_j, $j = 0, 1, ..., n-1$ with:

$$r_j = 1 + e^a(j+1)^b \qquad (21)$$

where the parameters to be estimated are only a and b. Notice that in our example, modelling r_j with (21) involves fitting only 2 parameters (a and b) and not individually r_0, r_1, r_2, r_3. Although (21) is not linear, we can transform it easily into a linear expression:

$$R_j = a + b \cdot \ln(j+1). \qquad (22)$$

where $R_j = \ln(r_j - 1)$.

Notice that the representative link ratio for the development year r_j (and so, R_j) can be interpreted as an uncertain quantity. So, in our case we can quantify in a natural way R_j as a TFN $\tilde{R}_j = \left(R_j, l_{R_j}, r_{R_j}\right)$ where:

$$R_j = \frac{\sum_{i=0}^{n-j-1} \ln(r_{i,j} - 1)}{n-j} \tag{23a}$$

$$l_{R_j} = R_j - \ln\left(\text{Min}\{r_{0,j}, r_{1,j}, \ldots, r_{n-j-1,j}\} - 1\right) \tag{23b}$$

$$r_{R_j} = \ln\left(\text{Max}\{r_{0,j}, r_{1,j}, \ldots, r_{n-j-1,j}\} - 1\right) - R_j \tag{23c}$$

Table 9 shows the values of TFNs $\tilde{R}_j = (R_j, l_{R_j}, r_{R_j})$ $j = 0$, 1, 2, 3, that we deduce from Table 8, (23a), (23b) and (23c):

So, considering R_j, $j = 0,1,2,3$ as crisp observations of dependent variable, with OLS we adjust $R_j = -0.103 - 1.829 \cdot \ln(j + 1)$. In this way, Andrés-Sánchez [29] fits the evolution of claiming growth with FR:

$$\tilde{R}_j^* = \tilde{a} + \tilde{b}\ln(j+1) \tag{24}$$

In the case of pure MFP model, the coefficients will be the STFNs $\tilde{a} = (a, s_a)$ and $\tilde{b} = (b, s_b)$, So, the model (24) turns into (25):

$$\left(R_j^*, s_{R_j^*}\right) = (a, s_a) + (b, s_b)\ln(j+1) = (a + b\ln(j+1), s_a + s_b\ln(j+1)) \tag{25}$$

and with the data in Table 9 we find:

$$\left(R_j^*, s_{R_j^*}\right) = (-0.129, 0.223) + (-1.795, 0)\ln(j + 1)$$

If we use FLS, the model to be estimated is:

$$\begin{aligned}\left(R_j^*, l_{R_j}, r_{R_j}\right) &= (a, l_a, r_a) + (b, l_b, r_b)\ln(j+1) \\ &= (a + b\ln(j+1), l_a + l_b\ln(j+1), r_a + r_b\ln(j+1))\end{aligned} \tag{26}$$

and with the data in Table 9 we find:

$$\left(R_j^*, l_{R_j^*}, r_{R_j^*}\right) = (-0.103, 0.057, 0.067) + (-1.819, 0, 0)\ln(j+1)$$

	j	$\tilde{R}_j = (R_j, l_{R_j}, r_{R_j})$
Table 9 Fuzzy observations $\tilde{R}_j = (R_j, l_{R_j}, r_{R_j})$, $j = 0, 1, 2,$ 3 in our example	0	$(-0.072, 0.074, 0.110)$
	1	$(-1.453, 0.063, 0.061)$
	2	$(-2.036, 0.094, 0.094)$
	3	$(-2.632, 0.000, 0.000)$

To use Savic and Pedrycz [13] we take for the centres the OLS estimes of the coefficients, i.e. $a = -0.103$ and $b = -1.819$ and so, the STFN formulation is:

$$\left(R_j^*, s_{R_j^*}\right) = (-0.103, \ 0.244) + (-1.819, \ 0)\ln(j+1)$$

Likewise, with Ishibuchi and Nii [12] we fit:

$$\left(R_j^*, l_{R_j^*}, r_{R_j^*}\right) = (-0.103, \ 0.241, \ 0.241) + (-1.819, 0, 0)\ln(j+1)$$

After adjusting parameters $\tilde{a}$ and $\tilde{b}$, we can obtain the fuzzy growth rate of claims between jth and $(j + 1)$th development period, $\tilde{r}_j$, by evaluating (21) with FNs, i.e.:

$$\tilde{r}_j = 1 + e^{\tilde{a}}(j+1)^{\tilde{b}} \tag{27a}$$

So, taking into account that r_j is an increasing function of a and b (see Andrés-Sánchez [29]), its α-cuts can be obtained following (3b) when $\tilde{a}$ and $\tilde{b}$ are TFNs as:

$$r_{j_\alpha} = \left[\underline{r_j}(\alpha), \overline{r_j}(\alpha)\right] = \left[1 + e^{a - l_a(1-\alpha)}(j+1)^{b - l_b(1-\alpha)}, 1 + e^{a + r_a(1-\alpha)}(j+1)^{b + r_b(1-\alpha)}\right]$$
$$\tag{27b}$$

In the particular case where if $\tilde{a}$ and $\tilde{b}$ are STFNs, then:

$$r_{j_\alpha} = \left[\underline{r_j}(\alpha), \overline{r_j}(\alpha)\right] = \left[1 + e^{a - s_a(1-\alpha)}(j+1)^{b - s_b(1-\alpha)}, 1 + e^{a + s_a(1-\alpha)}(j+1)^{b + s_b(1-\alpha)}\right]$$
$$\tag{27c}$$

So, Tables 10, 11, 12 and 13 show the smoothed link rations from the FR models fitted in this section.

Table 10 α-cuts of claim growth rates $\tilde{r}_j$, $j = 0, 1, 2, 3$ from MFP

	$\tilde{r}_0$		$\tilde{r}_1$		$\tilde{r}_2$		$\tilde{r}_3$	
α	$\underline{r_0}(\alpha)$	$\overline{r_0}(\alpha)$	$\underline{r_1}(\alpha)$	$\overline{r_1}(\alpha)$	$\underline{r_2}(\alpha)$	$\overline{r_2}(\alpha)$	$\underline{r_3}(\alpha)$	$\overline{r_3}(\alpha)$
1	1.879	1.879	1.253	1.253	1.122	1.122	1.073	1.073
0.5	1.786	1.983	1.227	1.283	1.109	1.137	1.065	1.082
0	1.703	2.099	1.203	1.317	1.098	1.153	1.058	1.091

Table 11 α-cuts of claim growth rates $\tilde{r}_j$, $j = 0, 1, 2, 3$ from FLS

	$\tilde{r}_0$		$\tilde{r}_1$		$\tilde{r}_2$		$\tilde{r}_3$	
α	$\underline{r_0}(\alpha)$	$\overline{r_0}(\alpha)$	$\underline{r_1}(\alpha)$	$\overline{r_1}(\alpha)$	$\underline{r_2}(\alpha)$	$\overline{r_2}(\alpha)$	$\underline{r_3}(\alpha)$	$\overline{r_3}(\alpha)$
1	1.902	1.902	1.256	1.256	1.122	1.122	1.072	1.072
0.5	1.877	1.933	1.248	1.264	1.119	1.126	1.070	1.075
0	1.852	1.965	1.241	1.273	1.115	1.131	1.068	1.077

Table 12 α-cuts of claim growth rates $\tilde{r}_j$, $j = 0, 1, 2, 3$ from Savic and Pedrycz [13] fuzzy regression model

	$\tilde{r}_0$		$\tilde{r}_1$		$\tilde{r}_2$		$\tilde{r}_3$	
α	$\underline{r_0}(\alpha)$	$\overline{r_0}(\alpha)$	$\underline{r_1}(\alpha)$	$\overline{r_1}(\alpha)$	$\underline{r_2}(\alpha)$	$\overline{r_2}(\alpha)$	$\underline{r_3}(\alpha)$	$\overline{r_3}(\alpha)$
1	1.902	1.902	1.256	1.256	1.122	1.122	1.072	1.072
0.5	1.798	2.019	1.226	1.289	1.108	1.138	1.064	1.082
0	1.707	2.151	1.200	1.326	1.096	1.156	1.057	1.092

Table 13 α-cuts of claim growth rates $\tilde{r}_j$, $j = 0, 1, 2, 3$ from Ishibuchi and Nii [12] fuzzy regression model

	$\tilde{r}_0$		$\tilde{r}_1$		$\tilde{r}_2$		$\tilde{r}_3$	
α	$\underline{r_0}(\alpha)$	$\overline{r_0}(\alpha)$	$\underline{r_1}(\alpha)$	$\overline{r_1}(\alpha)$	$\underline{r_2}(\alpha)$	$\overline{r_2}(\alpha)$	$\underline{r_3}(\alpha)$	$\overline{r_3}(\alpha)$
1	1.902	1.902	1.256	1.256	1.122	1.122	1.072	1.072
0.5	1.800	2.019	1.227	1.289	1.108	1.138	1.064	1.082
0	1.709	2.151	1.201	1.326	1.096	1.156	1.057	1.092

6 Conclusions

This chapter has described fuzzy regression methods. Fuzzy regression is one of the most important instruments in fuzzy data analysis and it has been used widely in empirical applications. Taking into account the cost function to be minimised in order to adjust the coefficients, three different fuzzy regression methods have been developed: minimum fuzziness principle (MFP) methodology, fuzzy least squares (FLS) and procedures that mix both pure principles. So, whereas MFP is very sensitive to outliers and, in general, to any little change in data, FLS is not able to explain uncertainty in the linear relation of variables and also, when there is presence of fuzziness in data, the estimates does not reflect all the uncertainty of them. So, mixed methods arise as an interesting alternative that solve both drawbacks.

We also had paid a special attention to the applications of fuzzy regression in several areas of actuarial science as well as we have developed detailed two of the most common uses: the estimation of temporal structure of interest rates and the prediction of future claim costs in a non-life insurance context.

References

1. Dubois, D., Prade, H.: Foreword. In: Kacprzyk, J., Fedrizzi, M. (eds.) Fuzzy Regression Analysis. Physica-Verlag, Heidelberg (1992)
2. McCauley-Bell, P., Wang, H.: Fuzzy linear regression models for assessing risks of cumulative trauma disorders. Fuzzy Sets Syst. **92**, 317–340 (1997)
3. Lee, H.T., Chen, S.H.: Fuzzy regression model with fuzzy input and output data for manpower forecasting. Fuzzy Sets Syst. **119**, 205–213 (2001)
4. Ramenazi, R., Duckstein, L.: Fuzzy regression analysis of the effect of university research on regional technologies. In: Kacprzyk, J., Fedrizzi, M. (eds.) Fuzzy Regression Analysis, pp. 237–263. Physica-Verlag, Heidelberg (1992)
5. Watada, J.: Fuzzy time-series analysis and forecasting of sales volume. In: Kacprzyk, J., Fedrizzi, M. (eds.) Fuzzy Regression Analysis, pp. 211–227. Physica-Verlag, Heidelberg (1992)
6. Profillidis, V.A., Papadopoulos, B.K., Botzoris, G.N.: Similarities in fuzzy regression models and application on transportation. Fuzzy Econ. Rev. **4**(1), 83–98 (1999)
7. Tseng, F.-M., Tzeng, G.-H., Yu, H.-C., Yuan, B.J.-C.: Fuzzy ARIMA model for forecasting the foreign exchange market. Fuzzy Sets Syst. **118**, 9–19 (2001)
8. Chang, Y.-H.O., Ayyub, B.M.: Fuzzy regression methods-a comparative assessment. Fuzzy Sets Syst. **119**, 187–203 (2001)
9. Tanaka, H.: Fuzzy data analysis by possibility linear models. Fuzzy Sets Syst. **24**, 363–375 (1987)
10. Tanaka, H., Ishibuchi, H.: A possibilistic regression analysis based on linear programming. In: Kacprzyk, J., Fedrizzi, M. (eds.) Fuzzy Regression Analysis, pp. 47–60. Physica-verlag, Heidelberg (1992)
11. Chang, Y.-H.O.: Hybrid fuzzy least-squares regression analysis and its reliability measures. Fuzzy Sets Syst. **119**, 225–246 (2001)
12. Ishibuchi, H., Nii, M.: Fuzzy regression using asymmetric fuzzy coefficients and fuzzified neural networks. Fuzzy Sets Syst. **119**, 273–290 (2001)
13. Savic, D., Predrycz, W.: Fuzzy linear models: construction and evaluation. In: Kacprzyk, J., Fedrizzi, M. (eds.) Fuzzy Regression Analysis, pp. 91–100. Physica-Verlag, Heidelberg (1992)
14. Buckley, J.J., Qu, Y.: On using & α-cuts to evaluate fuzzy equations. Fuzzy Sets Syst. **38**, 309–312 (1990)
15. Diamond, P.: Fuzzy least squares. Inf. Sci. **46**, 141–157 (1988)
16. Andrés, J., Terceño, A.: Applications of fuzzy regression in actuarial analysis. J. Risk Insur. **70**(4), 665–699 (2003)
17. Berry-Stölze, T.R., Koissi, M.-C., Shapiro, A.F.: Detecting fuzzy relationships in regression models: the case of insurer surveillance in Germany. Insur.: Math. Econ. **46**(3), 554–567 (2010)
18. McCulloch, J.H.: The tax-adjusted yield curve. J. Finance **30**, 811–829 (1975)
19. Shapiro, A.F.: Fuzzy regression and the term structure of interest rates revisited. In: 14th Annual International AFIR Colloquium, Boston (2004b). http://afir2004.soa.org/afir_papers.htm
20. Hung, W.-L., Yan, M.-S.: An omission approach for detecting outliers in fuzzy regression. Fuzzy Sets Syst. **157**, 3109–3122 (2006)
21. Chen, Y.-S.: Outliers detection and confidence interval modification in fuzzy regression. Fuzzy Sets Syst. **119**, 259–272 (2001)
22. Hojati, M., Bector, C.R., Smimou, K.: A simple method for computation of fuzzy linear regresion. Eur. J. Oper. Res. **166**, 172–184 (2005)
23. Moskowitz, J.J., Kim, Y.: On assessing H-value on fuzzy linear regression. Fuzzy Sets Syst. **58**(3), 303–327 (1993)

24. Koissi, M.C., Shapiro, A.F.: The temporal structure of interest rates. A least squares approach. ARCH (2009)
25. Apaydin, A., Baser, F.: Hybrid fuzzy least-squares regression analysis in claims reserving with geometric separation method. Insur.: Math. Econ. **47**, 113–122 (2010)
26. Taylor, G.: Claims Reserving in Non-life Insurance. North-Holland, Amsterdam (1986)
27. de Andrés Sánchez, J., Gómez, A.T.: Estimating a term structure of interest rates for fuzzy financial pricing by using fuzzy regression methods. Fuzzy Sets Syst **139**(2), 399–424 (2003)
28. Koissi, M.C., Shapiro, A.F.: Fuzzy formulation of the lee-carter model for mortality forecasting. Insur.: Math. Econ. **39**, 287–309 (2006)
29. de Andrés, J.: Calculating insurance claim reserves with fuzzy regression. Fuzzy Sets Syst. **157**(23), 3091–3108 (2006)
30. de Andrés-Sánchez, J.: Claim reserving with fuzzy regression and the two ways of ANOVA. Applied Soft Computing, **12**(8), 2435–2441 (2012)
31. Kremer, E.: IBNR claims and the two way model of ANOVA. Scandinavian Actuarial J, **1**, 47–55 (1982)
32. Kaufmann, A.: Fuzzy subsets applications in O.R. and management. In: Jones, A., Kaufmann, A., Zimmermann, H.-J. (eds.) Fuzzy Set Theory and Applications, pp. 257–300. Reidel, Dordretch (1986)
33. Buckley, J.J.: The fuzzy mathematics of finance. Fuzzy Sets Syst. **21**, 57–73 (1987)
34. Li, M.: Calzi. towards a general setting for the fuzzy mathematics of finance. Fuzzy Sets Syst. **35**, 265–280 (1990)
35. Ostaszewski, K.: An Investigation into Possible Applications of Fuzzy Sets Methods in Actuarial Science. Society of Actuaries, Schaumburg (USA) (1993)
36. Cummins, J.D., Derrig, R.A.: Fuzzy financial pricing of property-liability insurance. North Am. Actuarial J. **1**, 21–44 (1997)
37. Echols, M.E., Elliott, J.W.: A Quantitative yield curve for estimating the term structure of interest rates. J. Financ. Quant. Anal. **11**, 87–114 (1976)
38. Straub, E.: Non-life insurance mathematics. Springerm, Berlin (1997)
39. Sherman, R.E.: Extrapolating, smoothing and Interpolating development factors. Proc. Casualty Actuarial Soc. **71**, 122–123 (1984)

Fuzzy Correlation and Fuzzy Non-linear Regression Analysis

Murat Alper Basaran, Biagio Simonetti and Luigi D'Ambra

Abstract In this chapter, we will deal with fuzzy correlation and fuzzy non-linear regression analyses. Both correlation and regression analyses that are useful and widely employed statistical tools have been redefined in the framework of fuzzy set theory in order to comprehend relation and to model observations of variables collected as either qualitative or approximately known quantities which are no longer being utilized directly in classical sense. When fuzzy correlation and fuzzy non-linear regression are concern, dealing with several computational complexities emerging due to the nature of fuzzy set theory is a challenge. It should be noted that there is no well-established formula or method in order to calculate fuzzy correlation coefficient or to estimate parameters of the fuzzy regression model. Therefore, a rich literature will accompany with the readers. While extension principle based methods are utilized in the computational procedures for fuzzy correlation coefficient, the distance based methods preferred rather than mathematical programming ones are employed in parameter estimation of fuzzy regression models. That extension principle combined with either fuzzy arithmetic or non-linear programming is two different methods proposed in the literature will be examined with small but illustrative examples in detail for fuzzy correlation analysis. Fuzzy non-linear regression has been a relatively new studied method when compared to fuzzy linear regression. However, both employ similar tools. S-curve fuzzy regression and two types of quadratic fuzzy regression models in the literature will be discussed.

M.A. Basaran (✉)
Department of Management Engineering, Alanya Alaaddin Keykubat University,
Antalya, Turkey
e-mail: muratalper@yahoo.com

B. Simonetti
Department of Economics Juridical and Social System Studies,
University of Sannio, Benevento, Italy
e-mail: simonetti@unisannio.edu.it

L. D'Ambra
Department of Economics Management and Institution,
University of Naples Federico II, Naples, Italy
e-mail: dambra@unina.edu.it

© Springer International Publishing Switzerland 2016
C. Kahraman and Ö. Kabak (eds.), *Fuzzy Statistical Decision-Making*,
Studies in Fuzziness and Soft Computing 343,
DOI 10.1007/978-3-319-39014-7_12

Keywords Fuzzy correlation analysis · Fuzzy non-linear regression analysis · S-curve regression · Quadratic regression

1 Introduction

Both correlation analysis and regression analysis are two of the most applied statistical tools in several disciplines due to its applicability and interpretability. They allow certain types of measurements to be used in classical statistical theory which means that observation are supposed to follow certain distributions. However, encountering observations either described by linguistic terms such as "bad", "good" and "very good", or approximately known quantities such as "around 2" is possible. With the introduction of fuzzy set theory, uncertainty different than one defined by probabilistic framework being modeled with possibility distribution for data collected as either qualitative or approximately known quantities has been a research area for data analysts.

Extending both methods to fuzzy framework gives rise to several proposed methods utilizing different aspects of fuzzy set theory.

2 Fuzzy Correlation Analysis

Correlation coefficient is a statistical measure which determines both the direction and strength of the linear relation between two variables which is defined by

$$r_{XY} = \frac{\sum_{i=1}^{n}(x_i - \bar{x})(y_i - \bar{y})}{\sqrt{\sum_{i=1}^{n}(x_i - \bar{x})^2 \sum_{i=1}^{n}(y_i - \bar{y})^2}} \tag{1}$$

where X and Y are variables whose values are denoted by (x_i, y_i), $i = 1, 2, \ldots, n$ and their corresponding arithmetic means are denoted by $\bar{x}$ and $\bar{y}$ respectively. Its range restricted in a closed interval $[-1, 1]$ tells how strong the linear dependence is between those variables with the knowledge of direction.

When the correlation coefficient is reconsidered in the fuzzy setting which means that observation values either are qualitative knowledge such as linguistic terms taking values of, for example, "bad" or "good" or "excellent", or are approximately known values, for instance, the value of the quantity can be defined around 2, measuring it is a need to quantify the relation. Both types of data are encountered when subjective or linguistic evaluations are provided by experts in the field of engineering, management or social sciences [1–3]. For example, the need for fuzzy correlation measure can arise when to quantify relation between the technology level and the management achievements of firms in management science or when to partition images to determine similarity or dissimilarity is concern in the field of engineering. Indeed, these types of exemplifications can easily be extended to any disciplines. Therefore, measuring correlation coefficient between two variables involving fuzziness is a need and computational procedures are challenging than that given in (1).

Computing fuzzy correlation employs basically two different methods. The first of which is to rely on Zadeh's extension principle, which aims at finding the membership function of fuzzy correlation. In order to determine membership function of fuzzy correlation, some methods are available providing with both analytical and numerical solutions, for example, using weakest t-norm and non-linear programming. Before explaining the details of the methods that are utilized in the computation of fuzzy correlation as well as fuzzy non-linear regression, some preliminary notions and definitions are needed which are fuzzy numbers, LR type fuzzy numbers, α-cuts of a fuzzy set and triangular norm, namely, t-norm, Zadeh's extension principle, fuzzy arithmetic. More detailed treatment of the subjects mentioned above can be found in variety of books pertinent to fuzzy set theory or fuzzy logic [4].

A fuzzy number is a convex subset of the real line R with a normalized membership function. For example, an asymmetric triangular fuzzy number $\tilde{x} = (x, \alpha, \beta)$ is defined by

$$\tilde{x}(t) = \begin{cases} 1 - \frac{x-t}{\alpha}, & \text{if } x - \alpha \leq t \leq x \\ 1 - \frac{t-x}{\beta}, & \text{if } x \leq t \leq x + \beta \\ 0, & \text{otherwise} \end{cases} \tag{2}$$

where the center value $x \in R$, left spread value $\alpha > 0$, and right spread value $\beta > 0$ are based on the definition of fuzzy number. When $\alpha = \beta$ is assumed, an asymmetric triangular fuzzy number is called a symmetric triangular fuzzy number and is denoted by $\tilde{x} = (x, \alpha)$. Other types of fuzzy numbers such as trapezoidal fuzzy number and Gaussian fuzzy number are also defined and utilized in various applications dependent upon the suitability, interpretability, and applicability.

A fuzzy number $\tilde{x} = (x, \alpha)_{LR}$ of type LR is a function from real numbers into the interval [0, 1] defined by

$$\tilde{x}(t) = \begin{cases} L\left(\frac{x-t}{\alpha}\right) & \text{for } x - \alpha \leq t \leq x \\ R\left(\frac{t-x}{\beta}\right) & \text{for } x \leq t \leq x + \beta \end{cases} \tag{3}$$

where L and R are non-increasing and continuous shape functions from [0,1] to [0,1] satisfying $L(0) = R(0) = 1$ and $L(1) = R(1) = 0$.

An α-cut of a fuzzy set is a crisp set defined by

$$A_\alpha = \{x \in A | \mu_A(x) \geq \alpha\} \tag{4}$$

A binary operation T on unit interval is said to be a triangular norm or t-norm if and only if T is associative, commutative, non-decreasing and $T(x, 1) = x$ for each $x \in [0, 1]$.

Extending ordinary arithmetic into fuzzy number setting is possible by employing Zadeh's extension principle defined by

$$\mu_B(y) = \underbrace{Sup}_{\substack{(x_1, \ldots, x_n) \in U_1 x \ldots x U_n \\ y = f(x_1, \ldots, x_n)}} \min(\mu_{A_1}(x_1), \ldots \mu_{A_n}(x_n)) \qquad (5)$$

where $A = A_1 x \ldots x A_n$ and $U = U_1 x \ldots x U_n$ are Cartesian product of the fuzzy sets $A_i, (i = 1, \ldots, n)$ and universal sets $U_i, (i = 1, \ldots, n)$ of fuzzy sets respectively.

2.1 Fuzzy Correlation Coefficient Based on the Weakest t-Norm (T_w) and Fuzzy Arithmetic

When Zadeh's extension principle is rewritten using one of union operators such as t-norm instead of minimization, the arithmetic operators are defined by

$$(\tilde{A} \oplus \tilde{B}) = \underbrace{Sup}_{x+y=z} T(\tilde{A}(x), \tilde{B}(y)) \qquad (6a)$$

$$(\tilde{A} \otimes \tilde{B}) = \underbrace{Sup}_{x \cdot y=z} T(\tilde{A}(x), \tilde{B}(y)) \qquad (6b)$$

$$(\tilde{A} \text{\o} \tilde{B}) = \underbrace{Sup}_{x/y=z} T(\tilde{A}(x), \tilde{B}(y)) \qquad (6c)$$

where $\tilde{A}$ and $\tilde{B}$ are fuzzy numbers and $\oplus$, $\otimes$, $\oslash$ are fuzzy arithmetic operators for addition, multiplication and division, respectively.

When fuzzy correlation is being computed, applying the extension principle based on the weakest t-norm denoted by T_w for a sample of n independent pairs of LR type fuzzy numbers is the method using the classical definition of the correlation coefficient given in (1) [5]. Instead of using the union operator Sup, T_w based fuzzy addition and multiplication are preferred in order to preserve the shape of the resultant LR type fuzzy numbers since it is the fact that fuzzy multiplication and division operators lead to resultant fuzzy numbers different than LR types except fuzzy addition and subtraction.

When the observations are fuzzy, the sample correlation coefficient given in (1) is rewritten.

$$\tilde{r}_{\widetilde{X,Y}} = \frac{\sum_{i=1}^{n} (\tilde{x}_i - \frac{1}{n}° \sum_{i=1}^{n} \tilde{x}_i)(\tilde{y}_i - \frac{1}{n}° \sum_{i=1}^{n} \tilde{y}_i)}{\sqrt{\sum_{i=1}^{n} (\tilde{x}_i - \frac{1}{n}° \sum_{i=1}^{n} \tilde{x}_i)^2 \sum_{i=1}^{n} (\tilde{y}_i - \frac{1}{n}° \sum_{i=1}^{n} \tilde{y}_i)^2}} \tag{7}$$

where $\tilde{x}_i = (x_i, \gamma_i)$ and $\tilde{y}_i = (y_i, \delta_i)$, $i = 1, 2, \ldots, n$, are symmetric triangular fuzzy numbers and $\bar{\tilde{x}} = \frac{1}{n}° \sum_{i=1}^{n} \tilde{x}_i$ and $\bar{\tilde{y}} = \frac{1}{n}° \sum_{i=1}^{n} \tilde{y}_i$ are the average values of fuzzy numbers $\tilde{X}$ and $\tilde{Y}$, respectively. Then the average values of fuzzy numbers $\tilde{X}$ and $\tilde{Y}$ are calculated based on T_w as follows:

$$\bar{\tilde{x}} = (\frac{1}{n} \sum_{i=1}^{n} x_i, \max_{1 \le i \le n} \gamma_i) \tag{8a}$$

$$\bar{\tilde{y}} = (\frac{1}{n} \sum_{i=1}^{n} y_i, \max_{1 \le i \le n} \delta_i) \tag{8b}$$

The expressions given in (8a) and (8b) can be written for just some observation using T_w as follows:

$$(\tilde{x} - \bar{\tilde{x}}_i) = (x_i - \frac{1}{n} \sum_{i=1}^{n} x_i, \max_{1 \le i \le n} \gamma_i)_L \tag{9a}$$

$$(\tilde{y} - \bar{\tilde{y}}_i) = (y_i - \frac{1}{n} \sum_{i=1}^{n} y_i, \max_{1 \le i \le n} \delta_i)_L \tag{9b}$$

Then the product of (9a) and (9b) is obtained as follows:

$$\left(\left(x_i - \frac{1}{n} \sum_{i=1}^{n} x_i \right) \left(y_i - \frac{1}{n} \sum_{i=1}^{n} y_i \right), \max \left| x_i - \frac{1}{n} \sum_{i=1}^{n} x_i \right| \max_{1 \le k \le n} \delta_k, \left(y_i - \frac{1}{n} \sum_{i=1}^{n} y_i \right) \max_{1 \le k \le n} \gamma_k \right)_L \tag{10}$$

The numerator of (7) is the summation of the product of (9a) and (9b) using T_w based fuzzy arithmetic denoted by

$$\left(\sum_{i=1}^{n} \left(x_i - \frac{1}{n} \sum_{k=1}^{n} x_k \right) \sum_{i=1}^{n} \left(y_i - \frac{1}{n} \sum_{k=1}^{n} y_k \right), \max_{1 \le i \le n} \left| x_i - \frac{1}{n} \sum_{k=1}^{n} x_k \right| \max_{1 \le k \le n} \delta_k, \left| x_i - \frac{1}{n} \sum_{k=1}^{n} x_k \right| \max_{1 \le k \le n} \gamma_k \right) \tag{11}$$

In order to compute the denominator of (7), we will follow the similar steps. The summation of the square of the differences between fuzzy observations and its fuzzy arithmetic mean for each variable is denoted using T_w based fuzzy arithmetic in (12) and (13).

$$\left(\sum_{i=1}^{n} \left(x_i - \frac{1}{n} \sum_{i=1}^{n} x_k \right)^2, \quad \max_{1 \le i \le n} \left| x_i - \frac{1}{n} \sum_{i=1}^{n} x_k \right| \max_{1 \le k \le n} \gamma_k \right)_L \quad (12)$$

$$\left(\sum_{i=1}^{n} \left(y_i - \frac{1}{n} \sum_{i=1}^{n} y_k \right)^2, \quad \max_{1 \le i \le n} \left| y_i - \frac{1}{n} \sum_{i=1}^{n} x_k \right| \max_{1 \le k \le n} \delta_k \right)_L \quad (13)$$

The product of (12) and (13) yields (14) and (15)

$$\sqrt{ \sum_{i=1}^{n} \left(x_i - \frac{1}{n} \sum_{i=1}^{n} x_k \right)^2 \sum_{i=1}^{n} \left(y_i - \frac{1}{n} \sum_{i=1}^{n} y_k \right)^2 }, \quad (14)$$

$$\frac{ \max\left\{ \sum_{i=1}^{n} \left(x_i - \frac{1}{n} \sum_{i=1}^{n} x_k \right)^2 \max_{1 \le i \le n} \left| y_i - \frac{1}{n} \sum_{i=1}^{n} y_k \right| \max_{1 \le k \le n} \delta_k, \right. }{ \left. \sum_{i=1}^{n} \left(y_i - \frac{1}{n} \sum_{i=1}^{n} y_k \right)^2 \max_{1 \le i \le n} \left| x_i - \frac{1}{n} \sum_{i=1}^{n} x_k \right| \max_{1 \le k \le n} \gamma_k, \right\} }{ \sum_{i=1}^{n} \left(x_i - \frac{1}{n} \sum_{i=1}^{n} x_k \right)^2 \sum_{i=1}^{n} \left(y_i - \frac{1}{n} \sum_{i=1}^{n} y_k \right)^2 } \quad (15)$$

where expressions in (14) and (15) are center and the spread part of the fuzzy number in denominator of (7), respectively.

Hence, both numerator and denominator are obtained. The last step is to divide those two fuzzy numbers. Its division is simply based on the implementation of the expression given in (6c). It is denoted by

$$(\tilde{A} \ø \tilde{B})(z) = \begin{cases} L\left[\frac{\left(\frac{a}{b}-z\right)}{\left(\left(\frac{1}{b}\right)\max(\alpha,z\beta)\right)} \right], & z \ge \min\left\{ \left(\frac{a-\alpha}{b}, \frac{a}{b+\beta} \right) \right\} \\ R\left[\frac{\left(z-\frac{a}{b}\right)}{\left(\frac{1}{b}\right)\max(\alpha,z\beta)} \right], & z \le \max\left\{ \frac{(a+\alpha)}{b}, \frac{a}{(b-\beta)} \right\} \end{cases} \quad (16)$$

where $a, b > 0$ and it is assumed that $L = R$, also $\tilde{A} = (a, \alpha)_{LL}$ and $\tilde{B} = (b, \beta)_{LL}$ are fuzzy numbers. Also, other cases including the different signs of two fuzzy numbers are easily defined and given in [5]. It should be noted that expression given in (16) holds for LL types fuzzy numbers.

A small data set presented in Table 1 will be used in order to exemplify calculations.

Table 1 Data set for both fuzzy numbers written in the form of symmetric triangular fuzzy numbers	$\tilde{x}_i = (x_i, \gamma_i)$	$\tilde{y}_i = (y_i, \delta_i)$
	(2.5,0.10)	(2.0,0.2)
	(3.0,0.4)	(2.6,0.3)
	(3.2,0.3)	(2.9,0.5)
	(3.5,0.2)	(3.8,0.4)
	(4.1,0.5)	(6.0,0.60)

Fuzzy arithmetic for both variables are obtained as $\bar{\bar{X}} = (3.26, 0.5)$ and $\bar{\bar{Y}} = (3.46, 0.6)$. Then using expression (11) results in (3.582, 1.27) which is the enumerator of (7). The denominator is calculated using (14) and (15) leading to (3.72, 2.71). When former value one is divided by the latter one, the membership function for correlation coefficient is denoted by

$$\tilde{r}_{\underset{XY}{\sim}} == \frac{(3.58, 1.27)}{(3.72, 2.71)} = \begin{cases} 1 - \frac{0.96-z}{\max(0.341, 0.728z)} & \text{if } 0.341 \le z \le 0.96 \\ 1 - \frac{z-0.96}{\max(0.341, 0.728z)} & \text{if } 0.96 \le z \le 1.304 \end{cases}$$

2.2 Fuzzy Correlation Based on Zadeh's Extension Principle

Another approach in the computation of fuzzy correlation coefficient is to use the α-cuts of fuzzy numbers in order to derive the membership function proposed by [6]. This method relies on the application of the extension principle aiming at finding the α-cuts of $\tilde{r}_{\underset{X,Y}{\sim}}$. The α-cuts of $\tilde{X}_i$ and $\tilde{Y}_i$ are denoted by

$$(X_i)_\alpha = \left[(X_i)_\alpha^L, (X_i)_\alpha^L \right] = [\min_x \{ x_i | \mu_{\tilde{x}_i}(x_i) \ge \alpha \}, \max_x \{ x_i | \mu_{\tilde{x}_i}(x_i) \ge \alpha \}] \tag{17a}$$

$$(Y_i)_\alpha = \left[(Y_i)_\alpha^L, (Y_i)_\alpha^L \right] = [\min_y \{ y_i | \mu_{\tilde{y}_i}(y_i) \ge \alpha \}, \max_y \{ y_i | \mu_{\tilde{y}_i}(y_i) \ge \alpha \}] \tag{17b}$$

Also, its interval form containing the values of both variables are denoted by

$$\left[(X_i)_\alpha, (Y_i)_\alpha \right] = [\min_x \{ x_i | \mu_{\tilde{x}_i}(x_i) \ge \alpha \}, \max_x \{ x_i | \mu_{\tilde{x}_i}(x_i) \ge \alpha \}] \tag{18}$$

where the α-cuts of $\tilde{X}_i$ and $\tilde{Y}_i$ are both crisp sets.

Then as mentioned in [6], a pair of non-linear mathematical programs are introduced in order to find the lower and upper bounds of the α-cuts of $\tilde{r}_{\underset{X,Y}{\sim}}$. Those are denoted as follows:

$$(r_{XY})_\alpha^L = \min \left[\left(\sum_{i=1}^n (x_i - \bar{x})(y_i - \bar{y}) \right) \bigg/ \sqrt{\sum_{i=1}^n (x_i - \bar{x})^2 \sum_{i=1}^n (y_i - \bar{y})^2} \right]$$
$$s.t \quad (X_i)_\alpha^L \le x_i \le (X_i)_\alpha^U, \quad \forall i$$
$$(Y_i)_\alpha^L \le x_i \le (Y_i)_\alpha^U, \quad \forall i$$

$$\tag{19a}$$

$$(r_{XY})_\alpha^U = \max\left[\left(\sum_{i=1}^n (x_i - \bar{x})(y_i - \bar{y})\Big/\sqrt{\sum_{i=1}^n (x_i - \bar{x})^2 \sum_{i=1}^n (y_i - \bar{y})^2}\right)\right]$$

$$s.t \quad (X_i)_\alpha^L \le x_i \le (X_i)_\alpha^U, \quad \forall i$$
$$(Y_i)_\alpha^L \le x_i \le (Y_i)_\alpha^U, \quad \forall i$$

$$(19b)$$

In the case of nonexistence of analytic solutions of non-linear programming problems, it is possible to obtain the numeric solutions for $(r_{XY})_\alpha^L$ and $(r_{XY})_\alpha^U$ at different α levels, which leads to the approximate shape of $L(r)$ and $R(r)$. A small data set which is given in Table 1 will be used to exemplify.

In order to work with a pair of non-linear programming problems, the α-cuts of variables for specified values ($\alpha = 0.0, 0.1, \ldots, 0.9, 1.0$) are tabulated in Tables 2 and 3, respectively.

For each α value, while the first column shows the left end point, the second column denotes the right end point. Similar construction is made for the fuzzy $\tilde{Y}$ variable in Table 3.

Based on those values presented in Tables 2 and 3, a pair of non-linear programming problem is solved in order to calculate correlation values for each corresponding α-cut values. Those are tabulated in Table 4.

Table 2 The α-cuts values for $\tilde{X}$

$\alpha = 0.0$		$\alpha = 0.1$		$\alpha = 0.2$		$\alpha = 0.3$		$\alpha = 0.4$		$\alpha = 0.5$	
2.4	2.6	2.41	2.59	2.42	2.58	2.4	2.6	2.41	2.59	2.42	2.58
2.6	3.4	2.64	3.36	2.68	3.32	2.6	3.4	2.64	3.36	2.68	3.32
2.9	3.5	2.93	3.47	2.96	3.44	2.9	3.5	2.93	3.47	2.96	3.44
3.3	3.7	3.32	3.68	3.34	3.66	3.3	3.7	3.32	3.68	3.34	3.66
3.6	4.6	3.65	4.55	3.70	4.50	3.6	4.6	3.65	4.55	3.70	4.50
$\alpha = 0.6$		$\alpha = 0.7$		$\alpha = 0.8$		$\alpha = 0.9$		$\alpha = 1.0$			
2.46	2.54	2.47	2.53	2.48	2.46	2.54	2.47	2.53	2.48		
2.84	3.16	2.88	3.12	2.92	2.84	3.16	2.88	3.12	2.92		
3.08	3.32	3.11	3.29	3.14	3.08	3.32	3.11	3.29	3.14		
3.42	3.58	3.44	3.56	3.46	3.42	3.58	3.44	3.56	3.46		
3.90	4.30	3.95	4.25	4.00	3.90	4.30	3.95	4.25	4.00		

Table 3 The α-cuts values for $\tilde{Y}$

$\alpha = 0.0$		$\alpha = 0.1$		$\alpha = 0.2$		$\alpha = 0.3$		$\alpha = 0.4$		$\alpha = 0.5$	
1.8	2.2	1.82	2.18	1.84	2.16	1.86	2.14	1.88	2.12	1.9	2.1
2.3	2.9	2.33	2.87	2.36	2.84	2.39	2.81	2.42	2.78	2.45	2.75
2.4	3.4	2.45	3.35	2.5	3.30	2.55	3.25	2.6	3.2	2.65	3.15
3.4	4.2	3.44	4.16	3.48	4.12	3.52	4.08	3.56	4.04	3.6	4.0
5.4	6.6	5.46	6.54	5.52	6.48	5.58	6.42	5.64	6.36	5.7	6.3
$\alpha = 0.6$		$\alpha = 0.7$		$\alpha = 0.8$		$\alpha = 0.9$		$\alpha = 1.0$			
1.92	2.08	1.94	2.06	1.96	2.04	1.98	2.02	2.0	2.0		
2.48	2.72	2.51	2.69	2.54	2.66	2.57	2.63	2.6	2.6		
2.7	3.1	2.75	3.05	2.8	3	2.85	2.95	2.9	2.9		
3.64	3.96	3.68	3.92	3.72	3.88	3.76	3.84	3.8	3.8		
5.76	6.24	5.82	6.18	5.88	6.12	5.94	6.06	6.0	6.0		

3 Fuzzy Non-linear Regressions

Fuzzy linear regression has been utilized as a modeling technique since the first introduction by [7] when one encounters different settings such as linguistically defined values, small data sets, unknown structure between dependent variable and independent variables, approximate measurements like intervals. Modeling endeavor covers several applications in many disciplines ranging from quality function deployment to determining claiming reserves [8, 9]. Also, it allows crisp numbers to be utilized in the modeling. Therefore, several types of fuzzy linear regression models and their parameter estimation methods have been proposed. The generic form of it can be denoted by

$$\tilde{Y} = \tilde{A}_0 + \tilde{A}_1 \otimes \tilde{X}_1 + \cdots + \tilde{A}_n \otimes \tilde{X}_k, \quad i = 1, \ldots, n \qquad (20)$$

where parameters, dependent and independent variables are all fuzzy numbers represented as one of the types such as triangular, trapezoidal and Gaussian fuzzy numbers.

Despite of the fact that several estimation methods have been defined, they are actually being grouped into two different methods that have been utilized and evolved during the research. The first of which is based on mathematical programming methods such as linear programming, goal programming, and non-linear programming and so on. The second one is to rely on the minimization of distance between two fuzzy sets so-called fuzzy least squares, which are the squares of the differences between the observed and estimated values of dependent variable.

When fuzzy non-linear regression is concern, the same variety pertinent to model types and their estimation methods are encountered. In this chapter, two different types of fuzzy non-linear regression models that are available in the literature will be presented with small but illustrative examples. The first one is called S-shaped curve fuzzy regression whose crisp version is widely utilized in the modeling of complex

Table 4 Upper and lower correlation values for each corresponding α-cuts

α-cuts	$\alpha = 0.0$	$\alpha = 0.1$	$\alpha = 0.2$	$\alpha = 0.3$	$\alpha = 0.4$	$\alpha = 0.5$	$\alpha = 0.6$	$\alpha = 0.7$	$\alpha = 0.8$	$\alpha = 0.9$	$\alpha = 1$
Min	0.4977	0.5663	0.6288	0.6891	0.7460	0.8024	0.8457	0.8865	0.9113	0.9418	0.9653
Max	1.00	1.00	1.00	1.00	1.00	1.00	0.9999	0.9980	0.9926	0.9821	0.9653

systems such as biology, agriculture and social economy. Both input variable and output variable in this model are fuzzy numbers. Its parameter estimation method is based upon minimizing the distance between fuzzy observed values and fuzzy estimated values which are represented by the pre-defined model. The parameter estimates are obtained as crisp values. The second one is called quadratic fuzzy regression model which appears to be two different types. While the first quadratic fuzzy regression allows quadratic term to be included in the model, the second one has interaction terms. While input variables have crisp values, the output variable and the parameters are fuzzy values. Its parameter estimation method uses the distance based methods aiming at minimizing the difference between observed values and estimated values proposed by [10, 12, 13].

3.1 S-Shaped Curve Fuzzy Regression

S-shaped curve fuzzy regression was proposed by [11] in order to model observations that are encountered in complex systems such as biology, social economy and agricultural sciences where the trend of growing is experienced slowly at the beginning, rapid increments are observed during process and it finishes with the saturation at the last phase.

Suppose that $\tilde{x}_i$ and $\tilde{y}_i$, $(i = 1, \ldots, n)$ are observations that are tried to be modeled defined by

$$\tilde{y} = (a + b \cdot \exp(-\tilde{x}))^{-1}, \quad a, b \in R \tag{21}$$

It is assumed that least squares based metrics between fuzzy numbers has better estimation ability when parameter estimation in fuzzy non-linear regression is concern. Therefore, metric defined in (22) will be utilized to determine parameters of the model given in (21).

$$\tilde{d}(\tilde{A}, \tilde{B}) = \left[\int_0^1 w^2(\alpha) d^2(A_\alpha, B_\alpha) dt \right]^{\frac{1}{2}}, \quad \tilde{A}, \tilde{B} \in F(R) \tag{22}$$

where $w^2(\alpha)$ should be chosen as a monotone increasing function in [0, 1], and $\tilde{A}$ and $\tilde{B}$ are fuzzy numbers defined on real line denoted by $F(R)$.

The motivation behind choosing monotone increasing function is based on the desire of having higher degree of membership level set when determining the distance between fuzzy numbers.

The distance based on the α-cuts of $\tilde{A}$ and $\tilde{B}$ given in (22) is denoted by

$$d^2 = (\tilde{A}_\alpha, \tilde{B}_\alpha) = [l(\alpha) - p(\alpha)]^2 + [r(\alpha) - q(\alpha)]^2 \tag{23}$$

where $\tilde{A}_\alpha = [l(\alpha), r(\alpha)]$ and $\tilde{B}_\alpha = [p(\alpha), q(\alpha)]$.

Utilizing the metric and the model given in (21) and (22) respectively, the least squares optimization problem is written in (24).

$$\text{Minimize } M(a,b) = \sum_{i=1}^{n} \tilde{d}^2 \left(a + b \exp(-\tilde{x}_i), \frac{1}{\tilde{y}_i} \right) \qquad (24)$$

The α-cuts of functions of $\tilde{X}$ and $\tilde{Y}$ are represented as follows:

$$(\tilde{Y}_i)_\alpha = [f_i(\alpha),\, g_i(\alpha)], \quad (\tilde{X}_i)_\alpha = [u_i(\alpha),\, v_i(\alpha)],$$
$$\left(\frac{1}{\tilde{Y}_i} \right)_\alpha = \left[\frac{1}{g_i(\alpha)}, \frac{1}{f_i(\alpha)} \right] \quad (\exp(-\tilde{x}_i))_\alpha = [\exp(-v_i), \exp(-u_i)], \qquad (25)$$
$$(\exp(\tilde{x}_i))_\alpha = [\exp(u_i), \exp(v_i)], \quad \alpha \in (0, 1]$$

where expression in (25) holds for positive fuzzy numbers.

Two different minimization functions are defined with respect to the sign of b, which are for $b \geq 0$ and $b < 0$, respectively.

In order to simplify the notations, the α-cut in parenthesis is removed. Also, w^2 is adapted instead of using $w^2(\alpha)$ in (22) and (25).

For $b \geq 0$, the α-cut of $(a + b \exp(-\tilde{x}_i))_\alpha$ is denoted by

$$[a + b \exp(-v_i), a + b \exp(-u_i)], \quad (i = 1, \ldots, n) \qquad (26)$$

Then, its least squares optimization function given in (24) is rewritten

$$\min_{a,b} M_+(a,b) = \sum_{i=1}^{n} \tilde{d}^2 \left(a + b\, exp(-x_i), \frac{1}{\tilde{y}_i} \right)$$
$$= \int_0^1 w^2 \left[\sum_{i=1}^{n} \left(a + b\, exp(-v_i) - \frac{1}{g_i} \right)^2 + \left(a + b \exp(-u_i) - \frac{1}{f_i} \right)^2 \right] d\alpha \qquad (27)$$

By taking derivatives of optimization function given in (27) with respect to parameters a and b, an equation system consisting two equations are obtained. The first equation system is denoted by ES1

$$ES1 = \begin{cases} 2na \int_0^1 w^2 d\alpha + b \int_0^1 w^2 \sum_{i=1}^{n} (\exp(-v_i) + \exp(-u_i)) d\alpha \\ \quad = \int_0^1 w^2 \sum_{i=1}^{n} \left(\frac{1}{f_i} + \frac{1}{g_i} \right) d\alpha \\ a \int_0^1 w^2 \sum_{i=1}^{n} (\exp(-v_i) + \exp(-u_i)) d\alpha + b \int_0^1 w^2 \sum_{i=1}^{n} (\exp(-2v_i) + \exp(-2u_i)) d\alpha \\ \quad = \int_0^1 w^2 \sum_{i=1}^{n} \left(\frac{1}{f_i} \exp(-2v_i) + \frac{1}{g_i} \exp(-2u_i) \right) d\alpha \end{cases}$$

$$(28)$$

For $b < 0$, the α-cut of $(a + b\exp(-\tilde{x}_i))_\alpha$ is denoted by

$$[a + b\exp(-u_i), a + b\exp(-v_i)], \qquad i = 1, \ldots, n) \tag{29}$$

Then, its least squares optimization function given in (24) is rewritten for this case.

$$
\min_{a,b} M_-(a,b) = \sum_{i=1}^{n} \tilde{d}^2(a + b\,exp(-x_i), \frac{1}{\tilde{y}_i})
$$

$$
= \int_0^1 w^2 \sum_{i=1}^{n} [\left(a + b\ exp(-u_i) - \frac{1}{g_i}\right)^2 + \left(a + b\exp(-v_i) - \frac{1}{f_i}\right)^2]d\alpha
$$

$$\tag{30}$$

By taking derivatives of optimization function given in (30) with respect to parameters a and b, an equation system consisting two equations are obtained. The second equation system is denoted by ES2

$$
ES2 = \begin{cases}
2na \int_0^1 w^2 d\alpha + b \int_0^1 w^2 \sum_{i=1}^{n} (\exp(-v_i) + \exp(-u_i)) d\alpha \\[2mm]
\quad = \int_0^1 w^2(\alpha) \sum_{i=1}^{n} \left(\frac{1}{f_i(\alpha)} + \frac{1}{g_i(\alpha)}\right) d\alpha \\[4mm]
a \int_0^1 w^2 \sum_{i=1}^{n} (\exp(-v_i) + \exp(-u_i)) d\alpha + b \int_0^1 w^2 \sum_{i=1}^{n} (\exp(-2v_i) + \exp(-2u_i)) d\alpha \\[2mm]
\quad = \int_0^1 w^2 \sum_{i=1}^{n} \left(\frac{1}{f_i(\alpha)}\exp(-v_i + \frac{1}{g_i(\alpha)}\exp(-2u_i)\right) d\alpha
\end{cases}
$$

$$\tag{31}$$

In order to find the parameters of fuzzy non-linear regression defined in the form of S-curve fuzzy model, ES1 and ES2 needs to be solved. For this purpose, criterion is defined by [11], which is denoted by (32) and (33) are utilized.

$$
D_b = 2n \int_0^1 w^2 d\alpha \int_0^1 \left(\frac{1}{f_i}\exp(u_i) + \frac{1}{g_i}\exp(v_i)\right) d\alpha
$$
$$
- \int_0^1 w^2 \sum_{i=1}^{n} (\exp(-v_i) + \exp(u_i)) d\alpha \int_0^1 w^2 \sum_{i=1}^{n} \left(\frac{1}{f_i} + \frac{1}{g_i}\right) d\alpha \tag{32}
$$

$$
D_{b_-} = 2n \int_0^1 w^2 d\alpha \int_0^1 w^2 \sum_{i=1}^{n} (\frac{1}{g_i}\exp(u_i) + \frac{1}{f_i}\exp(v_i)) d\alpha
$$
$$
- \int_0^1 w^2 \sum_{i=1}^{n} (\exp(-u_i) + \exp(-v_i)) d\alpha \int_0^1 w^2 \sum_{i=1}^{n} \left(\frac{1}{f_i} + \frac{1}{g_i}\right) d\alpha \tag{33}
$$

It is proved by [11] that $D_b \geq D_{b_-}$.

Based on values of D_b, The solution set of (28) or (31) is searched using the computational procedure

If $D_b \geq 0$, expression (28) has unique solution which is denoted by the form of parameter estimates

$$a = \frac{p_1}{D} \text{ and } b = \frac{D_b}{D} \tag{34}$$

where p_1 and D are determinant values which are defined by

$$p_1 = \begin{vmatrix} \int_0^1 w^2 \sum_{i=1}^{n} \left(\frac{1}{f_i} + \frac{1}{g_i}\right) d\alpha & \int_0^1 w^2 \sum_{i=1}^{n} (\exp(-v_i) + \exp(-u_i)) d\alpha \\ \int_0^1 w^2 \sum_{i=1}^{n} \left(\frac{1}{f_i}\exp(u_i) + \frac{1}{g_i}\exp(v_i)\right) d\alpha & \int_0^1 w^2 \sum_{i=1}^{n} (\exp(-2v_i) + \exp(-2u_i)) d\alpha \end{vmatrix} \tag{35}$$

$$D = \begin{vmatrix} 2n \int_0^1 w^2 d\alpha & \int_0^1 w^2 \sum_{i=1}^{n} (\exp(-v_i) + \exp(-u_i)) d\alpha \\ \int_0^1 w^2 \sum_{i=1}^{n} (\exp(-v_i) + \exp(-u_i)) d\alpha & \int_0^1 w^2 \sum_{i=1}^{n} (\exp(-2v_i) + \exp(-2u_i)) d\alpha \end{vmatrix} \tag{36}$$

If $D_b < 0$, then $D_{b-} \leq D_b \leq 0$. Hence expression (31) has a unique solution which is expressed in the form of parameter estimates

$$a = \frac{p_2}{D} \text{ and } b = \frac{D_{b-}}{D} \tag{37}$$

where p_2 is a determinant value which are defined by

$$p_2 = \begin{vmatrix} \int_0^1 w^2 d\alpha \sum_{i=1}^{n} \left(\frac{1}{f_i} + \frac{1}{g_i}\right) d\alpha & \int_0^1 w^2 \sum_{i=1}^{n} (\exp(-v_i) + \exp(-u_i)) d\alpha \\ \int_0^1 w^2 \sum_{i=1}^{n} \left(\frac{1}{g_i}\exp(u_i) + \frac{1}{f_i}\exp(v_i)\right) d\alpha & \int_0^1 w^2 \sum_{i=1}^{n} (\exp(-2v_i) + \exp(-2u_i)) d\alpha \end{vmatrix} \tag{38}$$

A small data set is tabulated in Table 5. Our aim is to determine parameters of fuzzy non-linear regression model defined in (21).

$$a = \frac{p_1}{D} = \frac{-6.53}{1.41} = -4.63 \text{ and } b = \frac{D_b}{D} = \frac{56.46}{1.41} = 40.04$$

Table 5 Data set for S-curve fuzzy regression

$\tilde{x}_i = (x_i, \gamma_i)$	$\tilde{y} = (y_i, \delta_i)$
(1.0,0.3)	(3.0,0.2)
(2.0,0.7)	(8.0,0.4)
(3.0,0.4)	(22.0,0.6)
(4.0,1.9)	(25.0,1.3)
(5.0,1.3)	(30.0,1.5)

The model given in (21) is denoted by

$$\tilde{y} = -4.63 + 40.04 \exp(-\tilde{x}) \tag{39}$$

3.2 Quadratic Fuzzy Regression

The second type of fuzzy non-linear regression model is quadratic fuzzy regression expressed in two different models. While the first of which is the one including a quadratic term, the second one contains a term consisting of the interaction of the independent variables. They are proposed by [12] and denoted by (40) and (41).

$$\tilde{Y}_i = A_0 + A_1 X_{i1} + A_2 X_{i1}^2, \quad i = 1, 2, \ldots, n \tag{40}$$

$$\tilde{Y}_i = A_0 + A_1 X_{i1} + A_2 X_{j1} + A_3 X_{i1} X_{j1}, \quad i, j = 1, 2, \ldots, n \tag{41}$$

In both models, it is assumed that input variables are non-negative crisp values and output variable is normal and convex fuzzy numbers with either symmetric or non-symmetric triangular membership functions. The parameter estimation method so-called fuzzy least squares which aim to minimizing the squares of the differences between the observed fuzzy dependent variable and the estimated fuzzy outputs are widely applied to estimation of the parameters. In order to define the difference between the observed and the estimated fuzzy numbers, some methods transforming those fuzzy numbers into crisp numbers are proposed in [10, 12, 13]. One of the methods called Overall Existence Ranking Index (OERI) was proposed in [13]. It is based on the usage of the inverse membership function which is simply a ranking method developed for fuzzy sets. For a given existence level w, the inverse image in terms of membership function, $\mu(x)$, is defined as

$$\mu^{-1}(w) = \{x : \mu(x) = w\} \tag{42}$$

Then for any two arbitrary fuzzy numbers A and B, if A is said to be larger than B at w where $w \in (0, 1], \{\mu_A^{-1}(w)\} > \{\mu_B^{-1}(w)\}$ holds. The inverse is not generally true. The OERI for a fuzzy number $A = (x, \alpha, \beta)$ is a crisp number defined as

$$OM(A) = x - \frac{1}{2}X_1(w)\alpha + \frac{1 - X_1(w)}{2}\beta \qquad (43)$$

where $X_1(w)$ is a weighting function determined by decision makers subjectively. The more realistic weighting function is the linear one mentioned in [13]. Then for fuzzy numbers A and B, the distance is defined as

$$D(A, B) = OM(A) - OM(B) \qquad (44)$$

When OERI is adapted into regression problem, its distance function can be written as follows:

$$MIN \sum_{i=1}^{n} [Y_i - (\widehat{Y}_i)]^2 = \sum_{i=1}^{n} [D(Y_i, \widehat{Y}_i)]^2 = \sum_{i=1}^{n} [OM(Y_i) - OM(\widehat{Y}_i)]^2 \qquad (45)$$

The minimization function and its constraints employing OERI can be written as

$$MIN \sum_{i=1}^{n} \left\{ \left[y_i^m - ax_i - \frac{X_1(w)}{2}(y_i^L - \gamma x_i) + \frac{1 - X_1(w)}{2}(y_i^R - \delta x_i) \right]^2 + (y_i^L - \gamma x_i)^2 + (y_i^R - \delta x_i)^2 \right\} \qquad (46)$$

$$y_i^m - (1 - \alpha)y_i^L \geq ax_i - (1 - \alpha)\gamma x_i$$

$$y_i^m + (1 - \alpha)y_i^R \leq ax_i + (1 - \alpha)\delta x_i$$

$$\gamma, \delta \geq 0$$

where fuzzy number $A = (a, \gamma, \delta)$ is the parameter of the regression and fuzzy number $Y = (y^m, y^L, y^R)$ is the observed dependent variable and $0 < \alpha \leq 1$. The formulation given in (46) is the case having one independent variables and its formulation can be easily extended to multiple cases of independent variables easily.

Similarly, Diamond [10] proposed another distance function defined as

$$d^2(A, B) = (x - y)^2 + [(x - y) - (\alpha - \gamma)]^2 + [(x - y) - (\beta - \delta)]^2 \qquad (47)$$

The formulations denoted based on the method proposed by Diamond is given as follows:

$$(X^T X)\alpha_L = X^T Y_L \qquad (48a)$$

$$(X^T X)\alpha_U = X^T Y_R \qquad (48b)$$

Table 6 Data set for quadratic regression models

$\tilde{y}_i$	x_1	x_2
(4,0.5)	1	2.1
(3,0.3)	2	3.3
(2,0.2)	3	4.6
(4,0.7)	4	1.8
(6,0.9)	5	3.7

where Y_L and Y_R are vectors denoting the left end points and the right end points of the response values and α_L and α_R are vectors denoting the left end points and the right end points of the center values of the predicted parameters.

where X is the data matrix denoted by

$$X = \begin{bmatrix} 1 & X_{11} & X_{11}^2 \\ \vdots & \vdots & \vdots \\ 1 & X_{n1} & X_{n1}^2 \end{bmatrix} \tag{49}$$

Similar construction was also proposed by [12] for the quadratic fuzzy regression containing the interaction term of independent variables.

The generic data matrix for the model given in (41) is given as follows:

$$X = \begin{bmatrix} 1 & X_{11} & X_{12} & X_{11}X_{12} \\ \vdots & \vdots & \vdots & \vdots \\ 1 & X_{n1} & X_{m2} & X_{n1}X_{m2} \end{bmatrix} \tag{50}$$

A small data set is used in order to illustrate the models given in (40) and (41) (Table 6).

The parameter estimates for model (40) and (41) are denoted respectively by

$$\tilde{y} = (6.37, 0.86) + (-3.12, 0.48)x_1 + (0.59, 0.10)x_1^2$$
$$\tilde{y} = (10.05, 1.53) + (-1.62, 0.18)x_1 + (-2.90, 0.51)x_2 + (0.76, 0.12)x_1x_2$$

4 Conclusion

Fuzzy correlation measure is an important fuzzy statistics that helps comprehend the relation between two variables that are collected as either linguistically defined values or approximately known quantities. In classical statistical theory, the correlation of these types of variables can no longer be calculated without losing information included. With the help of fuzzy set theory providing mathematical tools allowing to model uncertainty different than one defined by probabilistic approach, the relation between those variables can be quantified using fuzzy correlation measure. Despite of the fact several methods are available in the literature, two different methods are chosen due to having utilized fuzzy concepts directly in computational procedures and their reliable results. Both of them using basically Zadeh's extension principle with the combination of either fuzzy arithmetic and the weakest t-norm or non-linear programming problem are employed. Both methods with same small data set are run.

Fuzzy non-linear regression is a method fully benefiting from methodological developments used in fuzzy linear regression when it is defined in a form different than linear structure distance. They are called S-curve regression and quadratic fuzzy regression. It is a fact that distance based parameter estimation methods has better ability than mathematical programming ones do when parameter estimation is concern in fuzzy non-linear regression. Two data sets for S-curve fuzzy regression and quadratic fuzzy regression are employed respectively.

References

1. Ngan, S.-L.: Correlation coefficient of linguistic variables and its applications to quantifying relations in imprecise management data. Eng. Appl. Artif. Intell. **26**, 347–356 (2013)
2. Tang, Y., Mu, W., Zhao, L., Zhao, G.: An image segmentation method based on maximizing fuzzy correlation and its fast recursive algorithm. Comput. Electr. Eng. **40**, 833–843 (2014)
3. Huyn, V.-N., Kreinovich, V., Sriboonchitta, S., Suriya, K. (Eds): Econometrics of Risk. Springer (2015)
4. Zimmermann, H.J.: Fuzzy set theory and its applications, 3rd edn. Kluwer-Nijhoff, Boston (1996)
5. Hong, D.H.: Fuzzy measure for a correlation coefficient of fuzzy numbers under T_w (the weakest t-norm) based fuzzy arithmetic operation. Inf. Sci. **176**, 150–160 (2006)
6. Lin, S.-T., Kao, L.: Fuzzy measures for correlation coefficient of fuzzy numbers. Fuzzy Sets Syst. **128**, 267–275 (2002)
7. Tanaka, H., Uejima, S., and Asai, K.: Linear regression analysis with fuzzy model. IEEE Trans. Syst. Man Cyber. **12**, 903–907 (1982)
8. Liu, Y., Zhou, J., Chen, Y.: Using fuzzy non-linear regression to identify degree of compensation among customer requirements in QFD. Neuro Comput. **142**, 115–124 (2014)
9. Sanches, J.A.: Calculating insurance claim reserves with fuzzy regression. Fuzzy Sets Syst. **157**(23), 3091–3108 (2006)
10. Diamond, P.: Fuzzy least squares. Inf. Sci. **46**, 141–157 (1988)
11. Ruoning, X.: S-curve regression model in fuzzy environment. Fuzzy Sets Syst. **97**, 317–326 (1990)
12. Chen, Y.-S.: Fuzzy ranking and quadratic fuzzy regression. Comput. Math. Appl. **38**, 265–279 (1999)
13. Chang, P.T., Lee, E.S.: Ranking of fuzzy sets based on the concept of existence. Comput. Math. Appl. **27**(9/10), 1–21 (1994)

Fuzzy Decision Trees

Ayca Altay and Didem Cinar

Abstract Decision trees are one of the most widely used classification techniques because of their easily understandable representation. In the literature, various methods have been developed to generate useful decision trees. ID3 and SLIQ algorithms are two of the important algorithms generating decision trees. Although they have been applied for various real life problems, they are inadequate to represent ambiguity and vagueness of human thinking and perception. In this study, fuzzy ID3 and fuzzy SLIQ algorithms, which generate fuzzy decision trees, are discussed as well as their enhanced versions. Their performances are also tested using simple training sets from the literature.

Keywords Fuzzy decision trees · Induction algorithms · Classification

1 Introduction to Decision Trees

Decision trees are predictive models designed for supervised data mining that analyze data in a multi-variate and tree-like fashion [25]. They achieve object classification through splitting the branches of a tree where each split presents a test through an *attribute* or a *criterion*. Each split is called a *node* and the first split is called the *root* of the tree. When the splitting or the branching process is terminated, each of the last node is called a *terminal node* or a *leaf* of the tree. Each branch sequence provides a rule for classification of objects [26]. An example of a decision tree is provided in Fig. 1 [26]. In Fig. 1, the decision of playing tennis is achieved through considering weather related factors such as the outlook of the weather, the humidity and wind. The first split is achieved through the outlook which constitutes

A. Altay (✉) · D. Cinar
Department of Industrial Engineering, İstanbul Technical University,
34367 İstanbul, Turkey
e-mail: altaya@itu.edu.tr

D. Cinar
e-mail: cinard@itu.edu.tr

© Springer International Publishing Switzerland 2016
C. Kahraman and Ö. Kabak (eds.), *Fuzzy Statistical Decision-Making*,
Studies in Fuzziness and Soft Computing 343,
DOI 10.1007/978-3-319-39014-7_13

Fig. 1 A sample decision
tree [26]

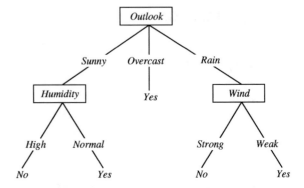

the root of the tree. If the sky is overcast, then the decision to play tennis is reached making this the final split of a certain conditional outlook. The decision to play tennis is the leaf or the terminal node of the tree. However, if the weather is sunny or rainy, a direct decision cannot be reached without considering humidity or wind conditions. Hence, in such cases, the terminal node cannot be reached solely on the outlook.

Decision trees split a complex decision process into a set of simpler decisions to classify the given objects with an easily understandable representation [27]. This is the reason that decision trees are important tools in data mining literature [31]. Many algorithms have been proposed to construct decision trees. Although these methods have generated useful decision trees for classification problems, they are inadequate to represent ambiguity and vagueness of human thinking and perception [46].

In classical set theory, an element either belongs to a certain set or not. For example, in the decision tree given in Fig. 1, it is assumed that *humidity* of an object is known precisely as it is either *high* or *normal*. Assume that there is a certain limit, 20 %, which is used to determine humidity of an object, i.e. if *humidity* is over that limit, then it belongs to *high*, otherwise it belongs to *normal*. If an object has 21 % *humidity*, should we classify the object as *not play tennis*?

Crisp sets may not be realistic for the real world problems including vagueness and subjectivity. Therefore, fuzzy sets have been integrated to the decision trees to enhance the uncertainty handling capability. In this chapter, classical induction algorithms to construct fuzzy decision trees are discussed.

The reminder of this chapter is organized as follows. Well-known crisp decision tree induction algorithms from the literature are explained in the next section. Induction algorithms for fuzzy decision trees and corresponding literature review are given in Sect. 3. The algorithms are applied to a sample training set from the literature and the results are discussed in Sect. 4. Finally, a brief summary on recent approaches is presented in Sect. 5 and concluding remarks are presented in Sect. 6.

2 Decision Trees and Classification Problems

A formal definition of a typical classification problem can be described as follows. S is the set of objects in a training set in which each object is described by attributes $A = \{A_1, \ldots, A_N\}$. Domain of each attribute A_i is represented by a set of discrete linguistic terms $L(A_i) = \{A_i^1, \ldots, A_i^{n_i}\}$. Each object $s \in S$ is classified by a set of classes $C = \{C_1, \ldots, C_K\}$. Let n denotes the number of objects where $<x_1, y_1>, \ldots, <x_n, y_n>$ be the objects of the data, x_s and y_s being the inputs and output of object s, respectively. The set of classes can be numerical values, ordered or unordered factors. Construction of a decision tree involves following decisions: (*i*) which attribute to split, (*ii*) when to stop splitting, and (*iii*) how to assign terminal nodes to a class.

A small training set about credit risk assessment procedure [23] is given in Table 1. Each customer is evaluated according to risk potential. As a classification problem, each customer is an object of S. There are three possible decisions for *risk* evaluation: *high*, *moderate*, and *low*. *Credit history*, *debt*, *collateral* and *income* are the attributes which are used to decide risk potential of a customer. Attribute evaluations for 14 customers are shown in Table 1. Attributes and their values are given as follows:

- *Credit history* = {*good, bad, unknown*}
- *Debt* = {*high, low*}
- *Collateral* = {*adequate, none*}
- *Income* = {*low* ($0 to $15 K), *moderate* ($15 to $35 K), *high* (over $35 K}

Table 1 A training data on risk assessment

#	Credit history	Debt	Collateral	Income	Risk
1	Bad	High	None	$0 to $15K	High
2	Unknown	High	None	$15 to $35K	High
3	Unknown	Low	None	$15 to $35K	Moderate
4	Unknown	Low	None	$0 to $15K	High
5	Unknown	Low	None	over $35K	Low
6	Unknown	Low	Adequate	over $35K	Low
7	Bad	Low	None	$0 to $15K	High
8	Bad	Low	Adequate	over $35K	Moderate
9	Good	Low	None	over $35K	Low
10	Good	High	Adequate	over $35K	Low
11	Good	High	None	$0 to $15K	High
12	Good	High	None	$15 to $35K	Moderate
13	Good	High	None	over $35K	Low
14	Bad	High	None	$15 to $35K	High

In this study, induction algorithms used to construct decision trees are investigated. Since constructing optimal binary decision trees is an NP-complete problem, efficient heuristics have been developed to generate near-optimal decision trees [20]. ID3 algorithm, CART and SLIQ algorithms are analyzed in the context of this chapter. Credit risk evaluation example given above will be used to explain ID3 and CART algorithms, whereas SLIQ algorithm will provide its own example.

2.1 ID3 Algorithm

ID3 (Interactive Dichotomizer 3) algorithm, which is developed by Quinlan [33, 34] in 1986, is one of the most well-known decision tree induction algorithms. Basically, it uses an information theoretic measure of entropy to evaluate the discriminatory power of each attribute. ID3 uses *information gain* of each attribute to build a decision tree. The attribute adding the greatest information about the decision is selected first [32]. The greatest information gain means the greatest decrease in entropy which is calculated for set S as follows:

$$E(S) = \sum_{k=1}^{K} -p(k)\log_2 p(k) = \sum_{k=1}^{K} -\frac{|C_k|}{|S|} log_2 \frac{|C_k|}{|S|} \tag{1}$$

where $E(S)$ represents the entropy and $p(k)$ is relative frequency of class k in set S, i.e. $p(k)$ is the ratio of the objects in class k to the whole set S. In all computations, $0 \cdot \log_2 0$ is assumed as 0. If decisions for all objects are the same, then entropy will be zero. This means there is no need to split the node on the corresponding decision level. Let T_i be the set of subsets created from splitting set S by attribute A_i. Information gain for each attribute is computed as follows:

$$IG(S, A_i) = E(S) - \sum_{t \in T_i} p(t)E(t) = E(S) - \sum_{t \in T_i} \frac{|t|}{|S|} E(t) \tag{2}$$

$$E(t) = \sum_{k=1}^{K} -p(t_k)\log_2 p(t_k) = \sum_{k=1}^{K} -\frac{|t_k|}{|t|} log_2 \frac{|t_k|}{|t|} \tag{3}$$

where $t = \cup_{k=1}^{K} t_k$. At each iteration of ID3 algorithm, the attribute having the greatest gain of information is selected as the decision level. A pseudocode including main steps of ID3 algorithm is given in Algorithm 1.

Algorithm 1: ID3 Algorithm

Input: Classification data
1 Calculate $p(k), E(S)$
2 Compute $IG(S, A_j) = max_{i \in S} IG(S, A_i)$
3 Branch for attribute A_j
4 Update S
5 **if** $E(S) = 0$ **or** no attribute remains to split **then** terminate this branch;
6 **if** all branches terminated **then** terminate the algorithm;
7 **else**
8 | Go to Step 1

Output: Decision tree

A decision tree for the credit risk assessment data can be generated by ID3 algorithm as follows. In the data set, there are 6 customers classified as *high risk*, 3 customers as *moderate risk* and 5 customers as *low risk*. The entropy for set S is computed as follows:

$$E(S) = -\frac{6}{14} log_2 \left(\frac{6}{14}\right) - \frac{3}{14} log_2 \left(\frac{3}{14}\right) - \frac{5}{14} log_2 \left(\frac{5}{14}\right) = 1.531$$

To compute the information gain of an attribute, entropy value for each subset t of that attribute should be calculated. For attribute *credit history*, entropy values for *good*, *bad* and *unknown* should be computed. Among the 14 customers, 5 of them has *good* credit history with 1 *high risk*, 1 *moderate risk* and 3 *low risk* levels. Entropy value for *good* credit history can be obtained as follows:

$$E(good) = -\frac{1}{5} log_2 \left(\frac{1}{5}\right) - \frac{1}{5} log_2 \left(\frac{1}{5}\right) - \frac{3}{5} log_2 \left(\frac{3}{5}\right) = 1.371$$

Similarly, entropy values for other features are found as $E(bad) = 0.811$ and $E(unknown) = 1.522$. In data set, there are customers with 5 *good*, 4 *bad* and 5 *unknown* credit history. So, the information gain for attribute *credit history* can be obtained as follows:

$$IG(S, \text{ credit history}) = 1.531 - \frac{5}{14} 1.317 - \frac{4}{14} 0.811 - \frac{5}{14} 1.522 = 0.266$$

Information gain for other attributes are found as $IG(S, debt) = 0.063$, $IG(S, collateral) = 0.207$ and $IG(S, income) = 0.967$. Since attribute *income* is the greatest one with 0.967 information gain, it becomes the root node of decision tree with three branches as *high*, *moderate* and *low*.

Firstly, branch *high* will be handled to construct the next level of decision tree. Customers having high income constitute the new set S_{high}. New information gain values are obtained for the rest of attributes given that the customers have high income. There are 6 high income customers in the data set of which 5 is *low risk* and 1 is *moderate risk*. No *high risk* one among the high income customers. Entropy for S_{high} is computed as follows:

$$E(S_{high}) = -\frac{5}{6}log_2\left(\frac{5}{6}\right) - \frac{1}{6}log_2\left(\frac{1}{6}\right) = 0.65$$

We will find information gain values for attributes *credit history*, *debt* and *collateral*, because they have not been placed on the decision tree yet. Let us compute the information gain for *credit history*. Since all customers having high income and good credit history is classified as low risk, $E(good) = 0$. Because of the same reason $E(bad)$ and $E(unknown)$ are also zero. The information gain for attribute *credit history* is obtained as follows:

$$IG(S_{high}, \text{credit history}) = 0.65 - \frac{3}{6}0 - \frac{1}{6}0 - \frac{2}{6}0 = 0.65$$

With similar computation we have $IG(S_{high}, debt) = 0.109$ and $IG(S_{high}, collateral) = 0.191$. Since the gain of information for *credit history* is the greatest one, it becomes next level of the corresponding branch. A decision node for *credit history* is added to the decision tree with three branches as *good*, *bad* and *unknown*. Since entropy for each branch is zero, the tree does not grow from this branch anymore. Computations proceed similarly for the rest of branches for node *income* (*moderate* and *low*) until the entropy for each branch is found zero or no attributes remains to be split. The whole decision tree obtained by ID3 algorithm is given at Fig. 2.

After the decision tree is constructed, classification rules are derived. For example one of the rules obtained from the decision tree given in Fig. 2 is

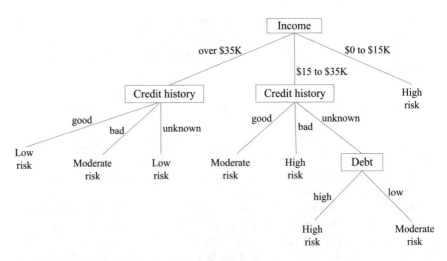

Fig. 2 Decision tree obtained using ID3 algorithm

"**if** *income* is *high* (over 35$) **and** *credit history* is *good* **then** object (customer) is in the *low risk* group". New objects can easily be classified with the rules obtained by decision tree.

Simplicity and comprehensibility are the most important features of a decision tree. In ID3 algorithm, an attribute appears only once on a decision path which is important to satisfy comprehensibility. On the other hand, it may result with the overlapping classes where ID3 cannot provide any information about the inter-section regions [27].

2.2 Other Information Gain Measures

The heterogeneity in an outcome by attribute classification is also called impurity [1]. In order to find the optimal split variable, ID3 algorithm uses the entropy which attempts to find the maximum information gain in the decreasing heterogeneity of the data. Depending on the outcome of the objects, different gain measures become suitable for the splitting process. Other information gain measures are presented below.

2.2.1 GINI Impurity

GINI impurity is mainly used in Classification Trees, where the outcome is binary or categorical. It is calculated as [1]

$$GINI(t) = 1 - \sum_{k=1}^{K} [p_t(k)]^2 \tag{4}$$

where $GINI(t)$ represents the GINI impurity of node t and $p_t(k)$ is relative frequency of class k in node t. The GINI impurity of a split is calculated as

$$GINI(S, A_i) = \sum_{t \in T_i} \frac{|t|}{|S|} GINI(t) \tag{5}$$

As for the aforementioned example, the credit history being *good* results in 3 *low*, 1 *moderate* and 1 *high* risk decision.

$$GINI(good) = 1 - [(\frac{3}{5})^2 + (\frac{1}{5})^2 + (\frac{1}{5})^2] = \frac{14}{25} = 0.560$$

In a similar fashion, the GINI impurity for the credit history being *unknown* and *bad* are calculated as 0.640 and 0.375, respectively. Total impurity of *credit history* is

$$GINI(S, \; credit \; history) = \frac{5}{14} \cdot GINI(good) + \frac{5}{14} \cdot GINI(unknown)$$
$$+ \frac{4}{14} \cdot GINI(bad) = 0.536$$

The GINI impurity for *debt*, *collateral* and *income* are calculated as 0.612, 0.563 and 0.262, respectively. A higher GINI impurity indicates a higher level of heterogeneity and a lower determination level on a decision; whereas a lower GINI impurity indicates a more determined rule on a decision. Since the lowest GINI impurity belongs to the *income* attribute, this attribute becomes the root of the tree with three branches that are *low*, *moderate* and *high*.

2.2.2 Misclassification Error

Misclassification error is also an impurity measure that evaluates the number of outcomes in different classes for binary and categorical outcomes [40]. The misclassification error rate of node t is calculated as [1]

$$ME(t) = 1 - max_k[p(t_k)] \tag{6}$$

For the credit risk example, branching the tree on *good* results in 3 low risks, 1 high risk and 1 moderate risk, making the fractions of answers 0.6, 0.2 and 0.2. Misclassification error for *good* credit history is computed as follows:

$$ME(good) = 1 - max(0.6, 0.2, 0.2) = 1 - 0.6 = 0.4$$

The maximum value of fraction of answers (the class that the majority of answers belong to) also provides the class for that node. In that sense, if a person has a *good* credit history, they are expected to be in the *low* risk class. Similarly, the misclassification error of an *unknown* credit history results in 2 low risks, 2 high risks and 1 moderate risk. Hence,

$$ME(unknown) = 1 - max(0.4, 0.4, 0.2) = 1 - 0.4 = 0.6$$

The class for *unknown* credit risk cannot be fully determined from this node, since the fractions or the frequency of low and high risks are the same. However, misclassification error remains the same whichever class is chosen. The misclasification error is calculated for *bad* credit history as

$$ME(bad) = 1 - max(0.75, 0.25, 0) = 0.25$$

Hence, the class besomes *high* risk, since most of the people with bad credit histories belong to the high risk group. The miscalculation error for this first split is calculated through all misclassified data. The misclassified number of objects for people with *good, unknown* and *bad* credit history is 2, 3 and 1, respectively. In a total of 14 objects, 6 of them are placed incorrectly. Hence, $ME(S, credit\ history) = 6/14 = 0.428$. The misclassification error rate are calculated as 0.5, 0.428 and 0.214 for the *debt, collateral* and *income* attributes. Hence, the first split is achieved through *income*.

The comparison of entropy, GINI impurity and misclassfication error is summarized in the Fig. 3. The maximum impurity level can be 0.5 for GINI impurity and misclassfication error and 1 for entropy which indicates the case of maximum heterogeneity in data. As can be seen from Fig. 3, entropy has a nonlinear and more accelarated increase as the heterogeneity increases.

2.2.3 Goodman and Kruskal Index

Generally used in unsupervised clustering, Goodman-Kruskal Index is a measure of misclassification that compares the distances between elements [14]. Goodman-Kruskal Index is based on distance comparisons of components from clusters. Let (p, q, r, s) be four different elements that are clustered and named a quadruple. In clustering, it is essential that the elements within a cluster are close to each other and the elements in different clusters are apart. In that manner, a quadruple is assigned concordant if it satisfies one of the conditions below:

- $d(p, q) > d(r, s)$, p and q are in different clusters, and r and s are in the same cluster.
- $d(p, q) < d(r, s)$, p and q are n same clusters, and r and s are in different clusters.

where d signifies distance. On the other hand, a quadruple is assigned disconcordant if it satisfies one of the conditions given below:

Fig. 3 The comparison between entropy, GINI impurity and misclassfication error

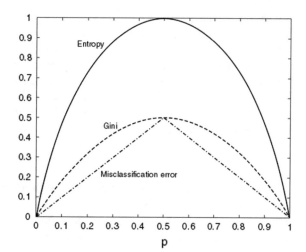

- $d(p,q) > d(r,s)$, p and q are in the same cluster, and r and s are in different clusters.
- $d(p,q) < d(r,s)$, p and q are in different cluster and r and s are in the same cluster.

All concordance and discordance conditions signify within two couples, the closer ones are allowed to be in the same cluster whereas further ones are to be in different clusters. The Goodman-Kruskal Index calculates the concordance ratio of all possible quadruples for clustering. It is formulated as in the following equation.

$$GK = (Q_C - Q_D)/(Q_C + Q_D) \tag{7}$$

where Q_C is the number of concordant quadruples and Q_D is the number of discordant quadruples. According to the formula, in case of many concordant quadruples and few discordant quadruples, the Goodman-Kruskal ratio increases. Hence, a large value of the index indicates a more robust classification. However, the distance metric requires numerical values for classification. Although categorical values can be presented in multi-level binary representation, the accuracy of this index requires numerical inputs and categorical outputs. Hence, this is not an appropriate index for the credit risk classification example mentioned above.

2.2.4 Deviance

Deviance is another measure of impurity which is calculated as

$$D(t) = -2 \sum_{k=1}^{K} [n_{tk}(ln(p_{tk}))] \tag{8}$$

where n_{tk} is the number of objects observed at node t in class k and p_{tk} is the probability of being in class k at node t [26]. Analogous to the standard deviation, the deviance is a measure of the deviation of objects in a class. Hence, for a class in order to be more homogeneous, the deviance should be as small as possible. For the credit risk example, the root node has 5 *low* risk, 3 *moderate* risk and 6 *high* risk values. Let $k = 1, 2, 3$ indicate classes *low, moderate* and *high* classes, respectively. Hence, the probability of a *low* risk object is $p(1) = 5/14 = 0.357$. Likely, the probability of a *moderate* risk object is $p(1) = 3/14 = 0.214$ and a *high* risk is $p(3) = 6/14 = 0.429$. The total deviance of the root node is calculated as

$$D(1) = -2[(5 \cdot ln(0.357)) + (3 \cdot ln(0.214)) + (6 \cdot ln(0.429))] = 29.71$$

If the first split is made by splitting through credit history, the split results in 4 bad objects, 5 unknown objects and 5 unknown objects. Out of 4 bad objects, 3 of them end in high risk and one of them ends in moderate risk. Out of 5 unknown objects, 2 of them end in low risk, 2 of them end in high risk, and one of them ends in moderate risk. Lastly, out of 5 good objects, 3 of them end in low risk, one of them ends in moderate risk, and one of them ends in high risk. Hence, the deviance of that split would be

$$D(2) = -2[(3 \cdot ln(\frac{3}{4})) + (1 \cdot ln(\frac{1}{4})] - 2[(2 \cdot ln(\frac{2}{5})) + 2 \cdot ln(\frac{2}{5}))$$
$$+ 1 \cdot ln(\frac{1}{5}))] - 2[(3 \cdot ln(\frac{3}{5})) + 1 \cdot ln(\frac{1}{5})) + 1 \cdot ln(\frac{1}{5}))] = 24.55$$

With such a split, the deviance would be decreased from 29.71 to 24.55, which is a decrease of 5.24 units of impurity. If the split attribute would be debt, the new deviance value of the split would be 28.49, which is a decrease of 1.22. The deviance value of the split would be 25.71, in case of collateral and 10.95 in case of income. Since the smallest deviance value is achieved by income, it is selected as the first attribute for the split of the tree.

2.3   Tree Pruning

In order to avoid overfitting problem, pruning is the process of avoiding abundant nodes of a tree. In this case, the mathematical definition of an overfitting is as follows:

$$Er(h, D) < Er(h', D) \text{ and } Er(h) > Er(h') \tag{9}$$

where h and h' are different subsets of a set of objects set D, H is the universal set and $Er(\cdot)$ is the error rate [19]. If the overall error rate of h is higher than h', yet, the error rate relation is reversed when the tree constructed on the data that D provides, it is said that the tree overfits D. In order to avoid overfitting, trees are eliminated from nodes which cause overfitting. Pruning is applied in two ways: (*i*) *Pre-pruning*: the node is eliminated before its addition to the tree, (*ii*) *Post-pruning*: the node is eliminated after its addition to the tree. Different pre-pruning methods are analysed in literature. In pre-pruning a threshold is determined for constructing the tree and when this threshold is reached, generation of new nodes are stopped. One threshold is to limit the number of maximum branches of the tree and cancel growth when this number is reached [21]. The main drawback of this approach is the tendency to collect less-relevant attributes on the tree and losing vital attributes due to the limited size of the tree. In pre-pruning the main challenge is to determine the optimum threshold for pruning.

In post-pruning, the tree is let to fully and perfectly grow. Once the tree is generated, another set of data than training are fed into the tree in order to find the "best pruned tree". Hence, post-pruning techniques require the data to be divided in two: training data and testing data. The training data are used for the tree generation and the testing data are used for the pruning process [21]. There are various post-pruning algorithms in literature, two of which are explained below.

2.3.1 Cost Complexity Pruning

Once a base tree (T_0) is generated, various sub-trees $(T_1, T_2, \ldots, T_t)$ of this tree are sequentially constructed in a way that minimizes the error rate per leaf node using the following formula [37].

$$\alpha = \frac{\varepsilon((pruned(T,t), S) - \varepsilon(T,S))}{|leaves(T)| - |leaves(pruned(T,t)|} \qquad (10)$$

where $\varepsilon(T, S)$ is the error rate of tree T over sample S. $|leaves(T)|$ is the number of leaves on tree T. $pruned(T, t)$ specifies any tree by removing node t and re-attaching remaining nodes. Hence, by error rate (α) minimization, the node t to be removed is selected. By a recursion of that approach, best pruned tree is obtained. The tree obtained in ID3 algorithm has 0 error rate due to small sized and clean data; hence, it is not appropriate for pruning.

2.3.2 Minimum Error Pruning

Proposed by Niblett [28], Minimum Error Pruning attempts to prune the tree using a proposed error rate measure that is compared between pruned and original tree. This comparison is achieved through comparing error rates which imply if the error rate decreases when each nonterminal route is pruned. The error rate formula for the pruned tree is given below:

$$Er_p = \frac{n - n_c + k - 1}{n + k} \qquad (11)$$

where n is the number of objects that satisfy the conditions of the related non-terminal node, n_c is the maximum number of elements that belong the same class out of n objects and k is the number of classes. Assume that the error rate is calculated for the *credit history* node in case of a *moderate* (between 15 and 35 K) income (Note that the error rate cannot be calculated for the *credit history* node under the *low* (0–15 K) income, since this is a terminal node). Let Er_p be the error rate of the pruned tree. If the tree is pruned from that node, that is, if the splits that originate from *credit history* are removed and *credit history* becomes a terminal node, a total of 4 objects are left (the objects which satisfy a *moderate* income). 2 of these objects belong to *high* risk class and 2 of them belong to *moderate* risk class. Hence, the number of objects in the most crowded class (n_c) is 2. There are 3 classes in total $(k = 3)$. The error rate is $Er_p = (4 - 2 + 3 - 1)/(4 + 3) = 4/7 = 0.571$.

When it comes to the error term for the unpruned tree, there is another node *debt* with three branches. One branch belongs to *high* debt which has 3 objects, 2 of which have *high* risk and one of which has *moderate* risk. Other branch belongs to *low* debt with one object which results in *moderate* risk. The error rate for an unpruned subtree is calculated as

$$Er_u = \sum_{k=1}^{K} [\frac{n_k}{n} \cdot (\frac{n - n_c + k - 1}{n + k})] \tag{12}$$

where n_k is the number of elements that satisfy the conditions on the branch that leads up to kth class. The error rate of the unpruned tree becomes

$$Er_u = \frac{3}{4} \cdot \frac{3 - 2 + 3 - 1}{3 + 3} + \frac{1}{4} \cdot \frac{1 - 1 + 3 - 1}{1 + 3} = 0.375$$

Since the error rate is lower in the unpruned tree, the tree should not be pruned. At this point, the importance of separate training and testing objects arises, since the tree is constructed using training objects with the objective of minimizing error.

2.4 C4.5 Algorithm

Although ID3 algorithm is an efficient method for classification of symbolic data, it requires a discretization procedure prior to attribute selection for nonsymbolic (numeric, continuous) data [27]. Classification and Regression Trees (CART) [4] and C4.5 [35] are decision tree induction algorithms which do not require prior partitioning. In these algorithms, thresholds are dynamically computed and an attribute may be used multiple time with different thresholds. In spite of an improvement in accuracy, these methods can result in a reduction of comprehensibility.

The C4.5 algorithm, also proposed by Quinlan [36], overcomes some disadvantages of the ID3 algorithm. The C4.5 algorithm

- can handle attributes both numeric and categorical values
- can handle data with missing attribute values
- prunes trees after training.

In order to handle numerical values of attributes, following steps are included:

- sort the numerical attribute (a_i) values in an ascending order
- determine adjacent values that the decision reverses or changes (say, b and c where $b, c \in \mathbb{R}$)
- calculate the mean of these values (say, $d = (a + b)/2$)
- categorize and reassign the attribute values by controlling if they are less than or more than or equal to the mean (the classes are "$<d$" and "$\geq d$").

The steps of the algorithm are listed below:

1. Training data and testing data are separated.
2. The minimum number of objects for a class is determined as a threshold.
3. Numerical values are categorized.
4. Missing attribute values are predicted in a way which provides the most information gain.
5. By selecting an appropriate impurity measure, the tree is generated as in the ID3 algorithm. Tree generation is ceased when the threshold number is reached.
6. The tree is pruned using one of the appropriate pruning techniques.

2.5 Classification and Regression Trees (CART)

CART is developed by Breimen et al. [4] for constructing binary decision trees which are called classification trees and numerical decision trees which are called regression trees [4]. Classification trees utilize GINI impurity for splitting whereas C4.5 algorithm is generally entitled to information gain (entropy) related measures. Another difference is that for the pruning process, CART uses cost-complexity error method whereas C4.5 does not use recursive methods. Additionally, in classification trees, each split is binary. Likely, the decision tree can be split using the same attribute more than once. The pseudocode including main steps of CART is given in Algorithm 2.

In case of the credit card risk example, the first split had been made on *income* by using the GINI impurity. The first split leads to a leaf in terms of *low* incomes. Hence, the first binary split involves if the *income* is *low* or not, and the first branch becomes a leaf. In CART, the second split does not have to be on income again.

Continuing with the credit risk example, we have 14 data for credit risk assessment. Although the size of data is small, we will use the first 10 data for training and the last 4 data for testing. The GINI impurities for all nodes are calculated as $GINI(good) = 0$, $GINI(Unknown) = 0.64$, $GINI(Bad) = 0.444$, $GINI(High) = 0.444$, $GINI(Low) = 0.653$, $GINI(None) = 0.571$, $GINI(Adequate) = 0.444$, $GINI(0 - 15K) = 0$, $GINI(15K - 35K) = 0.5$, $GINI(over\,35K) = 0.32$. The attribute GINI impurities are $G(S, credithistory) = 0.453$, $G(S, debt) = 0.59$, $GINI(S, collateral) = 0.533$, $GINI(S, income) = 0.26$. The GINI impurity for all objects is 0.62. Hence, the greatest impurity reduction is achieved by *income* by a decrease of $0.62 - 0.26 = 0.36$. Since the tree should be binary in CART, the nodes for *income* should be decided. The smallest GINI impurity belongs to $0 - 15K$, which decreases the impurity to 0. Hence, the first split is achieved through *income* by splitting the ones that are less than 15 K and that are more than 15 K. The impurity of the branch $0 - 15K$ is 0, hence no more splitting is made after this node, making this node a terminal one or a leaf. The other branch includes objects with income values over 15 K. The GINI impurity of that node is calculated as 0.571. To see if further splitting is required, we calculate the GINI impurities of all other nodes, which are calculated as $GINI(good) = 0$, $GINI(Unknown) = 0.625$, $GINI(Bad) = 0$, $GINI(High)$ $= 0.444$, $GINI(Low) = 0.5$, $GINI(None) = 0.565$, $GINI(Adequate) = 0.444$, $GINI$ $(15K - 35K) = 0.5$, $GINI(over\ 35K) = 0.32$. The attribute GINI impurities are $G(S, credithistory) = 0.358$, $G(S, debt) = 0.484$, $GINI(S, collateral), = 0.513$, $GINI(S, income) = 0.371$. The greatest contribution to decreasing GINI impurity is again achieved by income and the tree is split into two more nodes based on if the income is between 15 and 35K or higher than 35K.

Algorithm 2: CART Algorithm

Input: Training data
1 Calculate $GINI(t)$ for all nodes
2 Compute $GINI(S, A_j) = min_{i \in S} GINI(S, A_i)$
3 **if** attribute A_j has two nodes **then** branch for attribute A_j;
4 **else**
5   Compute $GINI(t_l) = min_{t \in A_j} IG(t)$
6   Create two branch as $(t_1, ..., t_l)$ and $(t_{l+1}, ..., t_n)$ where $t_1, ..., t_n \in A_j$
7 **if** $GINI$ of the branch is 0 **then** terminate this branch;
8 **if** all branches terminated **then** terminate the algorithm;
9 **else**
10   Update S
11   Go to Step 1
Output: Decision tree

If the income is between 15 and 35 K, the impurity of that node is 0.5. Since all attributes of income have been covered beforehand, no further splitting on *income* is feasible. The GINI impurities of nodes are $GINI(unknown) = 0.5$, $GINI(high) = 0$, $GINI(low) = 0$, $GINI(None) = 0.5$. The attribute GINI impurities are $GINI(S, credithistory) = 0.5$, $GINI(S, debt) = 0$, $GINI(S, collateral) = 0.5$. Hence, another split is achieved through debt being *high* or *low*. This split yields in two nodes with one object in each. Hence, the nodes are declared terminal.

If the *income* is higher than 35 K, the impurity of this node is 0.32. The impurities of all nodes are $GINI(good) = 0$, $GINI(unknown) = 0$, $GINI(bad) = 0$, $GINI(high) = 0$, $GINI(low) = 0.375$, $GINI(none) = 0$, $GINI(adequate) = 0.444$. The attribute GINI impurities are $GINI(S, credithistory) = 0$, $GINI(S, debt) = 0.3$, $GINI(S, collateral) = 0.267$. Hence, the split is achieved considering *credit history* which has three attribute values with all 0 impurities. Assuming that the split is achieved on if the *credithistory* is *bad* or not, the node with *bad* credit history is left with one object and therefore is a leaf. The other node that involves objects with *not bad* history has a GINI impurity of 0, and no more splitting is necessary for that node. In that manner, the decision tree is constructed as in Fig. 4. The rules are derived from the decision tree as follows:

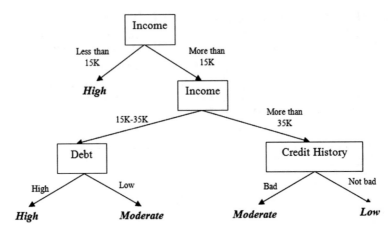

Fig. 4 Classification tree of the credit risk example

1. **if** the income is less than 15 K, **then** risk is *high*,
2. **if** the income is higher than 15 K **and** between 15 K and 35 K **and** the debt is *high*, **then** risk is *high*,
3. **if** the income is higher than 15 K **and** between 15 K and 35 K **and** the *debt* is *low*, **then** risk is *moderate*,
4. **if** the income is higher than 35 K **and** credit history is *bad*, **then** the risk is *moderate*.
5. **if** the income is higher than 35 K **and** the credit history is *not bad*, **then** the risk is *low*.

The tree is generated using 10 data, the rest of the data will be used for pruning, using Cost Complexity Pruning. First test object has an income less than 15 K, hence according to the tree, the object should indicate a *high* risk which it does. Likely, the other three data also fits to the tree and the error rate of the unpruned tree is 0. To see if we can benefit from pruning, it is required to check if any pruning also yields an error rate of 0, since it is an impossible task to surpass such error rate. For example, lets prune the *debt* node. In this case, there are 1 *moderate* and 1 *high* risk objects in the training data. Assigning any class to that node would lead to the same error rate, since testing data also contains 1 *moderate* and 1 *high* risk objects. This yields that if any pruning is done, error rate will increase. Hence, the tree is left unpruned in terms of *debt* node. Similarly, any pruning leads to a higher error rate than 0, which means no pruning should be done over the generated tree.

Similar to classification trees, regression trees use the same structure. However, in this case, numerical outputs are classified, that is, the rules result in a continuous number. The objective of the tree is to minimize the error between the actual output and the predicted class of the output in terms of an error measure (i.e. squared error) [10].

2.6   Supervised Learning in Quest (SLIQ)

This algorithm is proposed by Mehta et al. [24] for binary decision trees which involve numerical or categorical values. It uses the GINI impurity and a split limit with a breadth-first approach. During the tree building process, two main decisions have to be made: (*i*) evaluation of splits for each attribute and the selection of the best split and (*ii*) creation of partitions using the best split. Splitting through numerical attributes are in the form of $A < v$ where A is the attribute value and v is the split value and a real number. The pseudocode including main steps of SLIQ is given in Algorithm 3.

Algorithm 3: SLIQ Algorithm

Input: Numerical or categorical data
1 Sort data with respect to attribute A_i
2 Determine split points v_{ij} for A_i
3 Compute $GINI(S, A_i, v_{il}) = min_j GINI(S, A_i, v_{ij})$
4 Create two branch as $< v_{il}$ and $> v_{il}$
5 **if** $GINI$ of the branch is 0 **then** terminate this branch;
6 **if** all branches terminated **then** terminate the algorithm;
7 **else**
8 | Update S
9 | $i \leftarrow i + 1$
10 └   Go to Step 1

Output: Binary decision tree

The following example in Table 2 is to be used for SLIQ. In this example, all values are numerical and there are two output classes. The first step of this algorithm is to determine a split value for numerical attributes. In order to achieve that, the numerical values are sorted and the class outputs are listed. Sorting for the first attribute A_1, Table 3 is obtained.

The split points are determined as the points where the output changes. For example, observing the Table 3, the first change appears to occur when the attribute value of A_1 switches from 46 to 47, as the object with the value of 46 belongs to Class 2 and the object with the value of 47 belongs to Class 1. Their mean is selected as the split point. Hence, a split could occur in the form of $A_1 < 46.5$ and $A_1 > 46.5$. Other split point options include the switch from 51 to 52 in the form $A_1 < 51.5$ and $A_1 > 51.5$. Other split points are 53, 54.5, 55.5 and 58.5. Note that for the value of 55, two objects belong to Class 1 and one object belongs to Class 2. However, a certain split cannot be achieved when two different classes have the same attribute value; as a result, 55 is not chosen as a split value. In order to find the right split, the GINI impurity is used. The root node involves ten Class 1 objects and ten Class 2 objects, making the GINI impurity 0.5. The GINI for the first split involves two branches $GINI(A_1 < 46.5) = 0$ since all objects belong to the same

Table 2 Example table for SLIQ

#	A_1	A_2	A_3	Class
1	38	69	21	2
2	42	69	1	2
3	43	58	52	2
4	44	58	9	2
5	46	69	3	2
6	46	58	2	2
7	47	66	12	1
8	48	66	0	1
9	49	66	0	1
10	50	66	1	1
11	51	66	1	1
12	52	69	3	2
13	54	66	0	1
14	54	68	7	2
15	55	66	0	1
16	55	66	18	1
17	55	68	15	2
18	56	66	1	1
19	56	66	2	1
20	61	68	1	2

class. The GINI impurity for the branch $A_1 > 46.5$ is calculated as 0.408. The GINI impurity of the split is $G(S, A_1, 46.5) = 0.286$. Other split values are $GINI(S, A_1, 51.5) = 0.495$, $GINI(S, A_1, 53) = 0.490$, $GINI(S, A_1, 54.5) = 0.476$, $GINI(S, A_1, 55.5) = 0.490$ and $GINI(S, A_1, 58.5) = 0.474$. The maximum decrease in the GINI is achieved by the split point 46.5; that is, if the first split were to be chosen through A_1, the split point would be 46.5. However, a similar approach is to be carried out for all attributes. $GINI(S, A_1)$ becomes the smallest GINI impurity, thus, 0.286.

Similarly for A_2, the split points are 62 and 67 where the GINI impurities are $GINI(S, A_2, 62) = 0.412$ and $GINI(S, A_2, 67) = 0.231$. The second split point provides a decrease of 0.269 and $GINI(S, A_2) = 0.231$. In terms of A3, the split point options are 0.5, 2.5, 10.5, 13.5, 16.5 and 19.5 with GINI impurities 0.375, 0.374, 0.493, 0.469, 0.490 and 0.444. Hence, $GINI(S, A_3) = 0.374$ by the point 2.5. Considering A_1, A_2 and A_3, the maximum decrease on GINI impurity is provided by A2 by splitting from the value of 67. The left branch $A_2 > 67$ involves a GINI impurity of 0, hence all objects belong to the same class, making this node a leaf. However, the GINI impurity of the branch $A_2 < 67$ has a GINI of 0.355 and needs

Table 3 Sorted A_1 data for SLIQ

A_1	Class
38	2
42	2
43	2
44	2
46	2
46	2
47	1
48	1
49	1
50	1
51	1
52	2
54	1
54	2
55	1
55	1
55	2
56	1
56	1
61	2

further splitting over 13 objects. Further splitting provides two alternative trees. Over the split $A_2 < 67$, both $GINI(S, A_1, 46.5)$ and $GINI(S, A_2, 62)$ has values of 0. Two trees generated by SLIQ algorithm are given in Figs. 5 and 6.

Fig. 5 Second tree generated by the SLIQ algorithm

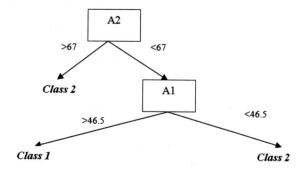

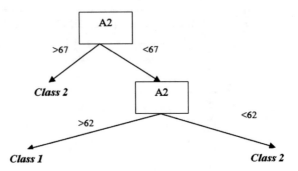

Fig. 6 First tree generated by the SLIQ algorithm

3 Fuzzy Decision Trees

Fuzzy sets are integrated to induction algorithms to improve their comprehensibility by combining with cognitive uncertainties [46]. A fuzzy decision tree can be considered as a generalized version of the crisp case [43]. Fuzzy decision trees were first mentioned by Chang and Pavlidis [8] in 1977. Since then, many fuzzy decision tree induction algorithms have been proposed in the literature. A comprehensive review on fuzzy decision trees can be found in Chiang and Hsu [9]. In this section, widely used fuzzy induction algorithms—fuzzy ID3 and fuzzy SLIQ algorithms— are discussed and a brief literature review on recent studies using these algorithms is given.

Before discussing fuzzy decision trees, some fundamental fuzzy operations related with fuzzy decision tree induction process are given as the following. Let $\tilde{S}$ be a fuzzy set of n objects in a training set. A fuzzy subset for class k can be represented with $\tilde{C}_k$. Relative frequency p_k for class k can be computed as follows:

$$p(k) = \frac{M(\tilde{C}_k)}{\sum_{k=1}^{K} M(\tilde{C}_k)} \tag{13}$$

$$M(\tilde{C}_k) = \sum_{x \in \tilde{S}} \mu_{\tilde{C}_k}(x) \tag{14}$$

where $M(\tilde{C}_k)$ is the cardinality of subset $\tilde{C}_k$ and $\mu_{\tilde{C}_k}(x)$ is the membership value of object x to class k. For each attribute A_i, there are n_i linguistic terms which are represented by $A_i^1, \ldots, A_i^{n_i}$. Let $\tilde{T}_i$ be the set of subsets created from splitting set $\tilde{S}$ by attribute A_i. Relative frequency of class k in subset $\tilde{i} \in \tilde{T}_i$ is

$$p(\tilde{t}_k) = \frac{M(\tilde{t}_k \cap \tilde{C}_k)}{M(\tilde{t}_k)}, \qquad \cup_{k=1}^{K} \tilde{t}_k = \tilde{t} \qquad (15)$$

$$M(\tilde{t}_k \cap \tilde{C}_k) = \sum_{x \in \tilde{S}} \min(\mu_{\tilde{t}_k}(x), \mu_{\tilde{C}_k}(x)) \qquad (16)$$

3.1   Input Representation

In this study, a triangular membership function is used to determine the membership values for all linguistic terms. Each linguistic value A_i^j is represented with three value (a_j, b_j, c_j) where a_j and c_j are the least possible values while b_j is the most occurred value for A_i^j. Let x be a numerical value for attribute A_i. Membership function for each linguistic term $A_i^j (j = 1, \ldots, n_i)$ can be defined as follows:

$$\mu_{A_i^1}(x) = \begin{cases} 1 & a_1 \le x \le b_1 \\ (c_1 - x)/(c_1 - b_1) & b_1 < x < c_1 \\ 0 & x \ge c_1 \end{cases} \qquad (17)$$

$$\mu_{A_i^{n_i}}(x) = \begin{cases} 0 & x \le a_{n_i} \\ (x - a_{n_i})/(b_{n_i} - a_{n_i}) & a_{n_i} < x < b_{n_i} \\ 1 & n_i \le x \le c_{n_i} \end{cases} \qquad (18)$$

$$\mu_{A_i^j}(x) = \begin{cases} 0 & x \le a_j \\ (x - a_j)/(b_j - a_j) & a_j < x < b_j \\ (c_j - x)/(c_j - b_j) & b_j < x < c_j \\ 0 & x \ge c_j \end{cases}, \quad 1 < j < n_i \qquad (19)$$

The selection of the membership function is of a crucial importance. However, the most common way to determine the membership function has come to be trial-and-error [2]. It is a prevalent approach that the selection and parameter tuning of the membership function is tuned until a more desirable behavior is reached [18]. For many other applications of fuzzy logic, metaheuristic approaches are utilized for tuning membership function parameters [13, 28, 38].

The graphical representation of membership function used in this study can be seen in Fig. 7. Parameters of each attribute for the credit risk assessment example are given in Table 4. We assume that each customer is evaluated using a scale between 0 and 10 for *credit history*, *debt* and *collateral* attributes and *risk* decision. Fuzzy membership values are given in Table 5.

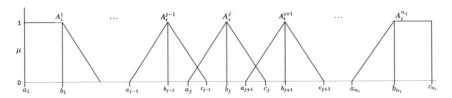

Fig. 7 Membership function for attribute A_i

Table 4 Triangular membership parameters for credit risk assessment problem

Attribute	Linguistic term	(a, b, c)
credit history	good	(4, 8, 10)
	bad	(0, 2, 6)
	unknown	(2, 5, 8)
debt	low	(0, 3, 8)
	high	(3, 8, 10)
collateral	none	(0, 3, 8)
	adequate	(3, 8, 10)
income	low	(0, 10K, 30K)
	moderate	(10K, 25K, 40K)
	high	(20K, 40K, 60K)
risk	low	(4, 8, 10)
	moderate	(2, 5, 8)
	high	(0, 2, 6)

3.2 Fuzzy ID3 Algorithm

Fuzzy ID3 algorithm and its variants are the most used induction algorithms in the literature because it does not require much computational effort to generate fuzzy decision trees and it is suitable for large-scale learning problems [43]. The main idea of fuzzy ID3 is the same with classical ID3. The main difference between two algorithms is the computation of entropy values as follows:

$$E(\tilde{S}) = \sum_{k=1}^{K} -p(k)\log_2 p(k) = \sum_{k=1}^{K} -\frac{M(\tilde{C}_k)}{\sum_{k=1}^{K} M(\tilde{C}_k)} log_2 \frac{M(\tilde{C}_k)}{\sum_{k=1}^{K} M(\tilde{C}_k)} \quad (20)$$

$$E(\tilde{t}) = \sum_{k=1}^{K} -p(\tilde{t}_k)\log_2 p(\tilde{t}_k) = \sum_{k=1}^{K} -\frac{M(\tilde{t}_k \cap \tilde{C}_k)}{\sum_{k=1}^{K} M(\tilde{t}_k \cap \tilde{C}_k)} log_2 \frac{M(\tilde{t}_k \cap \tilde{C}_k)}{\sum_{k=1}^{K} M(\tilde{t}_k \cap \tilde{C}_k)}$$

$$(21)$$

Table 5 Fuzzy membership values for credit risk assessment problem

#	Credit history			Debt		Collateral		Income			Risk		
	Good	Bad	Unknown	Low	High	None	Adequate	Low	Moderate	High	Low	Moderate	High
1	0	1	0	0.6	0.4	0.5	0.5	0.8	0.3	0	0	0.3	0.8
2	0	0.5	0.7	0.8	0.2	0.8	0.2	0.2	0.9	0.3	0	0.2	0.9
3	0.5	0	0.7	1	0	1	0	0	0.7	0.5	0.1	0.8	0.4
4	0.3	0.3	1	0.5	0.5	0.7	0.3	1	0	0	0	0.2	0.9
5	0.4	0.1	0.8	1	0	1	0	0	0.3	0.8	0.8	0.3	0
6	0	0.5	0.7	0.8	0.2	0.4	0.6	0	0	1	0.9	0.2	0
7	0	0.8	0.3	1	0	0.7	0.3	0	0.3	0	0	0	1
8	0	0.9	0.2	0.8	0.2	0.3	0.7	0	0.5	0.6	0.3	1	0.3
9	1	0	0	1	0	0.9	0.1	0	0	1	0.6	0.5	0
10	0.8	0	0.3	0	1	0.2	0.8	0	0	1	0.8	0.3	0
11	1	0	0	0.4	0.6	0.7	0.3	0.7	0.4	0	0	0	1
12	0.8	0	0.3	0.1	0.9	0.8	0.2	0.4	0.8	0.1	0.4	0.8	0.1
13	0.9	0	0.2	0	1	1	0	0	0.3	0.8	1	0	0
14	0	0.8	0.3	0.3	0.7	0.8	0.2	0.1	0.8	0.4	0	0.3	0.8

Morever, information gain is computed as follows:

$$IG(\tilde{S}, A_i) = E(\tilde{S}) - \sum_{\tilde{i} \in \tilde{T}_i} p(\tilde{i}) E(\tilde{i}) \tag{22}$$

At each iteration of fuzzy ID3, truth level of classifying objects within the branch into each class is calculated as follows [46]:

$$P(\tilde{i}, \tilde{C}_k) = \frac{M(\tilde{i} \cap \tilde{C}_k)}{M(\tilde{i})} \tag{23}$$

where $P(\tilde{i}, \tilde{C}_k)$ is the truth level of class k on the branch including set $\tilde{i}$. If the truth level for a branch is larger than a predetermined threshold parameter β, then the branch is terminated as a leaf. Otherwise, the next attribute having the greatest information gain is investigated to split the branch. All objects in a leaf are classified to the class with the highest truth level [46].

Fuzzy ID3 algorithm can be applied to the fuzzy credit risk assessment data in the following way. Let β be 0.7. Firstly, the attribute having the greatest entropy is selected as the main decision node using Eq. (20).

$$E(\tilde{S}) = -\frac{4.9}{16} log_2 \left(\frac{4.9}{16}\right) - \frac{4.9}{16} log_2 \left(\frac{4.9}{16}\right) - \frac{6.2}{16} log_2 \left(\frac{6.2}{16}\right) = 1.576$$

$$E(good) = -\frac{3.2}{5.7} log_2 \left(\frac{3.2}{5.7}\right) - \frac{2.6}{5.7} log_2 \left(\frac{2.6}{5.7}\right) - \frac{1.8}{5.7} log_2 \left(\frac{1.8}{5.7}\right) = 1.509$$

Similarly, entropy values for other features are found as $E(bad) = 1.314$ and $E(unknown) = 1.502$. The information gain for *credit history* is given in the following:

$$IG(\tilde{S}, credit\ history) = 1.576 - \frac{5.7}{16.1} 1.509 - \frac{4.9}{16.1} 1.314 - \frac{5.5}{16.1} 1.502 = 0.129$$

The information gain for the other attributes are found as $IG(S, debt) = 0.037$, $IG(S, collateral) = 0.052$ and $IG(S, income) = 0.301$. Since attribute *income* is the greatest one with 0.301 information gain, it becomes the root node of decision tree with three branches as *high*, *moderate* and *low*. The truth level of each class $k = \{low\_risk, moderate\_risk, high\_risk\}$ for branch *low* is computed as follows:

$$P(\tilde{S}_{low}, \tilde{C}_{low\_risk}) = \frac{0.4}{4} = 0.1$$

$$P(\tilde{S}_{low}, \tilde{C}_{moderate\_risk}) = \frac{0.3 + 0.2 + 0.2 + 0.4}{4} = 0.3$$

$$P(\tilde{S}_{low}, \tilde{C}_{high\_risk}) = \frac{0.8 + 0.2 + 0.9 + 0.8 + 0.7 + 0.1 + 0.1}{4} = 0.9$$

$P(\tilde{t}, \tilde{C}_k)$ represents the truth level of class k for set $\tilde{t}$. Since $P(\tilde{S}_{low}, \tilde{C}_{high\_risk}) = 0.9 > \beta = 0.7$, the branch *low* is terminated and all objects on this branch classified as *high_risk*. Truth level of classes for branch *moderate* in the first level are found as $P(\tilde{S}_{moderate}, \tilde{C}_{low\_risk}) = 0.264$, $P(\tilde{S}_{moderate}, \tilde{C}_{moderate\_risk}) = 0.585$, and $P(\tilde{S}_{moderate}, \tilde{C}_{high\_risk}) = 0.66$. Since none of them is larger than β parameter, splitting procedure is performed for branch *moderate* as follows. Firstly, the entropy of set $\tilde{S}_{moderate}$ is computed.

$$E(\tilde{S}_{moderate}) = -\frac{1.4}{8} log_2\left(\frac{1.4}{8}\right) - \frac{3.1}{8} log_2\left(\frac{3.1}{8}\right) - \frac{3.5}{8} log_2\left(\frac{3.5}{8}\right) = 1.492$$

The recursive computation performed to compute information gain of each attribute as $IG(\tilde{S}_{moderate}, credit\,history) = 0.236$, $IG(\tilde{S}_{moderate}, debt) = 0.249$, $IG(\tilde{S}_{moderate}, collateral) = 0.226$. Since the information gain *debt* is the greatest one, it becomes the second level decision node of this branch. The computations continue until the whole objects are classified. The decision tree obtained using fuzzy ID3 algorithm is given in Fig. 8. Risk assessment decisions are given at the end of each leaf with their corresponding truth level P.

Many enhancements have been proposed for fuzzy ID3 algorithm in the literature. The studies of Umana et al. [42] and Janikow [16] are the first ones extending ID3 algorithms with using fuzzy sets. Hayashi [15] proposed a fuzzy ID3 algorithm with adjusting mechanism of AND/OR operator. Chang et al. [7] hybridized fuzzy ID3 algorithm with genetic algorithms to optimize the rule parameters in the tuning process. Bartczuk and Rutkowska [3] developed a new version of fuzzy ID3 algorithm which allows to use more than one attribute value in leaves. By this way, decision trees contain less number of the nodes than the ones constructed by classical algorithms. Wang et al. [44] obtained several fuzzy attribute reducts, which are subset of attributes that are necessary and sufficient to represent the given data, and generated a fuzzy decision tree for each fuzzy attribute reduct by a fuzzy ID3 algorithm. Jin et al. [17] proposed a generalized fuzzy partition entropy-based ID3 algorithm which considers the impact of the non-linear characteristics of the membership degree of fuzzy sets. There are also various applications of fuzzy ID3 algorithm in the literature, such as performance evaluation [22], online purchasing behavior analysis [45], image processing [11] and medical diagnosis [12].

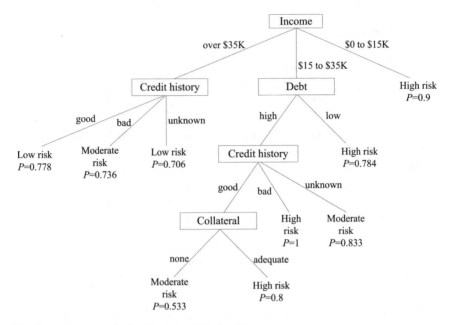

Fig. 8 Decision tree obtained by fuzzy ID3 algorithm

3.2.1 Generating the Classification Rules

Each branch from root to leaf can be converted to a rule. The decision nodes on a path gives the features of attributes which represent a given condition while the leaf at the end of a path is the final decision. The rules extracted from the decision tree given by Fig. 8 are as follows:

1. **if** *income* is *high* **and** *credit history* is *good* **then** *risk* is *low* $(P = 0.778)$
2. **if** *income* is *high* **and** *credit history* is *bad* **then** *risk* is *moderate* $(P = 0.736)$
3. **if** *income* is *high* **and** *credit history* is *unknown* **then** *risk* is *low* $(P = 0.706)$
4. **if** *income* is *moderate* **and** *debt* is *high* **and** *credit history* is *good* **and** *collateral* is *none* **then** *risk* is *moderate* $(P = 0.533)$
5. **if** *income* is *moderate* **and** *debt* is *high* **and** *credit history* is *good* **and** *collateral* is *adequate* **then** *risk* is *high* $(P = 0.8)$
6. **if** *income* is *moderate* **and** *debt* is *high* **and** *credit history* is *bad* **then** *risk* is *high* $(P = 1)$
7. **if** *income* is *moderate* **and** *debt* is *high* **and** *credit history* is *unknown* **then** *risk* is *moderate* $(P = 0.833)$
8. **if** *income* is *moderate* **and** *debt* is *low* **then** *risk* is *high* $(P = 0.784)$
9. **if** *income* is *low* **then** *risk* is *high* $(P = 0.9)$.

Quinlan [34] investigated the methods which are simplifying non-fuzzy decision trees without compromising their accuracy. Yuan and Shaw [46] applied rule simplification technique to fuzzy decision trees. In this technique, a rule is simplified by removing an attribute term from the condition (**if**) part. For a rule, an attribute is removed from the rule and the truth level is obtained in each time. The simplified rule having the greater truth level than the original rule is replaced with the original one. In the cresit risk assessment example, only rule 4 and rule 5 can be simplified. Simplified version of rules and their truth levels are given in the following:

4. **if** *income* is *moderate* **and** *credit history* is *good* **then** *risk* is *moderate* $(P = 0.696)$
5. **if** *income* is *moderate* **and** *debt* is *high* **and** *collateral* is *adequate* **then** *risk* is *high* $(P = 0.929)$.

It can be observed that truth levels of simplified rule 4 and 5 are greater than the rules obtained by fuzzy ID3.

3.2.2 Classification with Rules

The classification of an object with decision tree rules can be summarized as follows [46]:

1. Membership value of the object is computed for each rule.
2. If several rules results with the same class, maximum membership value among the rules are considered as the membership value of the object to the corresponding class.
3. If the object belongs to various classes with different membership values, it is assigned to the class having the largest membership value.

This procedure is applied for all objects in the data set. The classification result of training data set in credit risk assessment example is given at Table 6. Classification results obtained by fuzzy ID3 algorithm is exactly the same with the given classes in the training data.

3.3 Fuzzy SLIQ Algorithm

The SLIQ algorithm has been fuzzified by Chandra and Verghese [6]. In the crisp case, midpoints of the values where classes switch are determined as split points. According to Chandra and Verghese, the fuzziness lies in the choice of the split point, and the distance to the split point together with the standard deviation of the attribute values determine the output. Fuzzy membership values are calculated using standard deviation, the split point and user specified parameters β, α as given below:

Table 6 Results of fuzzy ID3 algorithm for credit risk assessment example

#	Real classification				Results of fuzzy ID3			
	Membership values			Decision	Membership values			Decision
	low	moderate	high		low	moderate	high	
1	0	0.3	0.8	high	0	0	0.8	high
2	0	0.2	0.9	high	0	0.2	0.8	high
3	0.1	0.8	0.4	moderate	0.1	0.5	0.4	moderate
4	0	0.2	0.9	high	0	0	0.9	high
5	0.8	0.3	0	low	0.8	0.3	0	low
6	0.9	0.2	0	low	0.7	0.2	0	low
7	0	0	1	high	0	0	0.8	high
8	0.3	1	0.3	moderate	0.2	0.6	0.3	moderate
9	0.6	0.5	0	low	0.6	0	0	low
10	0.8	0.3	0	low	0.8	0	0	low
11	0	0	1	high	0	0	0.7	high
12	0.4	0.8	0.1	moderate	0.1	0.8	0.1	moderate
13	1	0	0	low	0.8	0	0	low
14	0	0.3	0.8	high	0	0.3	0.7	high

$$
\mu_{val} = \begin{cases} \frac{lw}{lp + lw - val} & val < lp \\ 1 & lp \leq val \leq rp \\ \frac{rw}{val - rp + rw} & val > rp \end{cases}
$$

If the branch is in the form of $A < v$ or $A \leq v$, it is called a left split and if the branch is in the form of $A > v$ or $A \geq v$, it is called a right split. The calculation of parameters lp and rp are the same for left and right splits.

$$lp = split\_point - \beta \tag{24}$$

$$rp = split\_point + \beta \tag{25}$$

The calculation of lw and rw for the left split is given below:

$$lw = \alpha \cdot \sigma \tag{26}$$

$$rw = 0 \tag{27}$$

In case of a right split, the formula change as follows:

$$lw = 0 \tag{28}$$

$$rw = \alpha \cdot \sigma \tag{29}$$

where σ is the standard deviation of the attribute values. The parameters lw and rw control the slope of the membership functions and depend on the standard deviation, and in various applications $\beta \in [0, 1]$. However, for bigger spreads of the data, it is possible that $\beta > 1$. As given in the SLIQ algorithm in Sect. 2.6, the split points are determined and membership values are calculated using these split points. In this case, the GINI impurity uses fuzzy membership values; hence, the GINI impurity is fuzzified as follows:

$$GINI(x_j) = \sum_{v=1}^{V} \frac{N^{(v)}}{N^{(u)}} [1 - \sum_{k=1}^{K} (\frac{N_{w_k}^{(v)}}{N^{(v)}})^2] \tag{30}$$

where K is the total number of classes, V is the total number of partitions, $N^{(u)}$ is the sum of membership values of the objects in the dataset before split if x_j is chosen as the split point, $N^{(v)}$ is the sum of membership values in the vth partition and is $N_{w_k}^{(v)}$ the sum of the product of the fuzzy membership values of the attribute and the fuzzy membership values of the corresponding records for class w_k in the vth partition. For the example given in Table 2, let's assume that the split is about to be made for attribute A2 and the split point is 67, $\alpha = 1$ and $\beta = 0.5$. The standard deviation (σ) for this attribute is calculated as 3.54. For the left branch $A_2 < 67$, $lw = 3.54, rw = 0, lp = 66.5, rp = 67.5$. Since there are 2 classes, $K = 2$

Table 7 Example table for fuzzy SLIQ

A2	Membership values	Class
69	0.702	2
69	0.702	2
58	0.294	2
58	0.294	2
69	0.702	2
58	0.294	2
66	0.876	1
66	0.876	1
66	0.876	1
66	0.876	1
66	0.876	1
69	0.702	2
66	0.876	1
68	0.876	2
66	0.876	1
66	0.876	1
68	0.876	2
66	0.876	1
66	0.876	1
68	0.876	2

and any split over A2 would yield 2 branches $V = 2$, The values of fuzzy membership is given in the Table 7.

The sum of all membership values, that is $N^{(u)} = 15.078$. The sum of fuzzy membership values with partitions are $N^{(1)} = 0.876 \cdot 10 + 0.294 \cdot 3 = 9.642$ (for the left split) and $N^{(2)} = 0.702 \cdot 4 + 0.876 \cdot 3 = 5.436$ (for the right split). For the left split, 10 objects belong to Class 1, and their fuzzy membership value sums are $N_1^{(1)} = 0.876 \cdot 10 = 8.760$. Likely, 3 objects belong to Class 2; indicating $N_2^{(1)} = 0.294 \cdot 3 = 0.882$. The GINI impurity for that partition (branch) is

$$GINI(S, A2 < 67) = 1 - (\frac{8.760}{9.642})^2 - (\frac{0.882}{9.642})^2 = 0.166$$

As for the right split, that is, $A > 67$, all seven objects belongs to Class 2 making $N_1^{(1)} = 0$ and $N_2^{(1)} = 5.436$, Hence

$$GINI(S, A2 > 67) = 1 - (\frac{5.436}{5.436})^2 - (\frac{0}{5.436})^2 = 0$$

The total GINI over the A2 split is

$$GINI(S, A2) = \frac{9.642}{15.078} \cdot 0.166 + \frac{5.436}{15.078} \cdot 0 = 0.106$$

GINI impurities over all attributes and all split points are given in Table 8. The most decrease in GINI impurity is provided by the A2 attribute with a split at point 67.

Table 8 Fuzzy GINI impurities for different splits

Attribute	Split point	GINI
A1	46.5	0.220
A1	51.5	0.490
A1	53	0.479
A1	54.5	0.463
A1	55.5	0.463
A1	58.5	0.440
A2	62	0.374
A2	67	0.106
A3	0.5	0.362
A3	2.5	0.369
A3	10.5	0.500
A3	13.5	0.478
A3	16.5	0.500
A3	19.5	0.435

The left split is defined as $A < 67$ and the right split is defined as $A > 67$. 13 objects are lead to the left split. The right split has a GINI of 0, hence it is terminated a leaf node. However, the right side has 13 objects that are required to be further-partitioned. For the remaining 13 objects, the table of possible partitions and their GINI impurities are shown in Table 9. According to the table, there are two options on the second split. One split can be achieved through A1 being less or greater than 46.5 and another split can be achieved through A2 being less or greater than 62. Same trees as shown in Figs. 6 and 5 are obtained by fuzzy SLIQ algorithm.

4   Computational Results

The algorithms discussed in this study are evaluated by using a small training data set from Yuan and Shaw [46]. A sport activity is decided according to the weather condition of a given day. Membership values for weather data and activities are given in Table 10. Volleyball, swimming and weight lifting (w_lifting) are activities one of which is decided to play considering the outlook, temperature, humidity and wind. Decision tree obtained by fuzzy ID3 algorithm is given in Fig. 9. The rules obtained by fuzzy ID3 algorithm are as follows:

1. **if** *temperature* is *hot* **and** *outlook* is *sunny* **then** choose *swimming* ($P = 0.854$)
2. **if** *temperature* is *hot* **and** *outlook* is *cloudy* **then** choose *swimming* ($P = 0.722$)
3. **if** *temperature* is *hot* **and** *outlook* is *rain* **then** choose *w_lifting* ($P = 0.727$)
4. **if** *temperature* is *mild* **and** *wind* is *windy* **then** choose *w_lifting* ($P = 0.813$)
5. **if** *temperature* is *mild* **and** *wind* is *not_windy* **then** choose *volleyball* ($P = 0.784$)
6. **if** *temperature* is *cool* **then** choose *w_lifting* ($P = 0.884$).

Rule 3 can be simplified with rule simplification technique as follows: 3. **if** *outlook* is *rain* **then** choose *w_lifting* ($P = 0.889$). Classification results are given in Table 14. Objects 2 and 8 are classified into wrong classes and object 16 can be assigned one of two classes with the same membership value. Decision tree and classification results are obtained by using fuzzy entropy measure given in

Table 9 Fuzzy GINI impurities for the second split

Attribute	Split point	GINI
A1	46.5	0.000
A2	62	0.000
A3	1.5	0.185
A3	5.5	0.281
A3	10.5	0.341
A3	35	0.253

Table 10 A training set from Yuan and Shaw [46]

#	Outlook			Temperature			Humidity		Wind		Plan		
	Sunny	Cloudy	Rain	Hot	Mild	Cool	Humid	Normal	Windy	not_windy	Volleyball	Swimming	w_lifting
1	0.9	0.1	0	1	0	0	0.8	0.2	0.4	0.6	0	0.8	0.2
2	0.8	0.2	0	0.6	0.4	0	0	1	0	1	1	0.7	0
3	0	0.7	0.3	0.8	0.2	0	0.1	0.9	0.2	0.8	0.3	0.6	0.1
4	0.2	0.7	0.1	0.3	0.7	0	0.2	0.8	0.3	0.7	0.9	0.1	0
5	0	0.1	0.9	0.7	0.3	0	0.5	0.5	0.5	0.5	0	0	1
6	0	0.7	0.3	0	0.3	0.7	0.7	0.3	0.4	0.6	0.2	0	0.8
7	0	0.3	0.7	0	0	1	0	1	0.1	0.9	0	0	1
8	0	1	0	0	0.2	0.8	0.2	0.8	0	1	0.7	0	0.3
9	1	0	0	1	0	0	0.6	0.4	0.7	0.3	0.2	0.8	0
10	0.9	0.1	0	0	0.3	0.7	0	1	0.9	0.1	0	0.3	0.7
11	0.7	0.3	0	1	0	0	1	0	0.2	0.8	0.4	0.7	0
12	0.2	0.6	0.2	0	1	0	0.3	0.7	0.3	0.7	0.7	0.2	0.1
13	0.9	0.1	0	0.2	0.8	0	0.1	0.9	1	0	0	0	1
14	0	0.9	0.1	0	0.9	0.1	0.1	0.9	0.7	0.3	0	0	1
15	0	0	1	0	0	1	1	0	0.8	0.2	0	0	1
16	1	0	0	0.5	0.5	0	0	1	0	1	0.8	0.6	0

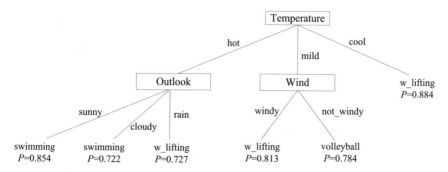

Fig. 9 Decision tree obtained by fuzzy ID3 algorithm

Sect. 3.2. Rules and classification results are the same with the ones obtained by Yuan and Shaw [46] where ambiguity measure was used instead of entropy.

Since the number of outcomes is three in Yuan and Shaw's example, fuzzy SLIQ algorithm is not able to tell which sport to play. However, based on the attribute membership values, it can decide if a person can play volleyball or not for different membership values of weather conditions. For generating such a tree, two modifications should be made on the data: The membership values of weather conditions can be assumed as a numerical value representing the related weather condition The outcome of this example is numerical, it should be converted to binary outcomes. In order to generate binary outcomes, it is assumed that if the membership value of the outcome is greater than 0.5, then the certain sport is played. Likely, if the membership value is less than 0.5, the sport is not played.

For volleyball, the attributes, split points and the fuzzy GINI impurity is determined as given in Table 11.

The result for *windy* and *not_windy* weather at splits 0.05 and 0.95 have the same GINI value. This result is expected since *windy* and *not_windy* are complementary attributes whose membership values sum to 1. One attribute can be chosen for branching. Assume that the first branching is made through weather being *windy* with a membership value that is less than 0.05. Branching results in 3 objects in "$\mu_{windy} < 0.05$" node(all having a membership value of 0) and 13 in the other node. The node "$\mu_{windy} < 0.05$" has a GINI value of 0 (all objects suggest that the person plays volleyball). Hence, this node is terminated as a leaf. The other node ($\mu_{windy} > 0.05$), has a GINI value of 0.26. Further split provides Table 12.

The minimum fuzzy GINI impurity is provided by the weather being *mild* with a membership value of 0.5. Hence, two branches are "$\mu_{mild} < 0.50$" and "$\mu_{mild} > 0.50$". If the membership value of the weather being *mild* is less than 0.5, then the decision is not to play volleyball for all object, hence this node is also terminated as a leaf. The other branch, however, has a GINI of 0.5 and can be further branched. The values for third split are given in Table 13.

Table 13 provides four alternatives with GINI impurities of 0. The decision can depend on the weather being *humid*, *normal*, *windy* or *not_windy*. Since *humid* and

Table 11 Fuzzy GINI impurities for Yuan and Shaw's first split

Attribute	Split point	GINI
Sunny	0.10	0.3661
Sunny	0.45	0.4264
Sunny	0.85	0.4182
Cloudy	0.05	0.4089
Cloudy	0.15	0.3942
Cloudy	0.25	0.4194
Cloudy	0.50	0.3875
Cloudy	0.65	0.4281
Cloudy	0.80	0.4276
Cloudy	0.95	0.3708
Rain	0.15	0.4281
Hot	0.25	0.4294
Hot	0.65	0.3229
Mild	0.25	0.3317
Mild	0.35	0.3801
Mild	0.75	0.4425
Mild	0.95	0.4066
Cool	0.75	0.4120
Cool	0.90	0.3693
Humid	0.05	0.4054
Humid	0.25	0.4214
Humid	0.40	0.3949
Humid	0.95	0.3795
Normal	0.60	0.3303
Normal	0.85	0.4386
Normal	0.95	0.4097
Windy	0.05	0.2356
Windy	0.25	0.4049
Windy	0.35	0.2411
Not_windy	0.65	0.2411
Not_windy	0.75	0.4049
Not_windy	0.95	0.2356

normal are complementary attributes just like *windy* and *not_windy*, the number of alternative trees can be decreased to two. In case of $\mu_{windy} > 0.05$ and $\mu_{mild} > 0.50$, if $\mu_{humid} < 0.15$, then the person is known not to play volleyball and if $\mu_{humid} > 0.15$, the person is known to play volleyball. Two decision trees are given in Figs. 10 and 11.

The same process applied to *swimming* and *w_lifting* yields to the trees shown in Figs. 12, 13 and 14. For the *w_lifting* decision, two alternative trees exist. The classes for each object obtained by fuzzy ID3 and fuzzy SLIQ algorithms are shown

Table 12 Second split to Fuzzy SLIQ for Yuan and Shaw

Attribute	Split point	GINI
Sunny	0.10	0.2009
Sunny	0.45	0.2326
Cloudy	0.50	0.1816
Rain	0.15	0.2907
Hot	0.50	0.2308
Mild	0.50	0.1538
Mild	0.90	0.1941
Cool	0.05	0.2626
Humid	0.15	0.2402
Humid	0.40	0.2272
Normal	0.60	0.2272
Normal	0.80	0.2402
Windy	0.25	0.2586
Windy	0.35	0.1918
Not_windy	0.65	0.1918
Not_windy	0.75	0.2586

Table 13 Third split to Fuzzy SLIQ for Yuan and Shaw

Attribute	Split point	GINI
Sunny	0.10	0.2480
Sunny	0.55	0.3163
Cloudy	0.35	0.3136
Cloudy	0.80	0.2639
Rain	0.05	0.3333
Rain	0.15	0.3333
Hot	0.10	0.5000
Hot	0.25	0.3333
Mild	0.75	0.3333
Mild	0.95	0.3333
Cool	0.05	0.3333
Humid	0.15	0.0000
Normal	0.85	0.0000
Windy	0.50	0.0000
Not_windy	0.50	0.0000

in Table 14. According to the results in Table 14, fuzzy SLIQ algorithm obtains better classification than fuzzy ID3 algorithm for sport decision problem.

For a detailed comparison of decision tree algorithms, we refer readers to the study of Niuniu and Yuxun [29]. This comparison is achieved for crisp cases; yet, the results obtain also holds for fuzzy decision trees. With a brief summary, it could be stated that the "No Free Lunch" theorem is valid for fuzzy decision tree cases.

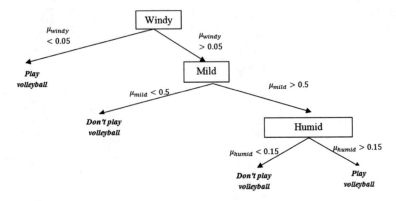

Fig. 10 Decision tree 1 obtained by fuzzy SLIQ algorithm for volleyball

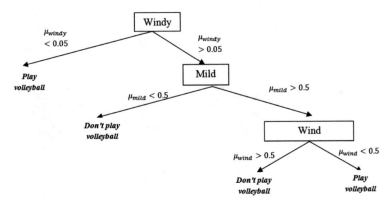

Fig. 11 Decision tree 2 obtained by fuzzy SLIQ algorithm for volleyball

Fig. 12 Decision tree obtained by fuzzy SLIQ algorithm for swimming

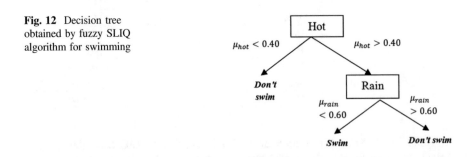

All algorithms have their advantages and disadvantages. The fuzzy ID3 algorithm is easy to apply to and has a strong learning capacity; yet, it is sensitive to noise, it cannot handle branching on multiple attributes at once and cannot handle continuous attribute values. It is also prone to attribute bias problem, that is, the algorithm

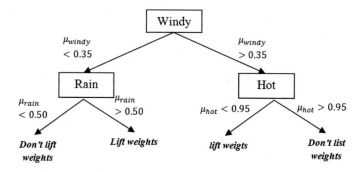

Fig. 13 Decision tree 1 obtained by fuzzy SLIQ algorithm for weight lifting

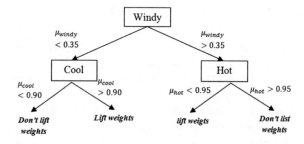

Fig. 14 Decision tree 2 obtained by fuzzy SLIQ algorithm for weight lifting

Table 14 Results of fuzzy ID3 algorithm for sport example

#	Real classification	Results of fuzzy ID3	Results of fuzzy SLIQ
1	swimming	swimming	swimming
2	volleyball	swimming	volleyball-swimming
3	swimming	swimming	swimming
4	volleyball	volleyball	volleyball
5	w_lifting	w_lifting	w_lifting
6	w_lifting	w_lifting	w_lifting
7	w_lifting	w_lifting	w_lifting
8	volleyball	w_lifting	volleyball
9	swimming	swimming	swimming
10	w_lifting	w_lifting	w_lifting
11	swimming	swimming	swimming
12	volleyball	volleyball	volleyball
13	w_lifting	w_lifting	w_lifting
14	w_lifting	w_lifting	w_lifting
15	w_lifting	w_lifting	w_lifting
16	volleyball	volleyball-swimming	volleyball-swimming

favors attributes with have a larger attribute value set, but is insignificant than others. The fuzzy C4.5 algorithm attempts to overcome this issue but using information gain and can handle continuous attribute values. Both algorithms do not consider multicollinearity, that is, the correlation among attributes. The efficiency of the fuzzy C4.5 is limited, since it conducts a linear search algorithm for determining threshold values and it cannot handle large datasets. Fuzzy CART algorithm can handle nonlinear data with noise with a higher accuracy; but for complex datasets with many attributes, the accuracy is reduced. The Fuzzy SLIQ algorithm can handle defaults attribute values which fuzzy ID3 cannot, and provides a higher implementation speed. However, for the sake of implementation speed needs a large memory to be allocated.

5   A Summary of Recent Classification Approaches

Given the most fundamental algorithms for fuzzy decision tress, some recent approaches are summarized in this section. One of the most recent approaches is developed by Tusor et al. [41] and utilizes greedy inference and complete inference mechanisms for the split decision.

The state-of-the art on fuzzy decision trees is the intuitionistic fuzzy decision trees; however, they stem their roots on the ID3 Algorithm [5]. Intuitionistic fuzzy sets offer an approach that considers a nonmembership value or function besides the conventional membership function. The difference between membership and non-membership values are defined as hesitation margin. The ID3 algorithm is applied for both membership and nonmembership values and the split is considered on the degree of a data point belonging to one class and not belonging to the others in a way that maximizes the fuzzy entropy reduction.

The aforementioned algorithms, especially ID3 algorithm, is with the advances of in Machine Learning algorithms, there is also a new branch for fuzzy decision trees where they are hybridized and tuned with Neural Network. This approach has its roots since early 2000s [30]. Yet, the latest approaches involves Neural Networks with more complicated structures [39].

6   Conclusion

The main points of this chapter are summarized as follows:

- Decision trees are elemental tools for classification problems. Various algorithms have been developed for sustaining the accuracy of a tree while avoiding abundance of branches.
- ID3 algorithm and its variants are considered as the most basic algorithms for constructing decision trees. However, ID algorithms are greatly capable of

processing categorical values, do not produce numerical outcomes and prone to overfit to the data.

- Numerous impurity measures that measure the information homogeneity are available. Most used ones are entropy, GINI impurity and misclassification error.
- Another method for overcoming the overfitting challenge, pruning trees is the most common approach. In order to prune trees, various methods have been proposed. Cost complexity and Error complexity methods are known to be widely used methods for pruning.
- Pruning methods are involved in algorithms such as C4.5, CART or SLIQ. In order to produce numerical outcomes as classes, regression trees are offered.
- In order to improve classification accuracy, trees are combined with cognitive uncertainties, leading to Fuzzy Decision Trees.
- As in the case of crisp decision trees, fuzzified versions of ID3 algorithms are widely exploited. Fuzzy ID3 algorithms offer membership values for inputs and outputs and attempt to explore the effects of vagueness.
- Fuzzy decision tree algorithms generally generate rules involving a *truth degree* related with each rule which indicates the level of generalizability of the related rule.
- In terms of impurity indices, their fuzzy versions use memberships instead of cardinalities.
- Fuzzy SLIQ algorithm use linguistic variables and attempt to find the optimal split value for numerical attributes.

Classification algorithms may have different performance depending on the input data. The relation between the methodology performance and input data can be investigated in further studies. Moreover, machine learning algorithms have become very important in classification literature. The methodologies mentioned in this chapter can be hybridized with machine learning algorithms and the effect of dynamic learning on the algorithms can be detected.

References

1. Aluja-Banet, T., Nafría E.: Generalized impurity measures and data diagnostics in decision trees. In: Blasius, J., Greenacre, M. (eds.) Visualization of Categorical Data, pp. 59–69. Academic Press, San Diego (1998)
2. Aranibar, L.A.Q.: Learning Fuzzy Logic from Examples. Master's thesis
3. Bartczuk, Ł., Rutkowska, D.: A new version of the fuzzy-id3 algorithm. In: Rutkowski, L., Tadeusiewicz, R., Zadeh, L., Żurada, J. (eds.) Artificial Intelligence and Soft Computing ICAISC 2006, volume 4029 of Lecture Notes in Computer Science, pp. 1060–1070. Springer, Berlin, Heidelberg (2006)
4. Breiman, L., Friedman, J., Olshen, R., Stone, C.: Classification and Regression Trees. Wadsworth and Brooks, Monterey, CA (1984)
5. Bujnowski, P., Szmidt, E., Kacprzyk, J.: Intuitionistic fuzzy decision tree: a new classifier. In: Angelov, P., Atanassov, K.T., Doukovska, L., Hadjiski, M., Jotsov, V., Kacprzyk, J.,

Kasabov, N., Sotirov, S., Szmidt, E., Zadrony, S. (eds.) Intelligent Systems'2014, volume 322 of Advances in Intelligent Systems and Computing, pp. 779–790. Springer International Publishing (2015)

6. Chandra, B., Varghese, P.P.: Fuzzy sliq decision tree algorithm. IEEE Trans. Systems, Man Cybern. Part B Cybern. **38**(5), 1294–1301 (2008)

7. Chang, J.-Y., Cho, C.-W., Hsieh, S.-H., Chen, S.-T.: Genetic algorithm based fuzzy id3 algorithm. In: Pal, N., Kasabov, N., Mudi, R.K., Pal, S.,Parui, S.K. (eds.) Neural Information Processing, volume 3316 of Lecture Notes in Computer Science, pp. 989–995. Springer, Berlin, Heidelberg (2004)

8. Chang, R.L.P., Pavlidis, T.: Fuzzy decision tree algorithms. IEEE Trans. Syst. Man Cybern. **7** (1), 28–35 (1977)

9. Chiang, I-J., Hsu, J.Y.: Fuzzy classification trees for data analysis. Fuzzy Sets Syst. **130**(1), 87–99 (2002)

10. Chou, P.A.: Optimal partitioning for classification and regression trees. IEEE Trans. Pattern Anal. Mach. Intell. **13**(4), 340–354 (1991)

11. Duan, R., Zhao, W., Huang, S., Hao, K.: High-speed corner detection based on fuzzy id3 decision tree. J. Central South Univ. **19**(9), 2528–2533 (2012)

12. Fan, C.-Y., Chang, P.-C., Lin, J.-J., Hsieh, J.C.: A hybrid model combining case-based reasoning and fuzzy decision tree for medical data classification. Appl. Soft Comput. **11**(1), 632–644 (2011)

13. Fang, G., Kwok, N.M., Ha, Q.: Automatic fuzzy membership function tuning using the particle swarm optimization. In: Pacific-Asia Workshop on Computational Intelligence and Industrial Application, 2008. PACIIA '08, vol. 2, pp. 324–328 (2008)

14. Goodman, L.A., Kruskal, W.H.: Measures of association for cross classifications. J. Am. Stat. Assoc. **49**(268), 732–764 (1954)

15. Hayashi, I., Maeda, T., Bastian, A., Jain, L.C.: Generation of fuzzy decision trees by fuzzy id3 with adjusting mechanism of and/or operators. In: The 1998 IEEE International Conference on Fuzzy Systems Proceedings, 1998. IEEE World Congress on Computational Intelligence, volume 1, pp. 681–685, May (1998)

16. Janikow, C.Z.: Fuzzy decision trees: issues and methods. IEEE Trans. Syst. Man Cybern. Part B Cybern. **28**(1), 1–14 (1998)

17. Jin, C., Li, F., Li, Y.: A generalized fuzzy id3 algorithm using generalized information entropy. Knowl.-Based Syst. **64**, 13–21 (2014)

18. Kadir, M.K.A., Hines, E.L., Qaddoum, K., Collier, R., Dowler, E., Grant, W., Leeson, M., Iliescu, D., Subramanian, A., Richards, K., Merali, Y., Napier, R.: Food security risk level assessment: a fuzzy logic-based approach. Appl. Artif. Intell. **27**(1), 50–61 (2013)

19. Kotsiantis, S.B.: Supervised machine learning: a review of classification techniques. In: Proceedings of the 2007 Conference on Emerging Artificial Intelligence Applications in Computer Engineering: Real Word AI Systems with Applications in eHealth, HCI, Information Retrieval and Pervasive Technologies, pp. 3–24, Amsterdam, The Netherlands, IOS Press (2007)

20. Kotsiantis, S.B.: Decision trees: a recent overview. Artif. Intell. Rev. **39**(4), 261–283 (2013)

21. Leong, L.K.: Analyzing Big Data With Decision Trees. Master's thesis, San Jose State University, USA (2014)

22. Li, Y., Jiang, D., Li, F.: The application of generating fuzzy id3 algorithm in performance evaluation. Procedia Engineering **29**, 229–234 (2012)

23. Luger, G.F.: Artificial Intelligence: Structures and Strategies for Complex Problem Solving. Addison-Wesley (2001)

24. Mehta, M., Agrawal, R., Rissanen, J.: Sliq: A Fast Scalable Classifier for Data Mining. pp. 18–32 (1996)

25. Mistikoglu, G., Gerek, I.H.l., Erdis, E., Mumtaz Usmen, P.E., Cakan, H., Kazan, E.E.: Decision tree analysis of construction fall accidents involving roofers. Expert Syst. Appl. **42** (4), 2256–2263 (2015)

26. Mitchell, T.M.: Machine Learning. McGraw-Hill Science/Engineering/Math (1997)

27. Mitra, S., Konwar, K.M., Pal, S.K.: Fuzzy decision tree, linguistic rules and fuzzy knowledge-based network: generation and evaluation. IEEE Trans. Syst. Man Cybern. Part C Appl. Rev. **32**(4), 328–339 (2002)
28. Nguyen, H.X., Huynh, T.M., Le, B.: A unified design for the membership functions in genetic fuzzy systems. Int. J. Comput. Sci. Issues **9**(3), 7–16 (2012)
29. Niblett, T.: Constructing decision trees in noisy domains. In: Second European Working Session on Learning, pp. 67–78 (1986)
30. Niuniu, X., Yuxun, L.: Notice of retraction review of decision trees. In: 2010 3rd IEEE International Conference on Computer Science and Information Technology (ICCSIT), vol. 5, pp. 105–109 (2010)
31. Olaru, C., Wehenkel, L.: On neurofuzzy and fuzzy decision tree approaches. In: Bernadette, B.-M., Ronald, R.Y., Lotfi A.Z. (eds.) Information, Uncertainty and Fusion, volume 516 of The Springer International Series in Engineering and Computer Science, pp. 131–145. Springer, US (2000)
32. Pach, F.P., Nemeth, J.A.O., Arva, P.: Supervised clustering and fuzzy decision tree induction for the identification of compact classifiers. In: 5th International Symposium of Hungarian Researchers on Computational Intelligence (2004)
33. Podgorelec, V., Kokol, P., Stiglic, B., Rozman, I.: Decision trees: an overview and their use in medicine. J. Med. Syst. **26**(5), 445–463 (2002)
34. Quinlan, J.R.: Induction of decision trees. Mach. Learn. **1**(1), 81–106 (1986)
35. Quinlan, J.R.: Simplifying decision trees. Int. J. Man-Mach. Stud. **27**:221234 (1987)
36. Quinlan, J.R.: C4.5: Programs for Machine Learning. Morgan Kaufmann Publishers Inc., San Francisco, CA, USA (1993)
37. Rokach, L., Maimon, O.: Data Mining with Decision Trees: Theroy and Applications. World Scientific Publishing Co., Inc, River Edge, NJ, USA (2008)
38. Shimojima, K., Fukuda, T., Hasegawa, Y.: Self-tuning fuzzy modeling with adaptive membership function, rules, and hierarchical structure based on genetic algorithm. Fuzzy Sets Syst. **71**(3), 295–309 (1995)
39. Singh, M.: Dynamic successive feed-forward neural network for learning fuzzy decision tree. In: Sergei, O.K., Dominik, I., Daryl, H.H., Boris, G.M. (eds.) Rough Sets, Fuzzy Sets, Data Mining and Granular Computing, volume 6743 of Lecture Notes in Computer Science, pp. 293–301. Springer, Berlin, Heidelberg (2011)
40. Tan, P.N., Steinbachand, M., Kumar, V.: Introduction to Data Mining. Addison-Wesley (2005)
41. Tusor, B., Varkonyi-Koczy, A.R., Takacs, M., Toth, J.T.: A fast fuzzy decision tree for color filtering. In: 2015 IEEE 9th International Symposium on Intelligent Signal Processing (WISP), pp. 1–6 (2015)
42. Umano, M., Okamoto, H., Hatono, I., Tamura, H., Kawachi, F., Umedzu, S., Kinoshita, J.: Fuzzy decision trees by fuzzy id3 algorithm and its application to diagnosis systems. In: Proceedings of the Third IEEE Conference on Fuzzy Systems, 1994. IEEE World Congress on Computational Intelligence, vol. 3, pp. 2113–2118, Jun (1994)
43. Wang, X., Chen, B., Qian, G., Ye, F.: On the optimization of fuzzy decision trees. Fuzzy Sets Syst. **112**(1), 117–125 (2000)
44. Wang, X.-Z., Zhai, J.-H., Lu, S.-X.: Induction of multiple fuzzy decision trees based on rough set technique. Inf. Sci. **178**(16), 3188–3202 (2008)
45. Xiaohu, W., Lele, W., Nianfeng, L.: An application of decision tree based on id3. Phys. Proc. **25**, 1017–1021 (2012)
46. Yuan, Y., Shaw, M.J.: Induction of fuzzy decision trees. Fuzzy Sets Syst. **69**(2), 125–139 (1995)

Fuzzy Shewhart Control Charts

Cengiz Kahraman, Murat Gülbay and Eda Boltürk

Abstract Process Control is the active correction of a process based on the results of process monitoring. Once the process monitoring tools have detected an assignable cause, this cause is removed to bring the process back into control. This chapter presents the process control techniques under fuzziness. Variable and attribute control charts are extended to their fuzzy versions.

Keywords Variable control charts · Attribute control charts · Fuzzy sets · p-chart · np-chart · c-chart · u-chart · $\bar{X}$ and R control charts · $\bar{X}$ and S control charts

1 Introduction

A process may either be classified as "in control" or "out of control". Based on the statistical methods, analytical decision-making tools which allow practitioners to measure, monitor, and control the process behavior are called *Statistical Process Control* (*SPC*). The most successful SPC tool is control charts, originally developed by Walter Shewhart in the early 1920s. Comparing with boundaries of a stable process with a graphical display, they enable online data tracing and abnormal conditions warning, which are an essential tool for continuous quality control. Basically, the control charts are the graphical display of a quality characteristic that has been measured or computed from a sample versus the sample number or time to monitor and show how the process is performing and how the capabilities are affected by changes to the process. This information is then used to make quality

C. Kahraman (✉) · E. Boltürk
Industrial Engineering Department, İstanbul Technical University,
İstanbul, Turkey
e-mail: kahramanc@itu.edu.tr

M. Gülbay
Industrial Engineering Department, Gaziantep University, Gaziantep, Turkey

© Springer International Publishing Switzerland 2016
C. Kahraman and Ö. Kabak (eds.), *Fuzzy Statistical Decision-Making*,
Studies in Fuzziness and Soft Computing 343,
DOI 10.1007/978-3-319-39014-7_14

improvements. The control charts attempt to distinguish between two types of process variation that impede peak performance. These variations are as common cause variations (random causes), which are intrinsic to the process and will always be present and special cause variations (assignable causes), which stem from external sources [1].

Based on the monitored quality characteristics in numerical or in "conforming" or "nonconforming" measurements, the control charts are categorized into two main groups, variables and attributes. This chapter deals with $\bar{X}$ and R, $\bar{X}$ and S control charts for variables and p, np, c, and u control charts for attributes. R is the range between x_{max} and x_{min} in a sample whereas S is the standard deviation of sample data [1].

The formulas for constructing the control limits on the $\bar{X}$ and R charts are tabulated in Table 1. These formulas need using the past data.

Where the constants A_2, D_3, and D_4 are the coefficients depending on the sample (observation) size and are tabulated for various sample sizes in Appendix.

When the sample size is variable and relatively large, say $n > 10$, the usage of $\bar{X}$ and S charts is advantageous. When σ is unknown, we can calculate the control limits of $\bar{X}$ and S control charts as given in the equations in Table 2.

where

$$s = \sqrt{\frac{\sum_{i=1}^{n}(x_i - \bar{x})^2}{n - 1}} \tag{1}$$

$$\bar{s} = \frac{s_1 + s_2 + \cdots + s_n}{n} \tag{2}$$

The coefficients A_3, B_3, and B_4 depend on the sample (observation) size and are tabulated for various sample sizes in Appendix.

Table 1 Control limits for $\bar{X}$ and R charts

	$\bar{X}$ chart	R chart
Center Line (CL)	$\bar{\bar{x}}$	$\bar{R}$
Lower Control Limit (LCL)	$\bar{\bar{x}} - A_2\bar{R}$	$D_3\bar{R}$
Upper Control Limit (UCL)	$\bar{\bar{x}} + A_2\bar{R}$	$D_4\bar{R}$

Table 2 Control limits for $\bar{X}$ and R charts

	$\bar{X}$ chart	S chart
CL	$\bar{\bar{x}}$	$\bar{s}$
LCL	$\bar{\bar{x}} - A_3\bar{s}$	$B_3\bar{s}$
UCL	$\bar{\bar{x}} + A_3\bar{s}$	$B_4\bar{s}$

2 Fuzzy Control Charts for Variables

For many problems, control limits could not be so precise. Uncertainty comes from the measurement system including operators and gauges, and environmental conditions [2]. A research work incorporating uncertainty into decision analysis is basically done through the probability theory and/or the fuzzy set theory. The former represents the stochastic nature of decision analysis while the latter captures the subjectivity of human behavior. A rational approach toward decision-making should take human subjectivity into account, rather than employing only objective probability measures. The fuzzy set theory is a perfect means for modeling uncertainty (or imprecision) arising from mental phenomena which is neither random nor stochastic. When human subjectivity plays an important role in defining the quality characteristics, the classical control charts may not be applicable since they require sharp information. The judgments in classical process control are either "in-control" or "out-of-control" while fuzzy control charts may yield several intermediate decisions. Fuzzy control charts are inevitable to use when the statistical data in consideration are imprecise and vague; or available information about the process is incomplete or includes human subjectivity [3]. In the fuzzy case, each sample, or subgroup, may be represented by a trapezoidal fuzzy number (a,b,c,d) or a triangular fuzzy number (a, b, b, d) or (a, c, c, d) with an α-cut (if necessary) as shown in Fig. 1.

2.1 Fuzzy $\bar{X}$ and R Control Charts

Let quality characteristic of a sample with a size of n be represented as fuzzy triangular numbers by $\tilde{X}_i(X_{ija}, X_{ijb}, X_{ijc},) \; i = 1, 2, \ldots, m; \; j = 1, 2, \ldots, n$. Using the

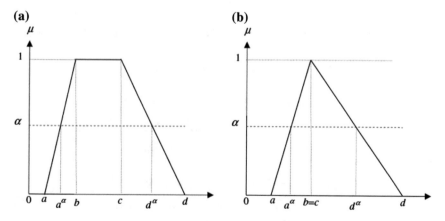

Fig. 1 Representation of a sample by trapezoidal and/or triangular fuzzy numbers: **a** Trapezoidal (a, b, c, d) and **b** triangular (a, b, b, d)

fuzzy arithmetic the mean of the each subgroup and grand average of the samples can be calculated by Eqs. (3) and (4).

$$\tilde{\bar{X}}_i = \left(\frac{\sum_{j=1}^n x_{ija}}{n}, \frac{\sum_{j=1}^n x_{ijb}}{n}, \frac{\sum_{j=1}^n x_{ijc}}{n} \right) = (\bar{x}_{ia}, \bar{x}_{ib}, \bar{x}_{ic}) \tag{3}$$

$$i = 1, 2, \ldots, m; \ j = 1, 2, \ldots, n$$

$$\tilde{\bar{X}} = \left(\frac{\sum_{i=1}^m x_{ia}}{m}, \frac{\sum_{i=1}^m x_{ib}}{m}, \frac{\sum_{i=1}^m x_{ic}}{m} \right) = (\bar{\bar{x}}_a, \bar{\bar{x}}_b, \bar{\bar{x}}_c) \tag{4}$$

The fuzzy range of each subgroup can be represented by the equation below.

$$\tilde{R}_i = \tilde{x}_{ij,max} - \tilde{x}_{ij,min} \quad i = 1, 2, \ldots, m; \ j = 1, 2, \ldots, n \tag{5}$$

Since fuzzy numbers cannot be easily compared to each other, we need a ranking method for fuzzy numbers in Eq. (17). There are many ranking methods for fuzzy numbers. An attempt to list most of the ranking methods was made in [4]. Once the maximum and minimum fuzzy observation is decided, the fuzzy range can be determined by the following equations.

$$\tilde{R}_i = \tilde{x}_{ij,max} - \tilde{x}_{ij,min} = (x_{ija}, x_{ijb}, x_{ijc})_{max} - (x_{ija}, x_{ijb}, x_{ijc})_{min} \tag{6}$$

$$\tilde{R}_i = (\tilde{x}_{ija,max} - \tilde{x}_{ijc,min}, \tilde{x}_{ijb,max} - \tilde{x}_{ijb,min}, \tilde{x}_{ijc,max} - \tilde{x}_{ija,min}) = (R_{ia}, R_{ib}, R_{ic}) \tag{7}$$

After calculating range of each subgroup, the fuzzy mean of the ranges can be defined as:

$$\tilde{\bar{R}} = \left(\frac{\sum_{i=1}^m R_{ia}}{m}, \frac{\sum_{i=1}^m R_{ib}}{m}, \frac{\sum_{i=1}^m R_{ic}}{m} \right) = (\bar{R}_a, \bar{R}_b, \bar{R}_c) \tag{8}$$

Control limits for the fuzzy $\tilde{\bar{X}}$ control charts are then formulized as follows:

$$\widetilde{CL} = \tilde{\bar{x}} = (\bar{\bar{x}}_a, \bar{\bar{x}}_b, \bar{\bar{x}}_c) = (CL_1, CL_2, CL_3) \tag{9}$$

$$\begin{aligned} \widetilde{UCL} = \tilde{\bar{x}} + A_2 \tilde{\bar{R}} &= (\bar{\bar{x}}_a, \bar{\bar{x}}_b, \bar{\bar{x}}_c) + A_2 (\bar{R}_a, \bar{R}_b, \bar{R}_c) \\ &= (\bar{\bar{x}}_a + A_2 \bar{R}_a, \bar{\bar{x}}_b + A_2 \bar{R}_b, \bar{\bar{x}}_c + A_2 \bar{R}_c) \\ &= (UCL_1, UCL_2, UCL_3) \end{aligned} \tag{10}$$

$$\begin{aligned} \widetilde{LCL} = \tilde{\bar{x}} - A_2 \tilde{\bar{R}} &= (\bar{\bar{x}}_a, \bar{\bar{x}}_b, \bar{\bar{x}}_c) - A_2 (\bar{R}_a, \bar{R}_b, \bar{R}_c) \\ &= (\bar{\bar{x}}_a - A_2 \bar{R}_c, \bar{\bar{x}}_b - A_2 \bar{R}_b, \bar{\bar{x}}_c - A_2 \bar{R}_a) \\ &= (LCL_1, LCL_2, LCL_3) \end{aligned} \tag{11}$$

Fuzzy control limits for the R charts can be derived in the same way.

$$\widetilde{CL} = \bar{\tilde{R}} = (\bar{R}_a, \bar{R}_b, \bar{R}_c) = (CL_1, CL_2, CL_3) \tag{12}$$

$$\begin{aligned}\widetilde{UCL} = D_4 \bar{\tilde{R}} &= D_4(\bar{R}_a, \bar{R}_b, \bar{R}_c) = (D_4\bar{R}_a, D_4\bar{R}_b, D_4\bar{R}_c) \\ &= (UCL_1, UCL_2, UCL_3)\end{aligned} \tag{13}$$

$$\widetilde{LCL} = D_3 \bar{\tilde{R}} = (D_3\bar{R}_a, D_3\bar{R}_b, D_3\bar{R}_c) = (LCL_1, LCL_2, LCL_3) \tag{14}$$

Example 1 From a production, the following 20 random samples whose sizes are 3 units are taken randomly. $\bar{\bar{X}}$ and R charts will be constructed. Table 3 gives the obtained data.

From Table 3, the center line is calculated as $\widetilde{CL} = \bar{\bar{\tilde{X}}} = (8.113, 8.540, 8.967)$ $n = 3$ and $A_2 = 1.023$. The UCL for $\bar{X}$ chart: $\widetilde{LCL} = (6.974, 7.834, 9.598)$.

The LCL for $\bar{X}$ chart: $\widetilde{LCL} = (7.482, 9.246, 10.106)$. Since all the sample means are between $\widetilde{LCL}$ and $\widetilde{UCL}$, the process is under control.

Table 3 Samples taken from the production

Sample no	1			2			3		
1	8.242	8.676	9.110	8.423	8.866	9.309	7.985	8.405	8.825
2	7.925	8.342	8.759	7.957	8.376	8.795	8.225	8.658	9.091
3	8.242	8.676	9.110	8.288	8.724	9.160	7.928	8.345	8.762
4	8.030	8.453	8.876	8.049	8.473	8.897	8.407	8.849	9.291
5	8.536	8.985	9.434	8.221	8.654	9.087	8.144	8.573	9.002
6	7.826	8.238	8.650	8.503	8.951	9.399	8.029	8.452	8.875
7	8.028	8.451	8.874	8.034	8.457	8.880	8.402	8.844	9.286
8	8.072	8.497	8.922	8.200	8.632	9.064	7.823	8.235	8.647
9	7.835	8.247	8.659	8.286	8.722	9.158	7.660	8.063	8.466
10	7.812	8.223	8.634	8.527	8.976	9.425	8.248	8.682	9.116
11	8.346	8.785	9.224	7.662	8.065	8.468	8.495	8.942	9.389
12	8.009	8.431	8.853	8.458	8.903	9.348	8.049	8.473	8.897
13	8.255	8.689	9.123	7.630	8.032	8.434	8.153	8.582	9.011
14	8.095	8.521	8.947	8.178	8.608	9.038	7.875	8.289	8.703
15	8.324	8.762	9.200	8.155	8.584	9.013	8.036	8.459	8.882
16	7.780	8.189	8.598	8.144	8.573	9.002	7.794	8.204	8.614
17	8.034	8.457	8.880	7.605	8.005	8.405	7.837	8.249	8.661
18	8.390	8.832	9.274	8.175	8.605	9.035	8.515	8.963	9.411
19	8.197	8.628	9.059	8.030	8.453	8.876	8.065	8.489	8.913
20	8.093	8.519	8.945	8.183	8.614	9.045	8.320	8.758	9.196

2.2 Fuzzy $\bar{X}$ and S Control Charts

Determination of the control limits for paired $\bar{X}$ and S charts are based on the standard deviation as mentioned in Introduction section. Hence, average standard deviation of the sample standard deviations need to be calculated. Let a quality characteristic of a sample of size n be represented by a triangular fuzzy number by $\tilde{X}_i\left(X_{ija}, X_{ijb}, X_{ijc}, \right) i = 1, 2, \ldots, m; j = 1, 2, \ldots, n$. Using the fuzzy arithmetic, the fuzzy standard deviation of each subgroup and the average of these standard deviations can be derived by Eqs. (15) and (16).

$$\tilde{\tilde{s}}_i = \sqrt{\frac{\sum_{j=1}^{n}\left(\tilde{x}_{ij} - \tilde{\tilde{x}}_i\right)^2}{n-1}} = \sqrt{\frac{\sum_{j=1}^{n}\left[\left(x_{ija}, x_{ijb}, x_{ijc}\right) - \left(\bar{x}_{ia}, \bar{x}_{ib}, \bar{x}_{ic}\right)\right]^2}{n-1}} \tag{15}$$
$$= \left(s_{ia}, s_{ib}, s_{ic}\right)$$

$$\tilde{\tilde{s}} = \frac{\sum_{i=1}^{m}\tilde{s}_i}{m} = \left(\frac{\sum_{i=1}^{m}s_{ia}}{m}, \frac{\sum_{i=1}^{m}s_{ib}}{m}, \frac{\sum_{i=1}^{m}s_{ic}}{m}\right) = \left(\bar{s}_a, \bar{s}_b, \bar{s}_c\right) \tag{16}$$

The control limits of fuzzy $\bar{X}$ control chart based on standard deviation are obtained as follows:

$$\widetilde{CL} = \tilde{\tilde{x}} = \left(\bar{\bar{x}}_a, \bar{\bar{x}}_b, \bar{\bar{x}}_c\right) = (CL_1, CL_2, CL_3) \tag{17}$$

$$\begin{aligned}
\widetilde{UCL} = \tilde{\tilde{x}} + A_3\,\tilde{\tilde{s}} &= \left(\bar{\bar{x}}_a, \bar{\bar{x}}_b, \bar{\bar{x}}_c\right) + A_3\left(\bar{s}_a, \bar{s}_b, \bar{s}_c\right) \\
&= \left(\bar{\bar{x}}_a + A_3\bar{s}_a, \bar{\bar{x}}_b + A_3\bar{s}_b, \bar{\bar{x}}_c + A_3\bar{s}_c\right) \\
&= (UCL_1, UCL_2, UCL_3)
\end{aligned} \tag{18}$$

$$\begin{aligned}
\widetilde{LCL} = \tilde{\tilde{x}} - A_3\,\tilde{\tilde{s}} &= \left(\bar{\bar{x}}_a, \bar{\bar{x}}_b, \bar{\bar{x}}_c\right) - A_3\left(\bar{s}_a, \bar{s}_b, s_c\right) \\
&= \left(\bar{\bar{x}}_a - A_3\bar{s}_c, \bar{\bar{x}}_b - A_3\bar{s}_b, \bar{\bar{x}}_c - A_3\bar{s}_a\right) \\
&= (LCL_1, LCL_2, LCL_3)
\end{aligned} \tag{19}$$

Similarly, the control limits of fuzzy s control chart are derived as follows:

$$\widetilde{CL} = \tilde{\tilde{s}} = \left(\bar{s}_a, \bar{s}_b, \bar{s}_c\right) = (CL_1, CL_2, CL_3) \tag{20}$$

$$\begin{aligned}
\widetilde{UCL} = B_4\,\tilde{\tilde{s}} &= B_4\left(\bar{s}_a, \bar{s}_b, \bar{s}_c\right) = \left(B_4\bar{s}_a, B_4\bar{s}_b, B_4\bar{s}_c\right) \\
&= (UCL_1, UCL_2, UCL_3)
\end{aligned}$$

$$\widetilde{LCL} = D_3\bar{s} = \left(B_3\bar{s}_a, B_3\bar{s}_b, B_3\bar{s}_c\right) = (LCL_1, LCL_2, LCL_3) \tag{21}$$

Example 2 Consider the data in Table 3. The sample standard deviiations are given in Table 4. The overall mean is $\bar{\bar{x}} = (8.113, 8.540, 8.967)$ and its defuzzified value is 8.540 when the defuzzification equation $(a + 2b + c)/4$ is used for a $\widetilde{TFN}$ = (a, b, c). The coefficient A_3 for n = 3 is 1.954. In this solution, the defuzzified values of samples are used in order to calculate their standard deviations.

From Table 4, the mean of standard deviations is obtained as 0.288. Then, $CL = (8.113, 8.540, 8.967)$

$$LCL = (8.113, 8.540, 8.967) - 1.954(0.288, 0.288, 0.288)$$
$$= (7.550, 7.977, 8.404)$$

$$UCL = (8.113, 8.540, 8.967) + 1.954(0.288, 0.288, 0.288)$$
$$= (8.676, 9.103, 9.530)$$

There is not enough evidence indicating that the process is certainly out of control. All the sample means are between $\widetilde{LCL}$ and $\widetilde{UCL}$ when defuzzified values are compared. However, it is possible that the process may be out of control since control limits and sample values have intersections. The detailed analyses can be found in [3, 5, 6].

Table 4 Defuzzified sample values

Sample no	Sample means	Defuzzified TFNs			Standard deviations
1	(8.217, 8.649, 9.081)	8.676	8.866	8.405	0.268
2	(8.036, 8.459, 8.882)	8.342	8.376	8.658	0.200
3	(8.153, 8.582, 9.011)	8.676	8.724	8.345	0.213
4	(8.162, 8.592, 9.021)	8.453	8.473	8.849	0.232
5	(8.300, 8.737, 9.174)	8.985	8.654	8.573	0.326
6	(8.119, 8.547, 8.975)	8.238	8.951	8.452	0.366
7	(8.155, 8.584, 9.013)	8.451	8.457	8.844	0.232
8	(8.032, 8.455, 8.878)	8.497	8.632	8.235	0.227
9	(7.927, 8.344, 8.761)	8.247	8.722	8.063	0.416
10	(8.196, 8.627, 9.058)	8.223	8.976	8.682	0.394
11	(8.168, 8.597, 9.027)	8.785	8.065	8.942	0.473
12	(8.172, 8.602, 9.033)	8.431	8.903	8.473	0.272
13	(8.013, 8.434, 8.856)	8.689	8.032	8.582	0.375
14	(8.049, 8.473, 8.896)	8.521	8.608	8.289	0.184
15	(8.172, 8.602, 9.032)	8.762	8.584	8.459	0.170
16	(7.906, 8.322, 8.738)	8.189	8.573	8.204	0.344
17	(7.825, 8.237, 8.649)	8.457	8.005	8.249	0.434
18	(8.360, 8.800, 9.240)	8.832	8.605	8.963	0.367
19	(8.097, 8.523, 8.949)	8.628	8.453	8.489	0.095
20	(8.199, 8.630, 9.062)	8.519	8.614	8.758	0.164

3   Fuzzy Approaches for Control Charts for Attributes

In this section, we give the fuzzy attribute charts for fraction nonconforming.

3.1   Fuzzy $\tilde{p}$ Control Chart

The fraction nonconforming is defined as the ratio of the number of nonconforming units in a population to the total number of units in that population. The units may have several quality characteristics that are examined simultaneously by the operator. If the unit does not conform to standard on one or more of these characteristics, the unit is classified as nonconforming [1].

The traditional p-control chart for known fraction nonconforming in the population would be as follows [1]:

$$UCL_p = p + 3\sqrt{\frac{p(1-p)}{n}} \tag{22}$$

$$CL_p = p \tag{23}$$

$$LCL_p = p - 3\sqrt{\frac{p(1-p)}{n}} \tag{24}$$

where; p is the fraction nonconforming in the population, n is the constant sample size.

If the fraction nonconforming of population is unknown, sample fraction nonconforming is used instead of it. The sample fraction nonconforming is defined as the ratio of the number of nonconforming units, that is:

$$p_j = \frac{d_j}{n} \tag{25}$$

$$\bar{p} = \frac{\sum_{j=1}^m d_j}{mn} = \frac{\sum_{j=1}^m p_j}{m} \tag{26}$$

where d_j: the number of nonconforming units in the jth sample, p_j: fraction nonconforming of jth sample, $\bar{p}$: the average of sample fractions nonconforming, m: the number of sample, $j = 1, 2, \ldots, m$.

The traditional p-control limits are computed from the average of sample fraction as [1]:

$$UCL_p = \bar{p} + 3\sqrt{\frac{\bar{p}(1-\bar{p})}{n}} \tag{27}$$

$$CL_p = \bar{p} \tag{28}$$

$$LCL_p = \bar{p} - 3\sqrt{\frac{\bar{p}(1-\bar{p})}{n}} \tag{29}$$

3.1.1 Fuzzy $\tilde{p}$-Control Chart Based on Constant Sample Size

In the fuzzy case, the number of nonconforming units is represented by the triangular fuzzy number $\left(d_{a_j}, d_{b_j}, d_{c_j}\right)$.

The fraction nonconforming is expressed by a triangular fuzzy number such as $\left(p_{a_j}, p_{b_j}, p_{c_j}\right)$. Here, $(\bar{p}_a, \bar{p}_b, \bar{p}_c)$ are the fuzzy averages of the fraction nonconforming, where $j = 1, 2, \ldots, m$:

$$p_{a_j} = \frac{d_{a_j}}{n} \tag{30}$$

$$p_{b_j} = \frac{d_{b_j}}{n} \tag{31}$$

$$p_{c_j} = \frac{d_{c_j}}{n} \tag{32}$$

$$\bar{p}_a = \frac{\sum p_{a_j}}{m} \tag{33}$$

$$\bar{p}_b = \frac{\sum p_{b_j}}{m} \tag{34}$$

$$\bar{p}_c = \frac{\sum p_{c_j}}{m} \tag{35}$$

Fuzzy center line, fuzzy upper and fuzzy lower limits of fuzzy $\tilde{p}$-control chart are obtained as follows:

$$U\tilde{C}L_p = \left(\bar{p}_a + 3\sqrt{\frac{\bar{p}_a(1-\bar{p}_a)}{n}}, \; \bar{p}_b + 3\sqrt{\frac{\bar{p}_b(1-\bar{p}_b)}{n}}, \; \bar{p}_c + 3\sqrt{\frac{\bar{p}_c(1-\bar{p}_c)}{n}} \right) \tag{36}$$

$$\tilde{CL}_p = (\bar{p}_a,\ \bar{p}_b,\ \bar{p}_c) \tag{37}$$

$$\tilde{LCL}_p = \left(\bar{p}_a - 3\sqrt{\frac{\bar{p}_c(1-\bar{p}_c)}{n}},\ \bar{p}_b - 3\sqrt{\frac{\bar{p}_b(1-\bar{p}_b)}{n}},\ \bar{p}_c - 3\sqrt{\frac{\bar{p}_a(1-\bar{p}_a)}{n}}\right) \tag{38}$$

3.1.2 α-Cut Fuzzy $\tilde{p}$-Control Chart Based on Constant Sample Size

The mean of α-cut is a set which includes all elements whose membership degrees are greater than equal to α. With α-cuts, the values of $\bar{p}_l^\alpha$ and $\bar{p}_r^\alpha$ are determined as follows:

$$\bar{p}_l^\alpha = \bar{p}_a + \alpha(\bar{p}_b - \bar{p}_a) \tag{39}$$

$$\bar{p}_r^\alpha = \bar{p}_c - \alpha(\bar{p}_c - \bar{p}_b) \tag{40}$$

α-cut fuzzy $\tilde{p}$-control chart is obtained by the following equations:

$$U\tilde{CL}_p^\alpha = \left(\bar{p}_l^\alpha + 3\sqrt{\frac{\bar{p}_l^\alpha(1-\bar{p}_l^\alpha)}{n}},\ \bar{p}_r + 3\sqrt{\frac{\bar{p}_r^\alpha(1-\bar{p}_r^\alpha)}{n}}\right) \tag{41}$$

$$L\tilde{CL}_p^\alpha = \left(\bar{p}_l^\alpha - 3\sqrt{\frac{\bar{p}_l^\alpha(1-\bar{p}_l^\alpha)}{n}},\ \bar{p}_r - 3\sqrt{\frac{\bar{p}_r^\alpha(1-\bar{p}_r^\alpha)}{n}}\right) \tag{42}$$

3.1.3 Fuzzy $\tilde{p}$-Control Chart Based on Variable Sample Size

The control limits of fuzzy $\tilde{p}$-control chart are calculated for each n_j by using triangular membership functions and fuzzy averages of sample fraction noncon-forming as follows:

$$\tilde{CL}_{p,j} = (\bar{p}_a,\ \bar{p}_b,\ \bar{p}_c) \tag{43}$$

$$U\tilde{CL}_{p,j} = \left(\bar{p}_a + 3\sqrt{\frac{\bar{p}_a(1-\bar{p}_a)}{n_j}},\ \bar{p}_b + 3\sqrt{\frac{\bar{p}_b(1-\bar{p}_b)}{n_j}},\ \bar{p}_c + 3\sqrt{\frac{\bar{p}_c(1-\bar{p}_c)}{n_j}}\right) \tag{44}$$

$$L\tilde{CL}_{p,j} = \left(\bar{p}_a - 3\sqrt{\frac{\bar{p}_a(1-\bar{p}_a)}{n_j}},\ \bar{p}_b - 3\sqrt{\frac{\bar{p}_b(1-\bar{p}_b)}{n_j}},\ \bar{p}_c - 3\sqrt{\frac{\bar{p}_c(1-\bar{p}_c)}{n_j}}\right) \tag{45}$$

3.2 Fuzzy $n\tilde{p}$ Control Chart

While p-control chart is related to the fraction of nonconforming, np-control chart is more convenient to deal with *the number of nonconforming units*. In many situations, observation of the number of nonconforming units is easier to interpret than the usual fraction nonconforming control chart [1].

In the conventional np-control chart for a known number of nonconforming units in the population is as follows [1]:

$$UCL_{np} = np + 3\sqrt{np(1-p)} \tag{46}$$

$$CL_{np} = np \tag{47}$$

$$LCL_{np} = np - 3\sqrt{np(1-p)} \tag{48}$$

where np is the number of nonconforming units in the population, n is a constant sample size.

If the number of nonconforming units in the population is unknown, then the average of the sample number of nonconforming units, $n\bar{p}$, is used. The number of nonconforming units in the jth sample is expressed as d_j, that is;

$$n\bar{p} = \frac{\sum_{j=1}^{m} d_j}{m} \tag{49}$$

The limits of the traditional np-control chart are given as follows [1]:

$$UCL_{np} = n\bar{p} + 3\sqrt{n\bar{p}(1-\bar{p})} \tag{50}$$

$$CL_{np} = n\bar{p} \tag{51}$$

$$LCL_{np} = n\bar{p} - 3\sqrt{n\bar{p}(1-\bar{p})} \tag{52}$$

In the fuzzy case, the number of nonconforming units for each sample is stated by a triangular fuzzy number $(d_{a_j}, d_{b_j}, d_{c_j})$. The average sample number of nonconforming units is expressed by a triangular fuzzy number $(n\bar{p}_a, n\bar{p}_b, n\bar{p}_c)$ as follows:

$$n\bar{p}_a = \frac{\sum_{j=1}^{m} d_{a_j}}{m} \tag{53}$$

$$n\bar{p}_b = \frac{\sum_{j=1}^{m} d_{b_j}}{m} \tag{54}$$

$$np_c = \frac{\sum_{j=1}^{m} d_{c_j}}{m} \tag{55}$$

The limits of fuzzy $n\tilde{p}$-control chart are calculated with the following equations;

$$U\tilde{C}L_{np} = \left(n\bar{p}_a + 3\sqrt{n\bar{p}_a(1-\bar{p}_a)},\ n\bar{p}_b + 3\sqrt{n\bar{p}_b(1-\bar{p}_b)},\ n\bar{p}_c + 3\sqrt{n\bar{p}_c(1-\bar{p}_c)}\right) \tag{56}$$

$$C\tilde{L}_{np} = (n\bar{p}_a,\ n\bar{p}_b,\ n\bar{p}_c) \tag{57}$$

$$L\tilde{C}L_{np} = \left(n\bar{p}_a - 3\sqrt{n\bar{p}_a(1-\bar{p}_a)},\ n\bar{p}_b - 3\sqrt{n\bar{p}_b(1-\bar{p}_b)},\ n\bar{p}_c - 3\sqrt{n\bar{p}_c(1-\bar{p}_c)}\right) \tag{58}$$

Example 3 200 products are randomly selected from a production and the defective units are recorded. The obtained results are as in Table 5.

From Table 5, CL = (0.009, 0.014, 0.019), LCL = (−0.011, −0.011, −0.010) → 0, and UCL = (0.029, 0.039, 0.048) is obtained. There is no sample mean completely above UCL or below LCL. The process might be out of control with some degree of possibility. The detailed analyses can be found in [3, 5, 6].

3.3 Fuzzy $\tilde{c}$ Control Chart

In the crisp case, control limits for number of nonconformities are calculated by

$$CL = \bar{c} \tag{59}$$

Table 5 Defective numbers in samples

Sample number	Defective number	TFN	Fuzzy sample $\tilde{p}$
1	Around 4	(3, 4, 5)	(0.015, 0.02, 0.025)
2	Around 3	(2, 3, 4)	(0.01, 0.015, 0.02)
3	Around 7	(6, 7, 8)	(0.03, 0.035, 0.04)
4	Around 3	(2, 3, 4)	(0.01, 0.015, 0.02)
5	Around 1	(0, 1, 2)	(0, 0.005, 0.01)
6	Around 1	(0, 1, 2)	(0, 0.005, 0.01)
7	Around 2	(1, 2, 3)	(0.005, 0.01, 0.015)
8	Around 4	(3, 4, 5)	(0.015, 0.02, 0.025)
9	Around 1	(0, 1, 2)	(0, 0.005, 0.01)
10	Around 2	(1, 2, 3)	(0.005, 0.01, 0.015)

$$LCL = \bar{c} - 3\sqrt{\bar{c}} \qquad (60)$$

$$UCL = \bar{c} + 3\sqrt{\bar{c}} \qquad (61)$$

where $\bar{c}$ is the mean of the nonconformities. In the fuzzy case, each sample, or subgroup, can be represented by a trapezoidal fuzzy number (a, b, c, d) or a triangular fuzzy number (a, b, b, d). Note that a trapezoidal fuzzy number becomes triangular when $b = c$. For the ease of representation and calculation, a triangular fuzzy number is also represented as a trapezoidal fuzzy number by (a, b, b, d) or (a, c, c, d). Center line, $\widehat{CL}$ is the mean of fuzzy samples, and it is represented by $(\bar{a}, \bar{b}, \bar{c}, \bar{d})$ where $\bar{a}, \bar{b}, \bar{c},$ and $\bar{d}$ are the arithmetic means of the values $a, b, c,$ and d, respectively. In the fuzzy case, it can be written as follows.

$$\widetilde{CL} = \left(\frac{\sum_{j=1}^{n} a_j}{n}, \frac{\sum_{j=1}^{n} b_j}{n}, \frac{\sum_{j=1}^{n} c_j}{n}, \frac{\sum_{j=1}^{n} d_j}{n} \right) = (\bar{a}, \bar{b}, \bar{c}, \bar{d}) \qquad (62)$$

$\widetilde{CL}$ can be represented by a fuzzy number whose fuzzy mode (multimodal) is the closed interval of $[\bar{b}, \bar{c}]$. $\widetilde{CL}, \widetilde{LCL},$ and $\widetilde{UCL}$ are calculated by [3, 5]:

$$\widetilde{CL} = (\bar{a}, \bar{b}, \bar{c}, \bar{d}) = (CL_1, CL_2, CL_3, CL_4) \qquad (63)$$

$$\widetilde{LCL} = \widetilde{CL} - 3\sqrt{\widetilde{CL}} = \left(\bar{a} - 3\sqrt{\bar{a}}, \bar{b} - 3\sqrt{\bar{b}}, \bar{c} - 3\sqrt{\bar{c}}, \bar{d} - 3\sqrt{\bar{d}} \right)$$
$$= (LCL_1, LCL_2, LCL_3, LCL_4) \qquad (64)$$

$$\widetilde{UCL} = \widetilde{CL} + 3\sqrt{\widetilde{CL}} = \left(\bar{a} + 3\sqrt{\bar{a}}, \bar{b} + 3\sqrt{\bar{b}}, \bar{c} + 3\sqrt{\bar{c}}, \bar{d} + 3\sqrt{\bar{d}} \right)$$
$$= (UCL_1, UCL_2, UCL_3, UCL_4) \qquad (65)$$

Example 4 A random sample of 15 products is taken from a production and the number of defects on each product is determined. The results are given in Table 6. We will calculate the fuzzy control limits for number of defects (nonconformities).
   Using Eqs. (63)–(65), we obtain the following control limits:
   CL = $(1.8, 2.867, 4.133, 4.867)$, LCL = 0, UCL = $(5.825, 7.946, 10.232, 11.485)$.

3.4 Fuzzy $\tilde{u}$ Control Chart

If we are related to the number of nonconformities on one product, c-control chart is used. When the sample size is not be constant due to the process constraints, u-control chart is preferred to monitor and evaluation of process. The classical u-control chart limits proposed by Shewhart are given the following equations:

Table 6 Number of defects for the example

Sample number	Number of defects
1	(2, 3, 4, 5)
2	(3, 4, 5, 6)
3	(1, 3, 5, 6)
4	(2, 3, 5, 6)
5	(1, 2, 3, 4)
6	(2, 4, 6, 6)
7	(1, 3, 5, 6)
8	(2, 3, 4, 5)
9	(3, 4, 5, 6)
10	(2, 3, 4, 4)
11	(1, 2, 4, 4)
12	(1, 1, 2, 3)
13	(2, 3, 4, 4)
14	(1, 2, 2, 3)
15	(3, 3, 4, 5)
Mean	(1.8, 2.867, 4.133, 4.867)

$$UCL_u = \bar{u} + 3\sqrt{\frac{\bar{u}}{n_j}} \tag{66}$$

$$CL_u = \bar{u} \tag{67}$$

$$LCL_u = \bar{u} - 3\sqrt{\frac{\bar{u}}{n_j}} \tag{68}$$

$$u_j = \frac{c_j}{n_j} \tag{69}$$

$$\bar{u} = \frac{\sum_{j=1}^m u_j}{m} \quad j = 1, 2, \ldots, m \tag{70}$$

where u_j is the number of nonconformities per inspection unit and $\bar{u}$ is the average number of nonconformities per inspection unit, n_j is the sample size, c_j is total nonconformities in a sample of n_j inspection units, and m is the number of sample.

In the fuzzycase, the number of nonconforming is expressed as a triangular fuzzy number $(u_{a_j}, u_{b_j}, u_{c_j})$. The fuzzy averages of nonconforming values are calculated by

$$\bar{u}_a = \frac{\sum u_{a_j}}{m} \tag{71}$$

$$\bar{u}_b = \frac{\sum u_{b_j}}{m} \tag{72}$$

$$\bar{u}_c = \frac{\sum u_{c_j}}{m} \tag{73}$$

The fuzzy $\tilde{u}$-control chart limits are given as follows:

$$U\tilde{C}L_u = \left(\bar{u}_a + 3\sqrt{\frac{\bar{u}_a}{n_j}}, \ \bar{u}_b + 3\sqrt{\frac{\bar{u}_b}{n_j}}, \ \bar{u}_c + 3\sqrt{\frac{\bar{u}_c}{n_j}} \right) \tag{74}$$

$$C\tilde{L}_u = (\bar{u}_a, \ \bar{u}_b, \ \bar{u}_c) \tag{75}$$

$$L\tilde{C}L_u = \left(\bar{u}_a - 3\sqrt{\frac{\bar{u}_c}{n_j}}, \ \bar{u}_b - 3\sqrt{\frac{\bar{u}_b}{n_j}}, \ \bar{u}_c - 3\sqrt{\frac{\bar{u}_a}{n_j}} \right) \tag{76}$$

3.4.1 α-Cut Fuzzy $\tilde{u}$-Control Chart

When α-cut is adapted to the fuzzy sets, the values of u_l^α and u_r^α are determined as follows:

$$\bar{u}_l^\alpha = \bar{u}_l + \alpha(\bar{u}_m - \bar{u}_l) \tag{77}$$

$$\bar{u}_r^\alpha = \bar{u}_r - \alpha(\bar{u}_r - \bar{u}_m) \tag{78}$$

α-cut fuzzy $\tilde{u}$-control chart is obtained by

$$U\tilde{C}L_u^\alpha = \left[\bar{u}_l^\alpha + 3\sqrt{\frac{\bar{u}_l^\alpha}{n_j}}, \ \bar{u}_r + 3\sqrt{\frac{\bar{u}_r^\alpha}{n_j}} \right] \tag{79}$$

$$C\tilde{L}_u^\alpha = \left[\bar{u}_l^\alpha, \ \bar{u}_r^\alpha \right] \tag{80}$$

$$L\tilde{C}L_u^\alpha = \left[\bar{u}_l^\alpha - 3\sqrt{\frac{\bar{u}_l^\alpha}{n_j}}, \ \bar{u}, \ \bar{u}_r - 3\sqrt{\frac{\bar{u}_r^\alpha}{n_j}} \right] \tag{81}$$

3.4.2 α-Level Fuzzy Median for α-Cut Fuzzy $\tilde{u}$-Control Chart

α-cut fuzzy $\tilde{u}$-control chart is transformed to crisp numbers via the fuzzy transformation techniques. α-level fuzzy midrange, fuzzy median, fuzzy average and fuzzy mode [7] are the transformation techniques. For a sample j, α-level fuzzy median value ($S^\alpha_{med-u,j}$) is calculated as follows:

$$S^\alpha_{med-u,j} = \frac{1}{3}\left[u^\alpha_{a,j}, u^\alpha_{c,j}\right] \tag{82}$$

By using these formulations, the fuzzy center line, fuzzy upper and fuzzy lower limits of α-level fuzzy median for α-cut fuzzy $\tilde{u}$-control chart is obtained by:

$$UCL^\alpha_{med-u} = CL^\alpha_{med-u} + 3\sqrt{\frac{CL^\alpha_{med-u}}{n_j}} \tag{83}$$

$$CL^\alpha_{med-u} = \frac{1}{3}\left[\bar{u}^\alpha_a, \bar{u}^\alpha_c\right] \tag{84}$$

$$LCL^\alpha_{med-u} = CL^\alpha_{med-u} - 3\sqrt{\frac{CL^\alpha_{med-u}}{n_j}} \tag{85}$$

The condition of process control for each sample is defined as:

$$Process\ control = \left\{ \begin{array}{ll} in\text{-}control, & for\ LCL^\alpha_{med-u} \leq S^\alpha_{med-u,j} \leq UCL^\alpha_{med-u} \\ out\text{-}of\ control, & for\ otherwise \end{array} \right\} \tag{86}$$

Example 5 The following defects have been observed on a textile product with the given sizes.

From Table 7, the following control limits (Table 8) are obtained.

Table 9 presents the u_i values calculated for each size. These values are examined if they are within the corresponding fuzzy control limits with respect to the sample sizes. All u_i values are between the control limits. Hence, the process is under control.

Table 7 Number of defects and variable sizes	Sample number	Sample size	Number of defects
	1	200	(3, 4, 5)
	2	400	(5, 6, 7)
	3	400	(6, 7, 8)
	4	200	(2, 3, 4)
	5	600	(7, 8, 9)
	6	400	(5, 6, 7)
	7	600	(9, 10, 11)
	8	300	(2, 3, 4)
	9	300	(3, 4, 5)
	10	600	(6, 7, 8)

Table 8 Fuzzy $\bar{u}$ control limits

| Sample size | $\bar{u} = (1.2, 1.45, 1.7)$ | |
	LCL	UCL
2	0	(3.524, 4.004, 4.466)
4	0	(2.843, 3.256, 3.656)
6	0	(2.542, 2.925, 3.297)
6	0	(2.542, 2.925, 3.297)
3	0	(3.097, 3.536, 3.958)

Table 9 u_i values for variable sizes

Sample no	Sample size	u_i values
1	2	(1.500, 2.000, 2.500)
2	4	(1.250, 1.500, 1.750)
3	4	(1.500, 1.750, 2.000)
4	2	(1.000, 1.500, 2.000)
5	6	(1.167, 1.333, 1.500)
6	4	(1.250, 1.500, 1.750)
7	6	(1.500, 1.667, 1.833)
8	3	(0.667, 1.000, 1.333)
9	3	(1.000, 1.333, 1.667)
10	6	(1.000, 1.167, 1.333)

4 Conclusion

Statistical process control is used when a large number of similar items are being produced. Every process is subject to variability. The variability when a process is running well is called inherent variability. The purpose of statistical process control is to give a signal when the process mean has moved away from the target. A second purpose is to give a signal when item to item variability has increased. Sometimes, variability cannot be measured with certainty. Some measurements can be vague enough to handle them with the fuzzy sets [8, 6]. In this chapter, we presented the fuzzy control charts for variables and attributes. With fuzzy control charts, a more flexible and informative evaluation of processes can be made. For further research, we suggest the extensions of fuzzy sets such as type-2 fuzzy sets, intuitionistic fuzzy sets, or hesitant fuzzy sets to be used in the development of fuzzy control charts.

Appendix

Table of coefficients for control charts for variables

n	A_2	A_3	c_4	B_3	B_4	B_5	B_6	d_2	d_3	D_1	D_2	D_3	D_4	E_2
2	1.880	2.659	0.798	0.000	3.267	0.000	2.606	1.128	0.853	0.000	3.686	0.000	3.267	2.660
3	1.023	1.954	0.886	0.000	2.568	0.000	2.276	1.693	0.888	0.000	4.358	0.000	2.574	1.772
4	0.729	1.628	0.921	0.000	2.266	0.000	2.088	2.059	0.880	0.000	4.698	0.000	2.282	1.457
5	0.577	1.427	0.940	0.000	2.089	0.000	1.964	2.326	0.864	0.000	4.918	0.000	2.114	1.290
6	0.483	1.287	0.952	0.030	1.970	0.029	1.874	2.534	0.848	0.000	5.078	0.000	2.004	1.184
7	0.419	1.182	0.959	0.118	1.882	0.113	1.806	2.704	0.833	0.204	5.204	0.076	1.924	1.109
8	0.373	1.099	0.965	0.185	1.815	0.179	1.751	2.847	0.820	0.388	5.306	0.136	1.864	1.054
9	0.337	1.032	0.969	0.239	1.761	0.232	1.707	2.970	0.808	0.547	5.393	0.184	1.816	1.010
10	0.308	0.975	0.973	0.284	1.716	0.276	1.669	3.078	0.797	0.687	5.469	0.223	1.777	0.975
11	0.285	0.927	0.975	0.321	1.679	0.313	1.637	3.173	0.787	0.811	5.535	0.256	1.744	0.945
12	0.266	0.886	0.978	0.354	1.646	0.346	1.610	3.258	0.778	0.922	5.594	0.283	1.717	0.921
13	0.249	0.850	0.979	0.382	1.618	0.374	1.585	3.336	0.770	1.025	5.647	0.307	1.693	0.899
14	0.235	0.817	0.981	0.406	1.594	0.399	1.563	3.407	0.763	1.118	5.696	0.328	1.672	0.881
15	0.223	0.789	0.982	0.428	1.572	0.421	1.544	3.472	0.756	1.203	5.741	0.347	1.653	0.864
16	0.212	0.763	0.984	0.448	1.552	0.440	1.526	3.532	0.750	1.282	5.782	0.363	1.637	0.849
17	0.203	0.739	0.985	0.466	1.534	0.458	1.511	3.588	0.744	1.356	5.820	0.378	1.622	0.836
18	0.194	0.718	0.985	0.482	1.518	0.475	1.496	3.640	0.739	1.424	5.856	0.391	1.608	0.824
19	0.187	0.698	0.986	0.497	1.503	0.490	1.483	3.689	0.734	1.487	5.891	0.403	1.597	0.813
20	0.180	0.680	0.987	0.510	1.490	0.504	1.470	3.735	0.729	1.549	5.921	0.415	1.585	0.803
21	0.173	0.663	0.988	0.523	1.477	0.516	1.459	3.778	0.724	1.605	5.951	0.425	1.575	0.794
22	0.167	0.647	0.988	0.534	1.466	0.528	1.448	3.819	0.720	1.659	5.979	0.434	1.566	0.786
23	0.162	0.633	0.989	0.545	1.455	0.539	1.438	3.858	0.716	1.710	6.006	0.443	1.557	0.778
24	0.157	0.619	0.989	0.555	1.445	0.549	1.429	3.895	0.712	1.759	6.031	0.451	1.548	0.770
25	0.153	0.606	0.990	0.565	1.435	0.559	1.420	3.931	0.708	1.806	6.056	0.459	1.541	0.763

References

1. Montgomery, D.C.: Introduction to Statistical Quality Control. Wiley
2. Senturk, S., Erginel, N.: Development of fuzzy $\tilde{\bar{X}}$ & $\tilde{R}$ and $\tilde{\bar{X}}$ & S control charts using α-cuts. Inf. Sci. **179**(10), 1542–1551 (2009)
3. Gülbay, M., Kahraman, C.: An alternative approach to fuzzy control charts: direct fuzzy approach. Inf. Sci. **177**(6), 1463–1480 (2007)
4. Rao, P.P.B., Shankar, N.R.: Ranking fuzzy numbers with a distance method using circumcenter of centroids and an index of modality. In: Advances in Fuzzy Systems, vol. 2011, Article ID 178308. doi:10.1155/2011/178308 (2011)
5. Gülbay, M., Kahraman, C.: Development of fuzzy process control charts and fuzzy unnatural pattern analyses. Comput. Stat. Data Anal. **51**, 434–451 (2006)
6. Gülbay, M., Kahraman, C., Ruan, D.: α-cut fuzzy control charts for linguistic data. Int. J. Intell. Syst. **19**, 1173–1196 (2004)
7. Wang, J.H., Raz, T.: On the construction of control charts using linguistic variables. Intell. J. Prod. Res. **28**, 477–487 (1990)
8. Şentürk, S., Erginel, N., Kaya, İ., Kahraman, C.: Fuzzy exponentially weighted moving average control chart for univariate data with a real case application. Appl. Soft Comput. **22**, 1–10 (2014)

Fuzzy EWMA and Fuzzy CUSUM Control Charts

Nihal Erginel and Sevil Şentürk

Abstract Exponentially Weighted Moving-Averages (EWMA) and Cumulative-Sum (CUSUM) control charts have the ability of detecting small shifts in the process mean. Classical EWMA and CUSUM charts are not capable to capture the uncertainty in case of incomplete data. Fuzzy EWMA and CUSUM control charts are developed in this chapter and numerical illustrations are given.

Keywords CUSUM · EWMA · Shewhart control charts · Fuzzy sets · Tabular CUSUM · V-mask

1 Introduction

Shewhart proposed the traditional control charts to detect if assignable causes exist in process. When data include uncertainty that comes from inherent of data collecting process or measurement system, the fuzzy set theory is a powerful tool to control processes. Fuzzy variable control charts and fuzzy attribute control charts are well documented in the literature. Firstly, fuzzy control charts have been introduced by Raz and Wang [14] and Wang and Raz [21]. After that, Kanagawa et al. [11], Gülbay et al. [10], Gülbay and Kahraman [8, 9], Faraz and Moghadam [7], Erginel [4], Şentürk and Erginel [17], Şentürk [16], Şentürk et al. [18], Erginel et al. [6]. Kaya and Kahraman [12], Erginel [5] have studied fuzzy control charts. But some special control charts such as exponentially weighted moving-average (EWMA) control charts and Cumulative–Sum Control Charts (CUSUM) are rarely handled in fuzzy environment.

N. Erginel (✉)
Industrial Engineering Department, Anadolu University, 26555 Eskişehir, Turkey
e-mail: nerginel@anadolu.edu.tr

S. Şentürk
Statistics Department, Anadolu University, 26470 Eskişehir, Turkey
e-mail: sdeligoz@anadolu.edu.tr

© Springer International Publishing Switzerland 2016
C. Kahraman and Ö. Kabak (eds.), *Fuzzy Statistical Decision-Making*,
Studies in Fuzziness and Soft Computing 343,
DOI 10.1007/978-3-319-39014-7_15

EWMA control chart is useful for detecting the small shifts in process mean. Therefore, a series of EWMA data on the chart tends to move slowly to the new level following a shift in the process, or will vary about the centerline with small fluctuations when the process is in control [3]. In traditional EWMA control charts, data are expressed with crisp value. But, if data include uncertainty or vagueness due to the measurement system and/or environmental conditions, traditional EWMA control chart is not sufficient to evaluate fuzzy data. In this case, these uncertainties are modeled by fuzzy EWMA control charts. Combining multivariate statistical quality control and the fuzzy set theory, fuzzy multivariate exponentially weighted moving average (F-MEWMA) control chart was proposed by Alipour and Noorossana [1]. Shu et al. [15] proposed fuzzy maximum generally weighted moving average (F-MaxGWMA) to detect outstanding diagnostic abilities for warning abnormal-manufacturing variation to model fuzziness of imprecise sample data. Şenturk et al. [19] proposed fuzzy EWMA (FEWMA) for univariate data under fuzzy environment for detecting small shifts in the data.

CUSUM control charts are also used for catching the small shifts in process mean. CUSUM was proposed firstly by Page in 1954. V-mask procedure that is helpful for determining whether the process is in control or out-of control was proposed by Barnhard [2]. The CUSUM chart directly incorporates all the information in the sequence of sample values by plotting the cumulative sum of the deviations of the sample values from a target value [13]. A CUSUM control chart for fuzzy quality data firstly was interpreted by Wang [20]. Wang proposed an optimal representative value for fuzzy quality data by means of a combination of a random variable with a measure of fuzziness. Applying the classical CUSUM chart for these representative values, a univariate CUSUM control chart concerning LR-fuzzy data under independent observations is constructed. Fuzzy Tabular CUSUM control chart is firstly proposed in the following section.

2   Fuzzy EWMA Control Chart

Exponentially weighted moving-average control chart is an effective tool for detecting the small shifts both in mean and in variance of the process. The theoretical structure of EWMA control chart is given as follows.

$$UCL_{EWMA} = \overline{\overline{X}} + A_2\bar{R}\sqrt{\frac{\lambda}{(2-\lambda)}} \tag{1}$$

$$CL_{EWMA} = \overline{\overline{X}} \tag{2}$$

$$LCL_{EWMA} = \overline{\overline{X}} - A_2\bar{R}\sqrt{\frac{\lambda}{(2-\lambda)}} \tag{3}$$

Table 1 Fuzzy averages and fuzzy ranges

t	$\tilde{\bar{X}}_t$	$\tilde{R}_j$
1	$(\bar{X}_{a,1}, \bar{X}_{b,1}, \bar{X}_{c,1})$	$(R_{a;1}, R_{b,1}, R_{c,1})$
.		
.		
...		
m	$(\bar{X}_{a,m}, \bar{X}_{b,m}, \bar{X}_{c,m})$	$(R_{a;m}, R_{b,m}, R_{c,m})$

where, $0 < \lambda \leq 1$ is a constant, $\overline{\overline{X}}$ is the overall mean, n is the sample size, $\bar{R}$ is the average of the R_i's while R_i is a range for each sample.

2.1 Theoretical Structure of FEWMA

Fuzzy EWMA control chart was proposed by Şenturk et al. [19] using the ranges calculated from a process in case of unknown $(\sigma_a, \sigma_b, \sigma_c)$. Fuzzy ranges $(R_{a,1}, R_{b,1}, R_{c,1})$ are computed from samples (Table 1).

Where, $\bar{R}_a$, $\bar{R}_b$, and $\bar{R}_c$ are the arithmetic means of the least possible values, the most possible values, and the largest possible values, respectively. $(R_{a,1}, R_{b,1}, R_{c,1})$ and $(\bar{R}_a, \bar{R}_b, \bar{R}_c)$ can be calculated similar to fuzzy $\tilde{\bar{X}} - \tilde{R}$ control charts. Fuzzy EWMA control limits for unknown $(\sigma_a, \sigma_b, \sigma_c)$ are given as follows;

$$U\tilde{C}L_{EWMA} = (\overline{\overline{X}}_a, \overline{\overline{X}}_b, \overline{\overline{X}}_c) + A_2(\bar{R}_a, \bar{R}_b, \bar{R}_c)\sqrt{\frac{\lambda}{(2-\lambda)}}$$

$$= \left(\overline{\overline{X}}_a + A_2\bar{R}_a\sqrt{\frac{\lambda}{(2-\lambda)}}, \overline{\overline{X}}_b + A_2\bar{R}_b\sqrt{\frac{\lambda}{(2-\lambda)}}, \overline{\overline{X}}_c + A_2\bar{R}_c\sqrt{\frac{\lambda}{(2-\lambda)}}\right)$$

$$(4)$$

$$C\tilde{L}_{EWMA} = (\overline{\overline{X}}_a, \overline{\overline{X}}_b, \overline{\overline{X}}_c) \tag{5}$$

$$L\tilde{C}L_{EWMA} = (\overline{\overline{X}}_a, \overline{\overline{X}}_b, \overline{\overline{X}}_c) - A_2(\bar{R}_a, \bar{R}_b, \bar{R}_c)\sqrt{\frac{\lambda}{(2-\lambda)}}$$

$$= \left(\overline{\overline{X}}_a - A_2\bar{R}_c\sqrt{\frac{\lambda}{(2-\lambda)}}, \overline{\overline{X}}_b - A_2\bar{R}_b\sqrt{\frac{\lambda}{(2-\lambda)}}, \overline{\overline{X}}_c - A_2\bar{R}_a\sqrt{\frac{\lambda}{(2-\lambda)}}\right)$$

$$(6)$$

α-cut fuzzy EWMA control limits for unknown $(\sigma_a, \sigma_b, \sigma_c)$ are represented by the following equations:

$$U\tilde{C}L_{EWMA}^{\alpha} = \left[\bar{\bar{X}}_l^{\alpha},\ \bar{\bar{X}}_r^{\alpha}\right] + A_2\left[\bar{R}_l^{\alpha},\ \bar{R}_r^{\alpha}\right]\sqrt{\frac{\lambda}{(2-\lambda)}}$$

$$= \left[\bar{\bar{X}}_l^{\alpha} + A_2\bar{R}_l^{\alpha}\sqrt{\frac{\lambda}{(2-\lambda)}},\ \bar{\bar{X}}_r^{\alpha} + A_2\bar{R}_r^{\alpha}\sqrt{\frac{\lambda}{(2-\lambda)}}\right] \quad (7)$$

$$C\tilde{L}_{EWMA}^{\alpha} = \left[\bar{\bar{X}}_l^{\alpha},\ \bar{\bar{X}}_r^{\alpha}\right] \quad (8)$$

$$L\tilde{C}L_{EWMA}^{\alpha} = \left[\bar{\bar{X}}_l^{\alpha},\ \bar{\bar{X}}_r^{\alpha}\right] - A_2\left[\bar{R}_l^{\alpha},\ \bar{R}_r^{\alpha}\right]\sqrt{\frac{\lambda}{(2-\lambda)}}$$

$$= \left[\bar{\bar{X}}_l^{\alpha} - A_2\bar{R}_r^{\alpha}\sqrt{\frac{\lambda}{(2-\lambda)}},\ \bar{\bar{X}}_r^{\alpha} - A_2\bar{R}_l^{\alpha}\sqrt{\frac{\lambda}{(2-\lambda)}}\right] \quad (9)$$

Fuzzy median transformation technique is integrated to the α-level fuzzy median for α-cut fuzzy EWMA control chart and unknown $(\sigma_a, \sigma_b, \sigma_c)$ as follows;

$$U\tilde{C}L_{med-EWMA}^{\alpha} = \tilde{C}L_{med-EWMA}^{\alpha} + \frac{1}{2}A_2\left[\bar{R}_l^{\alpha} + \bar{R}_r^{\alpha}\right]\sqrt{\frac{\lambda}{(2-\lambda)}} \quad (10)$$

$$\tilde{C}L_{med-EWMA}^{\alpha} = \frac{1}{2}\left(\bar{\bar{X}}_l^{\alpha} + \bar{\bar{X}}_r^{\alpha}\right) \quad (11)$$

$$L\tilde{C}L_{EWMA}^{\alpha} = \tilde{C}L_{med-EWMA}^{\alpha} - \frac{1}{2}A_2\left[\bar{R}_l^{\alpha} + \bar{R}_r^{\alpha}\right]\sqrt{\frac{\lambda}{(2-\lambda)}} \quad (12)$$

For a sample j, α-level fuzzy median value $(S_{med-EWMA,j}^{\alpha})$ is calculated as follows;

$$\tilde{S}_{med-EWMA,j}^{\alpha} = \frac{1}{2}\left(\bar{X}_{l,j}^{\alpha} + \bar{X}_{r,j}^{\alpha}\right) \quad (13)$$

The condition of process control for each sample is defined as;

Process control

$$= \begin{cases} in-control, & for\ L\tilde{C}L_{med-EWMA}^{\alpha} \leq \tilde{S}_{med-EWMA,j}^{\alpha} \leq U\tilde{C}L_{med-EWMA}^{\alpha} \\ out-of\ control, & for\ otherwise \end{cases} \quad (14)$$

2.2 Application on FEWMA

The fuzzy data in Table 2 are collected from a production process of plastic button. Fuzzy measurement values and their fuzzy averages and fuzzy ranges are given in Table 2.

In Table 2, $\bar{R}_a$, $\bar{R}_b$, and $\bar{R}_c$ are the arithmetic means of the least possible values, the most possible values, and the largest possible values, respectively. $(R_{a,1}, R_{b,1}, R_{c,1})$ and $(\bar{R}_a, \bar{R}_b, \bar{R}_c)$ are calculated similar to fuzzy $\tilde{\bar{X}} - \tilde{R}$ control charts. Fuzzy EWMA control limits for unknown $(\sigma_a, \sigma_b, \sigma_c)$ are given as follows where $\lambda = 0.2$;

$$
\begin{aligned}
U\tilde{C}L_{EWMA} &= \left(X_a + A_2 \bar{R}_a \sqrt{\frac{\lambda}{(2-\lambda)}}, \, \overline{\overline{X}}_b + A_2 \bar{R}_b \sqrt{\frac{\lambda}{(2-\lambda)}}, \, \overline{\overline{X}}_c + A_2 \bar{R}_c \sqrt{\frac{\lambda}{(2-\lambda)}} \right) \\
&= \left(3.98 + 0.577(0.01)\sqrt{\frac{0.2}{(2-0.2)}}, \, 4.00 + 0.577(0.03)\sqrt{\frac{0.2}{(2-0.2)}}, \right. \\
&\qquad \left. 4.02 + 0.577(0.09)\sqrt{\frac{0.2}{(2-0.2)}} \right) \\
&= (3.982, 4.005, 4.037)
\end{aligned}
$$

Table 2 Fuzzy measurement values, fuzzy averages and fuzzy ranges of plastic button

Sample	X_a	X_b	X_c	R
S1-1	3.99	4.00	4.02	
S1-2	3.96	3.99	4.01	$R_{a1} = 0.01$
S1-3	3.97	3.98	4.00	$R_{b1} = 0.03$
S1-4	3.95	3.97	3.98	$R_{c1} = 0.07$
S1-5	3.99	4.00	4.01	
S2-1	4.01	4.02	4.04	
S2-2	4.01			$R_{a2} = 0.00$
S2-3	3.98	4.00	4.01	$R_{b2} = 0.02$
S2-4	4.01	4.02	4.03	$R_{c2} = 0.06$
S2-5	4.00	4.01	4.02	
⋮	⋮	⋮	⋮	⋮
S10-1	3.97	3.98	3.99	
S10-2	3.98	3.99	4.00	$R_{a15} = 0.01$
S10-3	3.99	4.00	4.02	$R_{b15} = 0.04$
S10-4	3.99	4.01	4.03	$R_{c15} = 0.08$
S10-5	4.00	4.02	4.05	
	$\overline{\overline{X}}_a = 3.98$	$\overline{\overline{X}}_b = 4.00$	$\overline{\overline{X}}_c = 4.02$	$\bar{R}_a = 0.01$
				$\bar{R}_b = 0.03$
				$\bar{R}_c = 0.09$

$$C\tilde{L}_{EWMA} = (\overline{\overline{X}}_a, \overline{\overline{X}}_b, \overline{\overline{X}}_c) = (3.98, \ 4.00, \ 4.02)$$

$$L\tilde{C}L_{EWMA} = \left(\overline{\overline{X}}_a - A_2\bar{R}_c\sqrt{\frac{\lambda}{(2-\lambda)}}, \ \overline{\overline{X}}_b - A_2\bar{R}_b\sqrt{\frac{\lambda}{(2-\lambda)}}, \ \overline{\overline{X}}_c - A_2\bar{R}_a\sqrt{\frac{\lambda}{(2-\lambda)}} \right)$$

$$= \left(\begin{array}{c} 3.98 - 0.577(0.09)\sqrt{\frac{0.2}{(2-0.2)}}, \ 4.00 - 0.577(0.03)\sqrt{\frac{0.2}{(2-0.2)}}, \\ 4.02 - 0.577(0.01)\sqrt{\frac{0.2}{(2-0.2)}} \end{array} \right)$$

$$= (3.962, \ 3.994, \ 4.018)$$

α-cut fuzzy EWMA control limits for unknown $(\sigma_a, \ \sigma_b, \ \sigma_c)$ are represented by the following equations by using $\overline{\overline{X}}_l^\alpha, \overline{\overline{X}}_r^\alpha$ and $\bar{R}_l^\alpha, \bar{R}_r^\alpha$, where $\alpha = 0.65$.

$$\overline{\overline{X}}_l^\alpha = \overline{\overline{X}}_a + \alpha(\overline{\overline{X}}_b - \overline{\overline{X}}_a) = 3.98 + 0.65(4.00 - 3.98) = 3.993$$

$$\overline{\overline{X}}_r^\alpha = \overline{\overline{X}}_c - \alpha(\overline{\overline{X}}_c - \overline{\overline{X}}_b) = 4.02 - 0.65(4.02 - 4.00) = 4.007$$

$$\bar{R}_l^\alpha = \bar{R}_a + \alpha(\bar{R}_b - \bar{R}_a) = 0.01 + 0.65(0.03 - 0.01) = 0.023$$

$$\bar{R}_r^\alpha = \bar{R}_c - \alpha(\bar{R}_c - \bar{R}_b) = 0.09 - 0.65(0.09 - 0.03) = 0.051$$

The limits of α-cut fuzzy EWMA control chart are given as follows for plastic button:

$$U\tilde{C}L_{EWMA}^\alpha = \left(\overline{\overline{X}}_l^\alpha + A_2\bar{R}_l^\alpha\sqrt{\frac{\lambda}{(2-\lambda)}}, \ \overline{\overline{X}}_r^\alpha + A_2\bar{R}_r^\alpha\sqrt{\frac{\lambda}{(2-\lambda)}} \right)$$

$$= \left(3.993 + 0.577(0.023)\sqrt{\frac{0.2}{(2-0.2)}}, \ 4.007 + 0.577(0.051)\sqrt{\frac{0.2}{(2-0.2)}} \right)$$

$$= (3.997, \ 4.016)$$

$$C\tilde{L}_{EWMA}^\alpha = (\overline{\overline{X}}_l^\alpha, \ \overline{\overline{X}}_r^\alpha) = (3.993, \ 4.007)$$

$$L\tilde{C}L_{EWMA}^{\alpha} = \left(\overline{\overline{X}}_l^{\alpha} - A_2 \bar{R}_r^{\alpha} \sqrt{\frac{\lambda}{(2-\lambda)}}, \ \overline{\overline{X}}_r^{\alpha} - A_2 \bar{R}_l^{\alpha} \sqrt{\frac{\lambda}{(2-\lambda)}} \right)$$

$$= \left(3.993 - 0.577(0.051)\sqrt{\frac{0.2}{(2-0.2)}}, \ 4.007 - 0.577(0.023)\sqrt{\frac{0.2}{(2-0.2)}} \right)$$

$$= (3.983, \ 4.002)$$

The four transformation techniques that are well-known in descriptive statistics are introduced in the literature: fuzzy mode, α-level fuzzy midrange, fuzzy median and fuzzy average. It should be pointed out that there is no theoretical basis supporting any one specifically [21]. Fuzzy median transformation technique is integrated to the α-level fuzzy median for α-cut fuzzy EWMA control chart as follows:

$$UCL_{med-EWMA}^{a} = CL_{med-EWMA}^{a} + \frac{1}{2}A_2 \left[\bar{R}_l^{\alpha} + \bar{R}_r^{\alpha} \right] \sqrt{\frac{\lambda}{(2-\lambda)}}$$

$$= 4.00 + \frac{1}{2}0.577(0.023 + 0.051)\sqrt{\frac{0.2}{(2-0.2)}}$$

$$= 4.006$$

$$CL_{med-EWMA}^{\alpha} = \frac{1}{2}\left[\overline{\overline{X}}_l^{\alpha} + \overline{\overline{X}}_r^{\alpha} \right] = \frac{1}{2}(3.993 + 4.007) = 4.00$$

$$L\tilde{C}L_{EWMA}^{\alpha} = CL_{med-EWMA}^{\alpha} - \frac{1}{2}A_2 \left[\bar{R}_l^{\alpha} + \bar{R}_r^{\alpha} \right] \sqrt{\frac{\lambda}{(2-\lambda)}}$$

$$= 4.00 - \frac{1}{2}0.577(0.023 + 0.051)\sqrt{\frac{0.2}{(2-0.2)}}$$

$$= 3.993$$

Table 3 Control limits of fuzzy EWMA, α-level fuzzy median value and the process conditions	Sample no	$S_{med-EWMW,j}^{\alpha}$	$3.993 \leq S_{med-EWMA,j}^{\alpha} \leq 4.006$
	1	3.988	In control
	2	4.014	Out of control
	3	4.004	In control
	4	4.006	In control
	5	4.005	In control
	6	3.996	In control
	7	3.998	In control
	8	3.988	In control
	9	3.996	In control
	10	4.000	In control

As seen in Table 3, the production process of plastic button in clothing industry is "out of control" due to the second sample.

3 Fuzzy CUSUM Control Chart

CUSUM control charts use V-mask procedure. Typical V-mask control limits are given in Fig. 1.

In designing V-mask procedure, the following equations are used [13];

$$d = -2\frac{ln\Upsilon}{\delta^2} \tag{15}$$

$$\delta = \frac{\Delta}{\sigma_{\bar{X}}} \tag{16}$$

$$H = 2(d)(\sigma_{\bar{X}})\tan\theta \tag{17}$$

$$\tan\theta = \frac{\Delta}{2A} \tag{18}$$

where
d   lead distance
Υ the probability of incorrectly concluding that a shift has occurred (a false alarm)
Δ the shift in the process mean that it is desired to detect
$\sigma_{\bar{x}}$ the standard deviation of $\bar{x}$
H   The decision interval of the procedure, or the half height of the V-mask at point O

Fig. 1 The V-mask and scaling [13]

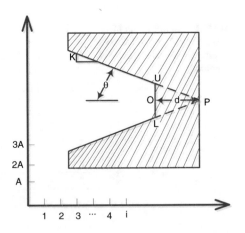

θ the angle of V-mask
A   a scale factor relating the vertical scale of V-mask

The application of V-mask procedure is complex in real life, so the tabular CUSUM is preferred by practitioners. The theoretical structure of tabular form of the CUSUM is given as follows [13]:

$$S_H(i) = \max[0, \bar{x}_i - (\mu_0 + K) + S_H(i-1)] \qquad (19)$$

and

$$S_L(i) = \max[0, (\mu_0 - K) - \bar{x}_i + S_L(i-1)] \qquad (20)$$

where $S_H(i)$ represents the upper one-sided tabular CUSUM for period i, $S_L(i)$ shows the lower one-sided tabular CUSUM for period i; μ_0 is the target value, $S_H(0) = S_L(0) = 0$ are the starting values. K is called the reference value and is usually chosen about halfway between the target μ_0 and the value of the mean corresponding to the out of control state, $\mu_1 = \mu_0 + \Delta$. That is, K is about one-half of the magnitude of the shift in process;

$$K = \frac{\Delta}{2} \qquad (21)$$

If either $S_H(i)$ or $S_L(i)$ exceeds the decision interval H, the process is out-of control otherwise the process is in control.

3.1   Fuzzy Tabular CUSUM

When manufacturing processes and/or measurement systems have some uncertainty and vagueness due to operators or gauges, the Fuzzy Tabular CUSUM control chart is more suitable than a traditional CUSUM control chart for analyzing the process.

If the data from a process are expressed as fuzzy numbers, they can be shown as triangular fuzzy numbers $x_i = (x_a, x_b, x_c)$ and fuzzy average $\bar{x}_i = (\bar{x}_a, \bar{x}_b, \bar{x}_c)$.

$S_H(i)$ and $S_L(i)$ are represented by the fuzzy numbers $(S_H(i)_a, S_H(i)_b, S_H(i)_c)$ and $(S_L(i)_a, S_L(i)_b, S_L(i)_c)$, respectively.

$$S_H(i)_a = \max\left[0, \bar{x}_a - (\mu_0 + K) + S_H(i-1)_a\right] \qquad (22)$$

$$S_H(i)_b = \max\left[0, \bar{x}_b - (\mu_0 + K) + S_H(i-1)_b\right] \qquad (23)$$

$$S_H(i)_c = \max\left[0, \bar{x}_c - (\mu_0 + K) + S_H(i-1)_c\right] \qquad (24)$$

$$S_L(i)_a = \max\left[0, \ (\mu_0 - K) - \bar{x}_c + S_L(i-1)_a\right] \tag{25}$$

$$S_L(i)_b = \max\left[0, \ (\mu_0 - K) - \bar{x}_b + S_L(i-1)_b\right] \tag{26}$$

$$S_L(i)_c = \max\left[0, \ (\mu_0 - K) - \bar{x}_a + S_L(i-1)_c\right] \tag{27}$$

$(S_H(i)_a, S_H(i)_b, S_H(i)_c)$ and $(S_L(i)_a, S_L(i)_b, S_L(i)_c)$ are transformed to crisp values by fuzzy transformation techniques for implementation fuzzy tabular CUSUM. α-level fuzzy midrange transformation technique is used in the theoretical structure of fuzzy tabular CUSUM.

Before applying transformation techniques, fuzzy tabular CUSUM based on α-cut is required. These α-cuts are composed of all elements whose membership degrees are greater than equal to α. The sets $A_\alpha = \{x \in X : \mu_A(x) \geq \alpha, \ 0 \leq \alpha \leq 1\}$ are the α-level sets of A. The α-level sets A_α called also the α-cut sets. α-cut fuzzy averages are obtained as follows:

$$\bar{x}_a^\alpha = \bar{x}_a + \alpha(\bar{x}_b - \bar{x}_a) \tag{28}$$

$$\bar{x}_c^\alpha = \bar{x}_c - \alpha(\bar{x}_c - \bar{x}_b) \tag{29}$$

Also, $S_H^\alpha(i)_a$, $S_H^\alpha(i)_c$ $S_L^\alpha(i)_a$ and $S_L^\alpha(i)_c$ based on α-cut for fuzzy tabular CUSUM are handled as follows, respectively:

$$S_H^\alpha(i)_a = \max\left[0, \ \bar{x}_a^\alpha - (\mu_0 + K) + S_H^\alpha(i-1)_a\right] \tag{30}$$

$$S_H^\alpha(i)_c = \max\left[0, \ \bar{x}_c^\alpha - (\mu_0 + K) + S_H^\alpha(i-1)_c\right] \tag{31}$$

and

$$S_L^\alpha(i)_a = \max\left[0, \ (\mu_0 - K) - \bar{x}_c^\alpha + S_L^\alpha(i-1)_a\right] \tag{32}$$

$$S_L^\alpha(i)_c = \max\left[0, \ (\mu_0 - K) - \bar{x}_a^\alpha + S_L^\alpha(i-1)_c\right] \tag{33}$$

The $S_H^\alpha(i)_a$, $S_H^\alpha(i)_c$ $S_L^\alpha(i)_a$ and $S_L^\alpha(i)_c$ are combined with α-level fuzzy midrange transformation technique for fuzzy tabular CUSUM, and given in the following equations:

$$S_{H-mr}^\alpha = \frac{S_H^\alpha(i)_a + S_H^\alpha(i)_c}{2} \tag{34}$$

$$S_{L-mr}^\alpha = \frac{S_L^\alpha(i)_a + S_L^\alpha(i)_c}{2} \tag{35}$$

Fuzzy control limits of fuzzy tabular CUSUM can be obtained in terms of triangular fuzzy numbers.

In designing fuzzy V-mask, the procedure is proposed as follows:

$$(H_a, H_b, H_c) = 2(d)(\sigma_{\overline{X}a}, \sigma_{\overline{X}b}, \sigma_{\overline{X}c}) \tan\theta \qquad (36)$$

where

$(\sigma_{\overline{X}a}, \sigma_{\overline{X}b}, \sigma_{\overline{X}c})$ fuzzy standard deviation of $(\overline{x}_a, \overline{x}_b, \overline{x}_c)$
(H_a, H_b, H_c) fuzzy decision interval of the procedure

Fuzzy control limits based on α-cuts for fuzzy tabular CUSUM are calculated as follows:

$$H_a^\alpha = H_a + \alpha(H_b - H_a) \qquad (37)$$

$$H_c^\alpha = H_c - \alpha(H_c - H_b) \qquad (38)$$

α-cut fuzzy control limits based on α-level fuzzy midrange transformation for fuzzy tabular CUSUM are handled as follows:

$$H_{mr}^\alpha = \frac{H_a^\alpha + H_c^\alpha}{2} \qquad (39)$$

If either fuzzy S_{H-mr}^α or S_{L-mr}^α exceeds the fuzzy decision interval H_{mr}^α, the process is out-of control otherwise the process is in control. The process control conditions are given as follows:

Process control:

$$\begin{cases} \text{in control;} & (S_{H-mr}^\alpha < H_{mr}^\alpha) \vee (S_{L-mr}^\alpha < H_{mr}^\alpha) \\ \text{out of control;} & \text{otherwise} \end{cases} \qquad (40)$$

3.2 Application on Fuzzy CUSUM

Fuzzy data from a manufacturing process are collected in terms of triangular fuzzy numbers and desired to detect the small shifts on process. The fuzzy mean of each sample and other values are given in Table 4.

For $\alpha = 0.65$, Table 5 is obtained.

$\Delta = 1$ (When there is 1 shift in μ_0, it is desired to detect)

$d = 10.5$ (This value is also handled from "Cumulative-sum control charts parameters" table by using ARL-Average Run length when process is in control [13]

A: 2 (the vertical scale of V-mask)

$$\tan\theta = 0.25$$

Table 4 Fuzzy measurement values and fuzzy upper one-sided and fuzzy lower one-sided tabular CUSUM for period i

Period	$\bar{x}_a$	$\bar{x}_b$	$\bar{x}_c$	$S_H(i)_a$	$S_H(i)_b$	$S_H(i)_c$	$S_L(i)_a$	$S_L(i)_b$	$S_L(i)_c$
1	8.100	8.250	8.750	0.000	0.000	0.000	0.750	1.250	1.400
2	8.900	9.300	9.900	0.000	0.000	0.000	0.350	1.450	2.000
3	10.200	10.800	11.200	0.000	0.300	0.700	0.000	0.150	1.300
4	9.800	10.300	10.800	0.000	0.100	1.000	0.000	0.000	1.000
5	8.9	9.400	9.600	0.000	0.000	0.100	0.000	0.100	1.600
6	8.850	9.370	9.900	0.000	0.000	0.000	0.000	0.230	2.250
7	9.450	10.080	10.600	0.000	0.000	0.100	0.000	0.000	2.300
8	11.290	11.790	11.990	0.790	1.290	1.590	0.000	0.000	0.510
9	10.550	11.000	11.500	0.840	1.790	2.590	0.000	0.000	0.000
10	9.350	9.850	10.200	0.000	1.140	2.290	0.000	0.000	0.150

Table 5 $S_H^\alpha(i)_a$, $S_H^\alpha(i)_c$ $S_L^\alpha(i)_a$ and $S_L^\alpha(i)_c$ based on α-cut for fuzzy tabular CUSUM

Period	$\bar{x}_a^\alpha$	$\bar{x}_c^\alpha$	$S_H^\alpha(i)_a$	$S_H^\alpha(i)_c$	$S_L^\alpha(i)_a$	$S_L^\alpha(i)_c$
1	8.198	8.425	0.000	0.000	1.075	1.303
2	9.160	9.510	0.000	0.000	1.065	1.643
3	10.590	10.940	0.090	0.440	0.000	0.553
4	10.125	10.475	0.000	0.415	0.000	0.000
5	9.225	9.470	0.000	0.000	0.030	0.275
6	9.188	9.556	0.000	0.000	0.000	0.587
7	9.860	10.262	0.000	0.000	0.000	0.228
8	11.615	11.860	1.115	1.360	0.000	0.000
9	10.843	11.175	1.458	2.035	0.000	0.000
10	9.675	9.973	0.633	1.508	0.000	0.000

$$(H_a, H_b, H_c) = 2(10.5)(\sigma_{\bar{X}_a}, \sigma_{\bar{X}_b}, \sigma_{\bar{X}_c})\tan\theta$$

$(\sigma_{\bar{X}_a}, \sigma_{\bar{X}_b}, \sigma_{\bar{X}_c}) = (0.94, 0.96, 0.98)$ are known from previous studies.

$$(H_a, H_b, H_c) = (4.935, 5.04, 5.145)$$

$$H_a^\alpha = H_a + \alpha(H_b - H_a) = 5.00325$$

$$H_c^\alpha = H_c - \alpha(H_c - H_b) = 5.07675$$

$$H_{mr}^\alpha = \frac{H_a^\alpha + H_c^\alpha}{2} = \frac{5.00325 + 5.07675}{2} = 5.04$$

Table 6 Control limits of fuzzy tabular CUSUM, S_{H-mr}^{α} values and the process conditions

Period	S_{H-mr}^{α}	H_{mr}^{α}	Process control
1	0.000	5.040	In control
2	0.000	5.040	In control
3	0.265	5.040	In control
4	0.208	5.040	In control
5	0.000	5.040	In control
6	0.000	5.040	In control
7	0.000	5.040	In control
8	1.238	5.040	In control
9	1.746	5.040	In control
10	1.070	5.040	In control

Table 7 Control limits of fuzzy tabular CUSUM, S_{L-mr}^{α} values and the process conditions

Period	S_{L-mr}^{α}	H_{mr}^{α}	Process control
1	1.189	5.040	In control
2	1.354	5.040	In control
3	0.276	5.040	In control
4	0.000	5.040	In control
5	0.153	5.040	In control
6	0.294	5.040	In control
7	0.114	5.040	In control
8	0.000	5.040	In control
9	0.000	5.040	In control
10	0.000	5.040	In control

$$S_{H-mr}^{0.65} = \frac{S_H^{0.65}(i)_a + S_H^{0.65}(i)_c}{2} = \frac{1.115 + 1.36}{2} = 1.2375$$

for period 8 and all $S_{H-mr}^{0.65}$ values are given in Table 6.

$$S_{L-mr}^{0.65} = \frac{S_L^{0.65}(i)_a + S_L^{0.65}(i)_c}{2} = \frac{1.075 + 1.3025}{2} = 1.18875$$

for period 1 and all $S_{L-mr}^{0.65}$ values are given in Table 7.

The manufacturing process is "in control" based on fuzzy tabular CUSUM.

4 Conclusions

EWMA and CUSUM control charts are very sensitive detect the small shifts whereas Shewhart control charts can detect larger shifts in process mean with respect to EWMA and CUSUM charts. Traditional control charts are not suitable to handle fuzzy data obtained from the process under fuzziness. Fuzzy data are caused from the incapability of measurement system, operators, gauges or methods. In this chapter, fuzzy EWMA and fuzzy tabular CUSUM control chart have been developed. Numerical examples on fuzzy control charts have also been given. Fuzzy multivariate CUSUM and fuzzy multivariate EWMA control charts can be developed to evaluate multivariate fuzzy data for further research. Besides, new extensions of fuzzy sets such as intuitionistic fuzzy sets or hesitant fuzzy sets can be used to develop fuzzy EWMA and CUSUM charts.

References

1. Alipour, H., Noorossana, R.: Fuzzy multivariate exponentially weighted moving average control chart. Int. J. Adv. Manufact. Technol. **48**, 1001–1007 (2010)
2. Barnhard G.A.: Control charts and stochastic processes. J. R. Stat. Soc. B **21** (1959)
3. Devor, R.E., Chang, T., Sutherland, J.W.: Statistical Quality Design and Control, Contemporary Concepts and Sutherland Methods. Pearson Education Inc., Upper Saddle River (2007)
4. Erginel, N.: Fuzzy individual and moving range control charts with α-cuts. J. Intell. Fuzzy Syst. **19**, 373–383 (2008)
5. Erginel, N.: Fuzzy rule-based $\tilde{p}$ and $n\tilde{p}$ control charts. J. Intell. Fuzzy Syst. **27**(1), 159–171 (2014)
6. Erginel, N., Şentürk, S., Kahraman, C., Kaya, İ.: Evaluating the packing process in food industry using $\tilde{\bar{X}}$ and $\tilde{S}$ control charts. Int. J. Comput. Intell. Syst. **4**(4), 509–520 (2011)
7. Faraz, A., Moghadam, M.B.: Fuzzy control chart a beter alternative for Shewhart Average Chart. Qual. Quant. **41**, 375–385 (2007)
8. Gülbay, M., Kahraman, C.: Development of fuzzy process control charts and fuzzy unnatural pattern analyses. Comput. Stat. Data Anal. **51**, 434–451 (2006)
9. Gülbay, M., Kahraman, C.: An alternative approach to fuzzy control chart: direct fuzzy approach. Inf. Sci. **77**(6), 1463–1480 (2006)
10. Gülbay, M., Kahraman, C., Ruan, D.: α-cut fuzzy control chart for linguistic data. Int. J. Intell. Syst. **19**, 1173–1196 (2004)
11. Kanagawa, A., Tamaki, F., Ohta, H.: Control charts for process average and variability based on linguistic data. Intell. J. Prod. Res. **31**(4), 913–922 (1993)
12. Kaya, İ., Kahraman, C.: Process capability analysis based on fuzzy measurements and fuzzy control charts. Expert Syst. Appl. **38**, 3172–3184 (2011)
13. Montgomery, D.C.: Introduction to Statistical Quality Control. Wiley, New York (1990)
14. Raz, T., Wang, J.H.: Probabilistic and memberships approaches in the construction of control chart for linguistic data. Prod. Planning Control **1**, 147 (1990)
15. Shu, H.M, Nguyen, T.L, Hsu. B.M.: Fuzzy MaxGWMA chart for identifying abnormal variations of on-line manufacturing processes with imprecise information. Expert Syst. Appl. (2013)
16. Şentürk, S.: Fuzzy regression control chart based on α-cut approximation. Int. J. Comput. Intell. Syst. **3**(1), 123–140 (2010)
17. Şentürk, S., Erginel, N.: Development of fuzzy $\tilde{\bar{X}} - \tilde{R}$ and $\tilde{\bar{X}} - \tilde{S}$ control charts using α-cuts. Inf. Sci. **179**, 1542–1551 (2009)
18. Şentürk, S., Erginel, N., Kaya, İ., Kahraman, C.: Design of Fuzzy $\tilde{u}$ Control Chart. J. Multiple Valued-Logic Soft Comput. **17**, 459–473 (2011)

19. Şentürk, S., Erginel, N., Kaya, İ., Kahraman, C.: Fuzzy exponentially weighted moving average control chart for univariate data with a real case application. Appl. Soft Comput. **22**, 1–10 (2014)
20. Wang, D.: A CUSUM control chart for fuzzy quality data. In: Lawry, J., Miranda, E., Bugarin, A., Li, S., Gil, M.A., Grzegorzewski, P., Hryniewicz, O. (eds.) Soft Methods for Integrated Uncertainty Modelling, pp. 357–364. Springer, Heidelberg (2006)
21. Wang, J.H., Raz, T.: On the construction of control charts using linguistic variables. Intell. J. Prod. Res. **28**, 477–487 (1990)

Linear Hypothesis Testing Based on Unbiased Fuzzy Estimators and Fuzzy Significance Level

Alireza Jiryaei and Mashaallah Mashinchi

Abstract A wide variety of applied problems of statistical hypothesis testing can be treated under a general setup of the linear models which includes analysis of variance. In this study, a new method is presented to test linear hypothesis using a fuzzy test statistic produced by a set of confidence intervals with non-equal tails. Also, a fuzzy significance level is used to evaluate the linear hypothesis. The method can be used to improve linear hypothesis testing when there is a sensitively in accepting or rejecting the null hypothesis. Also, as a simple case of linear hypothesis testing, one-way analysis of variance based on fuzzy test statistic and fuzzy significance level is investigated. Numerical examples are provided for illustration.

Keywords Analysis of variance · Confidence interval · Fuzzy critical value · Fuzzy test statistic · Fuzzy significance level · Linear hypothesis · Linear model

1 Introduction and Background

Analysis of Variance (ANOVA) is a common and popular method in the analysis of experimental designs. It includes important cases such as one-way and two-way ANOVA, and one-way and two-way analysis of covariance, and it has many useful applications in industry, agriculture and social sciences [8, 12, 13]. Various aspects of this topic have been considered in a fuzzy environment. One-way and two-way ANOVA using fuzzy unbiased estimators for variance parameter are discussed based on arithmetic operations on intervals by Buckley [3]. Wu [16] presented one-way ANOVA based on several notations of the α-cuts of fuzzy random variables, optimistic and pessimistic degrees and solving an optimization problem. An approach for one-way ANOVA has been carried out by Nourbakhsh et al. [10] for fuzzy data in which Zadeh's extension principle [9, 17] plays a key role for the applied

A. Jiryaei (✉) · M. Mashinchi
Department of Statistics, Faculty of Mathematics and Computer Sciences,
Shahid Bahonar University of Kerman, Kerman, Iran
e-mail: a.jiryae@gmail.com

© Springer International Publishing Switzerland 2016
C. Kahraman and Ö. Kabak (eds.), *Fuzzy Statistical Decision-Making*,
Studies in Fuzziness and Soft Computing 343,
DOI 10.1007/978-3-319-39014-7_16

computing operations. A statistical technique for testing the fuzzy hypothesis of one-way ANOVA is proposed by Kalpanapriya et al. [7] using the levels of pessimistic and optimistic of the triangular fuzzy data.

Linear hypothesis testing is an extension of analysis of variance. It can test hypotheses about the unknown parameters of the linear model, such as testing the equality of the means of several random variables [12]. Sometimes the observed value of test statistic is close to the related quantiles of statistical distributions, so there is uncertainty in accepting or rejecting the null hypothesis H_0. In this paper, a method is presented for linear hypothesis testing using a fuzzy test statistic and a fuzzy significance level. Moreover, the method can be used for modelling this uncertainty using fuzzy sets theory.

A method for testing statistical hypotheses in a fuzzy environment was introduced by Buckley [2, 3]. It considers a fuzzy test statistic and fuzzy critical values produced using confidence intervals with equal tails and arithmetic operations on intervals. In Buckley's method the fuzzy estimates are developed as fuzzy numbers, and their membership functions have been derived by Falsafain et al. [5]. In [2] the non-fuzzy hypotheses are tested, and in [14] and [1] the presented approach in [2] is generalized to the case where the statistical hypotheses and the observed data are also fuzzy. When dealing with non-symmetric statistical distributions, using confidence intervals with equal tails results in producing a fuzzy estimate where the membership degree for the unbiased point estimate of the required parameter is not equal to one [4]. While we expect that the unbiased point estimate has the highest importance in the fuzzy estimate, i.e. its membership degree should be equal to one. Solutions to overcome this problem using the confidence intervals with non-equal tails are provided by Buckley [3], and Falsafain and Taheri [4]. It has been shown that the solution presented by Falsafain and Taheri [4] is reduced to Buckley's method when dealing with symmetric statistical distributions. Moreover, it is possible to obtain the membership functions of the corrected fuzzy estimates. Therefore, we use this solution in this paper.

In order to discuss linear hypothesis testing based on fuzzy test statistic and fuzzy significance level, we first recall some basic concepts of fuzzy sets theory in Sect. 2. Section 3 contains a brief review of linear model and linear hypothesis. In Sect. 4, fuzzy test statistic and fuzzy critical value are discussed and decision rules are presented. Also, one-way ANOVA as a special case of linear hypothesis testing is discussed in Sect. 5. Two numerical examples are provided in Sect. 6 to show that our approach could perform quite well in practice. A conclusion is provided in Sect. 7.

2 Preliminaries

Some concepts of fuzzy sets theory, which will be referred to throughout this paper, are discussed in this section. Let U be a universal set and $F(U) = \left\{ \tilde{A} | \tilde{A} : U \to [0, 1] \right\}$. Any $\tilde{A} \in F(U)$ is called a fuzzy set on U. The α-cuts of $\tilde{A}$ is the crisp set $\tilde{A}_{\alpha} = \{u \in U | \tilde{A}(u) \geq \alpha\}$, for $0 < \alpha \leq 1$. Moreover, $\tilde{A}_0$ is separately defined [2] as the closure of the union of all the $\tilde{A}_{\alpha}$, for $0 < \alpha \leq 1$. The value $\tilde{A}(u)$ is interpreted as the membership degree of a point u. $\tilde{A} \in F(\mathbb{R})$ is called a fuzzy number, under the following conditions:

1. There is a unique $r_0 \in \mathbb{R}$ with $\tilde{A}(r_0) = 1$,
2. The α-cuts of $\tilde{A}$ are closed and bounded intervals on $\mathbb{R}$ for any $0 \leq \alpha \leq 1$,

where $\mathbb{R}$ is the set of all real numbers. In other words for every fuzzy number $\tilde{A}$ we have $\tilde{A}_{\alpha} = [a_1(\alpha), a_2(\alpha)]$ for all $\alpha \in [0, 1]$ which are the closed, bounded, intervals and their bounds are as functions of α.

To continue discussions, we need to clarify the concept of an unbiased fuzzy estimator, using the following definition. Similar to conventional statistics, a fuzzy estimator is a rule for calculating a fuzzy estimate of an unknown parameter based on observed data. Thus the rule and its result (the fuzzy estimate) are distinguished.

Definition 2.1 A fuzzy number $\tilde{\theta}$ is an unbiased fuzzy estimator for parameter θ from a statistical distribution if:

1. The α-cuts of $\tilde{\theta}$ are $(1 - \alpha)100\%$ confidence intervals for θ, with $\alpha \in [0.01, 1]$ and $\tilde{\theta}_{\alpha} = \tilde{\theta}_{0.01}$ for $\alpha \in [0, 0.01)$.
2. If $\hat{\theta}$ is an unbiased point estimator for θ then $\tilde{\theta}\left(\hat{\theta}\right) = 1$.

An explicit and unique membership function is given for a fuzzy estimate by the following theorem.

Theorem 2.1 [5] *Suppose that $X_1, X_2, \ldots, X_n$ is a random sample of size n from a distribution with unknown parameter θ. If, based on observations $x_1, x_2, \ldots, x_n$, we consider $\tilde{A}_{\alpha} = [\theta_1(\alpha), \theta_2(\alpha)]$ as a $(1 - \alpha)$ 100 % confidence interval for θ, then the fuzzy estimate of θ is a fuzzy set with the following unique membership function:*

$$\tilde{\theta}(u) = min\left\{\theta_1^{-1}(u), [-\theta_2]^{-1}(-u), 1\right\}.$$

To end this section, we give an introduction to interval arithmetic. Let $I = [a, b]$ and $J = [c, d]$ be two closed intervals. Then based on the interval arithmetic, we have

$$I + J = [a + c, b + d],$$

$$I - J = [a - d, b - c],$$

$$I \times J = [\alpha, \beta], \quad \alpha = min\{ac, ad, bc, bd\} \beta = max\{ac, ad, bc, bd\}$$

and

$$I/J = [a, b] \times [1/d, 1/c],$$

where zero does not belong to $J = [c, d]$ *in the last case.*

3   Linear Hypothesis Testing

In this section we give a brief review of linear hypothesis testing, for more details
see [12, 13]. The concepts of linear model and linear hypothesis are given in
Definition 3.1. The process of linear hypothesis testing is presented in Theorem 3.1.

Definition 3.1 Let $Y = (Y_1 Y_2 \ldots Y_n)'$ be a random column vector and $\mathbf{X}$ be a $n \times k$
matrix of full rank $k < n$ and known constants $x_{ij}, i = 1, 2, \ldots, n; j = 1, 2, \ldots, k$. It is
said that the distribution of Y satisfies a *linear model* if $E(Y) = \mathbf{X}\boldsymbol{\beta}$, where $\boldsymbol{\beta} =$
$(\beta_1 \beta_2 \ldots \beta_k)'$ is vector of unknown (scalar) parameters $\beta_1, \beta_2, \ldots, \beta_k$, where $\beta_j \in \mathbb{R}$
for $j = 1, 2, \ldots, k$. It is convenient to write $Y = \mathbf{X}\boldsymbol{\beta} + \boldsymbol{\epsilon}$, where $\boldsymbol{\epsilon} = (\epsilon_1 \epsilon_2 \ldots \epsilon_n)'$ is a
vector of non-observable independent normal random variables with common
variance σ^2 and $E(\epsilon_j) = 0$; $j = 1, 2, \ldots, n$. Relation $Y = \mathbf{X}\boldsymbol{\beta} + \boldsymbol{\epsilon}$ is known as a
linear model. The linear hypothesis concerns $\boldsymbol{\beta}$, such that $\boldsymbol{\beta}$ satisfies $H_0 : \mathbf{H}\boldsymbol{\beta} = \mathbf{0}$,
where $\mathbf{H}$ is a known $r \times k$ matrix of full rank $r \leq k$.

Theorem 3.1 *Consider the linear model* $Y = \mathbf{X}\boldsymbol{\beta} + \boldsymbol{\epsilon}$. *The generalized likelihood
ratio (GLR) test for testing the linear hypothesis* $H_0 : \mathbf{H}\boldsymbol{\beta} = \mathbf{0}$ *is to reject* H_0 *at
significance level* γ *if* $F \geq F_{1-\gamma, r, n-k}$, *where* $P_{H_0}(F < F_{1-\gamma, r, n-k}) = 1 - \gamma$ *and* F *is
the random variable given by*

$$F = \frac{SS^*/r}{SS/(n-k)}$$

where,

$$SS^* = \left(Y - \mathbf{X}\hat{\hat{\boldsymbol{\beta}}}\right)'\left(Y - \mathbf{X}\hat{\hat{\boldsymbol{\beta}}}\right) - \left(Y - \mathbf{X}\hat{\boldsymbol{\beta}}\right)'\left(Y - \mathbf{X}\hat{\boldsymbol{\beta}}\right)$$

and

$$SS = \left(Y - \mathbf{X}\hat{\boldsymbol{\beta}}\right)'\left(Y - \mathbf{X}\hat{\boldsymbol{\beta}}\right),$$

$\hat{\boldsymbol{\beta}}$ *is the maximum likelihood estimator (MLE) of* $\boldsymbol{\beta}$ *and* $\hat{\hat{\boldsymbol{\beta}}}$ *is the MLE of* $\boldsymbol{\beta}$ *under* H_0.
Moreover, under H_0 *the random variable* F *has the F-distribution with* r *and*
$(n - k)$ *degrees of freedom.*

Note 3.1 As the result of Theorem 3.1, it can be shown that the pivotal quantity SS/σ^2 has the distribution χ^2 with $(n - k)$ degrees of freedom and SS^*/σ^2 has the distribution χ^2 with r degrees of freedom, under the null hypothesis H_0. So both of these pivotal quantities can be used to produce confidence intervals for σ^2. It is clear that the statistics $SS/(n - k)$ and SS^*/r (under H_0) are the unbiased point estimators for the unknown parameter σ^2.

4 Linear Hypothesis Testing Based on Fuzzy Test Statistic

4.1 Testing at Precise Significance Level

In this section, taking into account Buckley's method in [2] and its modifications in [4], we consider testing the linear hypothesis based on a fuzzy test statistic and a fuzzy significance level. Because we could obtain a fuzzy test statistics to evaluate the linear hypothesis, we give several theorems sequentially. Also, we obtain a fuzzy critical value using α-cuts of a considered fuzzy significance level. Next we make two decision rules to the cases where the critical value is either crisp or fuzzy. In the rest of this paper, the symbols $\chi^2_{\xi,\upsilon}$ and $F_{\xi,\upsilon_1,\upsilon_2}$ will be used to represent the ξ'th quantile of the distribution χ^2 with υ degrees of freedom and the ξth quantile of the distribution F with υ_1 and υ_2 degrees of freedom, respectively.

Theorem 4.1.1 *In a linear model consider $SS/(n - k)$ as an unbiased point estimator for parameter σ^2. Then an unbiased fuzzy estimator for σ^2 is $\widetilde{\sigma^2}$ with the following α-cuts*

$$\left(\widetilde{\sigma^2}\right)_\alpha = \begin{cases} \left[SS/\chi^2_{1-\alpha+\alpha p',(n-k)}, SS/\chi^2_{\alpha p',(n-k)}\right] & 0.01 \leq \alpha \leq 1 \\ \left(\widetilde{\sigma^2}\right)_{0.01} & 0 \leq \alpha < 0.01 \end{cases}$$

in which p' is obtained from the relation $\chi^2_{p',(n-k)} = n - k$.

Proof Based on the pivotal quantity SS/σ^2, a $(1 - \alpha)\,100\,\%$ confidence interval for σ^2 is $\left[SS/\chi^2_{1-\alpha+\alpha p,(n-k)}, SS/\chi^2_{\alpha p,(n-k)}\right]$ for any $0 < \alpha < 1$ and $0 < p < 1$. When $\alpha = 1$ and $p = p'$, satisfying $\chi^2_{p',(n-k)} = n - k$, this interval becomes the point $SS/(n - k)$ which is unbiased point estimator for σ^2. Now fixing $p = p'$ and varying α from 0.01 to 1 we obtain nested intervals which are the α - cuts of a fuzzy number, say $\widetilde{\sigma^2}$. Finally, $\left(\widetilde{\sigma^2}\right)_\alpha = \left(\widetilde{\sigma^2}\right)_{0.01}$ for $0 \leq \alpha < 0.01$. So, we have the unbiased fuzzy estimator $\widetilde{\sigma^2}$ for σ^2. $\qquad\square$

Lemma 4.1.1 *The membership function of fuzzy estimator $\widetilde{\sigma^2}$ in Theorem 4.1.1 is as follows:*

$$\widetilde{\sigma^2}(u) = \begin{cases} \dfrac{1-G(SS/u)}{1-p'} & \dfrac{SS}{\chi^2_{0.99+0.01p',(n-k)}} \leq u \leq \dfrac{SS}{n-k} \\[3mm] \dfrac{G(SS/u)}{p'} & \dfrac{SS}{n-k} \leq u \leq \dfrac{SS}{\chi^2_{0.01p',(n-k)}} \\[3mm] 0 & otherwise, \end{cases}$$

where G is the cumulative distribution function of the χ^2 variable with $(n-k)$ degrees of freedom.

Proof By Theorem 4.1.1, we have $\theta_1(\alpha) = SS/\chi^2_{1-\alpha+\alpha p',(n-k)}$ for $0.01 \leq \alpha \leq 1$. Hence, $\theta_1^{-1}(u) = [1-G(ss/u)]/\left(1-p'\right)$. Also $\theta_2(\alpha) = SS/\chi^2_{\alpha p',(n-k)}$, therefore $[-\theta_2]^{-1}(-u) = [G(ss/u)]/p'$ for $0.01 \leq \alpha \leq 1$. Based on Theorem 2.1 $\widetilde{\sigma^2}(u) = \min\{\theta_1^{-1}(u), [-\theta_2]^{-1}(-u), 1\}$. So, the proof follows. □

Theorem 4.1.2 *Consider SS^*/r as an unbiased point estimator for parameter σ^2 under the null hypothesis $H_0 : \mathbf{H\beta} = \mathbf{0}$. Then, an unbiased fuzzy estimator for σ^2 is $\widetilde{\sigma^2_{H_0}}$ with α-cuts $\left(\widetilde{\sigma^2_{H_0}}\right)_\alpha$, where*

$$\left(\widetilde{\sigma^2_{H_0}}\right)_\alpha = \begin{cases} \left[SS^*/\chi^2_{1-\alpha+\alpha p'',r}, SS^*/\chi^2_{\alpha p'',r}\right] & 0.01 \leq \alpha \leq 1 \\[3mm] \left(\widetilde{\sigma^2_{H_0}}\right)_{0.01} & 0 \leq \alpha < 0.01 \end{cases}$$

in which p'' is obtained from the relation $\chi^2_{p'',r} = r$.

Proof Consider the pivotal quantity SS^*/σ^2. Now the proof is similar to that of Theorem 4.1.1. □

Notice that, similar to Lemma 4.1.1, one can derive the membership function of fuzzy estimator $\widetilde{\sigma^2_{H_0}}$, but under $H_0 : \mathbf{H\beta} = \mathbf{0}$.

Lemma 4.1.2 *The membership function of fuzzy estimator $\widetilde{\sigma^2_{H_0}}$ in Theorem 4.1.2 is as follows:*

$$\widetilde{\sigma^2_{H_0}}(u) = \begin{cases} \dfrac{1-G(SS^*/u)}{1-p''} & \dfrac{SS^*}{\chi^2_{0.99+0.01p'',r}} \leq u \leq \dfrac{SS^*}{r} \\[3mm] \dfrac{G(SS^*/u)}{p''} & \dfrac{SS^*}{r} \leq u \leq \dfrac{SS^*}{\chi^2_{0.01p'',r}} \\[3mm] 0 & otherwise, \end{cases}$$

where G is the cumulative distribution function of the χ^2 variable with r degrees of freedom.

Proof By Theorem 4.1.2, the proof is similar to that of Lemma 4.1.1. □

Remark 4.1.1 Theorems 4.1.1 and 4.1.2 define unbiased fuzzy estimators for σ^2 under null hypothesis H_0. Moreover, Lemmas 4.1.1 and 4.1.2 provide the membership functions of these two estimators.

Theorem 4.1.3 *The fuzzy test statistic for testing* $H_0 : \mathbf{H}\boldsymbol{\beta} = \mathbf{0}$ *is* $\tilde{F}$ *with* α-*cuts*

$$\tilde{F}_\alpha = \begin{cases} [f_1(\alpha)F, f_2(\alpha)F] & 0.01 \leq \alpha \leq 1 \\ \tilde{F}_{0.01} & 0 \leq \alpha < 0.01, \end{cases}$$

where

$$F = \frac{SS^*/r}{SS/(n-k)},$$

$$f_1(\alpha) = \left[(r)\chi^2_{\alpha p',(n-k)}\right] / \left[(n-k)\chi^2_{1-\alpha+\alpha p'',r}\right]$$

and

$$f_2(\alpha) = \left[(r)\chi^2_{1-\alpha+\alpha p',(n-k)}\right] / \left[(n-k)\chi^2_{\alpha p'',r}\right].$$

Proof Using the equality $\tilde{F}_\alpha = (\tilde{\sigma}^2_{H_0})_\alpha / (\tilde{\sigma}^2)_\alpha$ and interval arithmetic, the fuzzy test statistic follows from Buckley's method. □

Decision rule 4.1.1 After observing the data and crisp significance level γ, a typical method for rejecting or accepting the null hypothesis $H_0 : \mathbf{H}\boldsymbol{\beta} = \mathbf{0}$ can be made as follows. First we calculate the ratio $A_R/(A_R + A_L)$, where A_R (A_L) is area under the graph of the fuzzy test statistic $\tilde{F}$, but to the right (left) of the vertical line through $F_{1-\gamma,r,n-k}$ (see Fig. 1). Note that Fig. 1 just illustrates the sketch of A_R and A_L since the sides of $\tilde{F}$ are curves, not straight line segments. Next we choose a value for the credit level φ from $(0, 1]$, [1]. Finally, our decision rule at significance level γ is:

1. if $A_R/(A_R + A_L) \geq \varphi$, then reject the hypothesis $H_0 : \mathbf{H}\boldsymbol{\beta} = \mathbf{0}$,
2. if $A_R/(A_R + A_L) < \varphi$, then accept H_0.

Fig. 1 The areas A_R and A_L

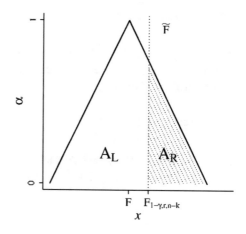

Remark 4.1.2 The presented decision rule 4.1.1 is reasonable since one can see that, by choosing any $\alpha \in [0, 1]$ and any $F \in \tilde{F}_\alpha$, this F is some value of the test statistic corresponding to this α which relates back to confidence intervals for σ^2. Therefore, if point (F, α) is in the region A_R then H_0 is rejected because $F \geq F_{1-\gamma,r,n-k}$, and if point (F, α) is in the region A_L then H_0 is accepted since $F < F_{1-\gamma,r,n-k}$.

Remark 4.1.3 In Decision rule 4.1.1, φ and γ are criterions which control possibilistic and probabilistic errors, respectively. Indeed they unify the concepts of randomness and fuzziness. The selected value of φ is more or less subjective and depends on the decision maker desire.

4.2 Testing at Fuzzy Significance Level

The approach for accepting or rejecting the null hypothesis in Subsection 4.1 is on the basis of comparing the observed fuzzy test statistic $\tilde{F}$ with the crisp critical value $F_{1-\gamma,r,n-k}$ at a crisp significance level γ. In practice it is more natural to consider the significance level as a fuzzy set since the test statistic is fuzzy. In fact, a fuzzy significance level is considered as a fuzzy number on $(0, 1)$, [6, 15]. Subsequently, we define a fuzzy significance level as a fuzzy number. We obtain a fuzzy critical value to evaluate the linear hypothesis using α-cuts of the defined fuzzy significance level. Finally, we provide a decision rule to decide whether to reject or accept the null hypothesis $H_0 : \mathbf{H\beta} = \mathbf{0}$.

Definition 4.2.1 A fuzzy significance level is a fuzzy number with the following α-cuts

$$\tilde{\gamma}_\alpha = \begin{cases} [\gamma_1 + (\gamma - \gamma_1)\alpha, \gamma_2 - (\gamma_2 - \gamma)\alpha] & 0.01 \leq \alpha \leq 1 \\ \tilde{\gamma}_{0.01} & 0 \leq \alpha < 0.01, \end{cases}$$

where $0 < \gamma_1 \leq \gamma \leq \gamma_2 < 1$.

Theorem 4.2.1 *In linear hypothesis testing based on fuzzy test statistics at the introduced fuzzy significance level in Definition 4.2.1, the critical value is a fuzzy number with the following α-cuts*

$$(\widetilde{cv})_\alpha = \begin{cases} [F_{(1-\gamma_2+(\gamma_2-\gamma)\alpha),r,n-k}, F_{(1-\gamma_1-(\gamma-\gamma_1)\alpha),r,n-k}] & 0.01 \leq \alpha \leq 1 \\ (\widetilde{cv})_{0.01} & 0 \leq \alpha < 0.01. \end{cases}$$

Proof The proof follows by substituting α-cuts of the fuzzy significance level $\tilde{\gamma}$ for crisp one γ in the crisp critical level and by using the interval arithmetic.   □

Example 4.2.1 Consider a linear hypothesis testing at the significance level $\gamma = 0.05$ where $n = 25$, $r = 3$ and $k = 4$. Assume that the significance level is a fuzzy number with the following α-cuts

$$\tilde{\gamma}_\alpha = \begin{cases} [0.03 + (0.05 - 0.03)\alpha, 0.07 - (0.07 - 0.05)\alpha] & 0.01 \le \alpha \le 1 \\ \tilde{\gamma}_{0.01} & 0 \le \alpha < 0.01. \end{cases}$$

Then based on Theorem 4.2.1 the fuzzy critical value is a fuzzy number with α-cuts $(\tilde{cv})_\alpha$ as follows

$$(\tilde{cv})_\alpha = \begin{cases} \left[F_{(0.93 + 0.02\alpha),3,21}, \quad F_{(0.97 - 0.02\alpha),3,21} \right] & 0.01 \le \alpha \le 1 \\ (\tilde{cv})_{0.01} & 0 \le \alpha < 0.01. \end{cases}$$

Figure 2 shows the fuzzy numbers $\tilde{\gamma}$ and $\tilde{cv}$.

Decision rule 4.2.1 After observing the data, the final decision rule is derived by comparing two fuzzy numbers $\tilde{cv}$ and $\tilde{F}$. Here a way is provided to decide whether to reject or accept the null hypothesis $H_0 : \mathbf{H}\boldsymbol{\beta} = \mathbf{0}$. First we calculate the ratio $A_R/(A_R + A_L)$, where A_R and A_L are depicted in Fig. 3. Note that Fig. 3 just illustrates the sketch of A_R and A_L since the sides of $\tilde{F}$ and $\tilde{cv}$ are curves, not straight line segments. Next we choose a value for the credit level φ from $(0, 1]$. Finally, our decision rule at significance level γ is as follows:

1. if $A_R/(A_R + A_L) \ge \varphi$, then reject the hypothesis $H_0 : \mathbf{H}\boldsymbol{\beta} = \mathbf{0}$,
2. if $A_R/(A_R + A_L) < \varphi$, then accept H_0.

Fig. 2 The fuzzy numbers $\tilde{\gamma}$ and $\tilde{cv}$

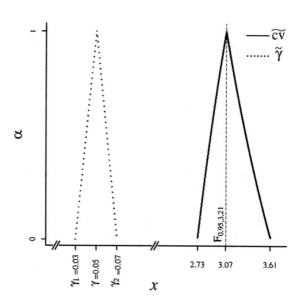

Fig. 3 The areas A_R and A_L

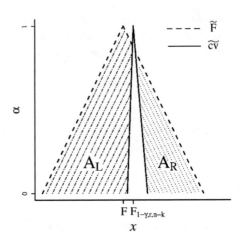

Remark 4.2.1 Decision rule 4.2.1 is reasonable since one can see that, by choosing any $\alpha \in [0, 1]$, any $F \in \tilde{F}_\alpha$ and any $F_{1-\gamma,r,n-k} \in (\tilde{cv})_\alpha$, $F_{1-\gamma,r,n-k}$ and F are some values of the test statistic and the critical level corresponding to this α which relates back to confidence intervals for σ^2 and α-cuts for the fuzzy significance level, respectively. Therefore, if point (F, α) is in the region A_R then we reject H_0 because $F \geq \max\{F_{1-\gamma,r,n-k} : F_{1-\gamma,r,n-k} \in (\tilde{cv})_\alpha\}$, if point (F, α) is in the region A_L then we decide to accept H_0 since $F < \min\{F_{1-\gamma,r,n-k} : F_{1-\gamma,r,n-k} \in (\tilde{cv})_\alpha\}$, and finally if point (F, α) is not in the region A_L or A_R then we do not make any decision on H_0. To this end, we have not shared point (F, α) in the final decision in Decision rule 4.2.1.

Remark 4.2.2 While, Buckley [2, 3] and Taheri et al. [1, 14] consider the problem of testing hypothesis based on a fuzzy test statistic and a crisp significance level, we assume that the significance level is fuzzy. Our method is, therefore, more convenient in real world studies.

5 One-Way ANOVA: A Simple Case of Linear Hypothesis Testing

In this section, one-way ANOVA is considered taking into account the method of linear hypothesis testing based on the fuzzy test statistic and fuzzy significance level. However, we must mention that the procedure proposed in this article is still applicable for any case of linear hypothesis testing. We now give a brief review of one-way ANOVA. For more detail refer to [8, 12]. Consider the linear model

$$Y_{ij} = \mu_i + \epsilon_{ij}, \quad j = 1, 2, \ldots, n_i; \quad i = 1, 2, \ldots, k,$$

where ϵ_{ij}'s have a normal distribution with an unknown variance σ^2, zero mean and μ_i's are unknown parameters. We are interested in testing the linear hypothesis $H_0 : \mu_1 = \mu_2 = \cdots = \mu_k$. To simplify the discussion we use the following notations

$$SSTr = \sum_{i=1}^{k} n_i(\bar{Y}_{i.} - \bar{Y}_{..})^2, \quad SSE = \sum_{i=1}^{k} \sum_{j=1}^{n_i} (Y_{ij} - \bar{Y}_{i.})^2, \quad n = \sum_{i=1}^{k} n_i \text{ and}$$

$$F1 = \frac{SSTr/(k-1)}{SSE/(n-k)},$$

where $\bar{Y}_{i.} = \sum_{j=1}^{n_i} Y_{ij}/n_i$ and $\bar{Y}_{..} = \sum_{i=1}^{k} \sum_{j=1}^{n_i} Y_{ij}/n$. By replacing $\boldsymbol{\beta} = (\mu_1\mu_2\ldots\mu_k)'$ in Theorem 3.1, it can be shown that $SS = SSE, SS^* = SSTr, r = (k-1)$, $F = F1$ and the null hypothesis $H_0 : \mu_1 = \mu_2 = \cdots = \mu_k$ is rejected if the observed value of $F1$ statistic is greater than or equal to $F_{1-\gamma,k-1,n-k}$. The case described above is referred to as a one-way analysis of variance which is a very simple case of linear hypothesis testing. One-way ANOVA has many applications in agricultural and engineering sciences.

6 Illustrative Examples

Example 6.1 An experiment is conducted to determine if there is a difference in the breaking strength of a monofilament fibre produced by four different machines for a textile company. Also it is known that all fibres are of equal thickness. A random sample is selected from each machine. The fibre strength y for each specimen is shown in Table 1. The one-way ANOVA model is $Y_{ij} = \mu_i + \epsilon_{ij}, j = 1, 2, \ldots, n_i;$ $i = 1, 2, 3, 4$. We are going to test the null hypothesis $H_0 : \mu_1 = \mu_2 = \mu_3 = \mu_4$. All computations are done by R software [11].

In the traditional statistics point of view and based on Theorem 3.1, we have $F1 = 2.789$ and $F_{0.95,3,32} = 2.901$. Therefore we accept H_0 at the crisp significance level $\gamma = 0.05$ because $F1 < F_{0.95,3.32}$. In other words, there is not any difference at significance level 0.05 in the breaking strength of a monofilament fibre produced by

Table 1 Breaking strength data where y is strength in pounds

Machine 1	$y_{11} = 37$	$y_{12} = 41$	$y_{13} = 40$	$y_{14} = 40$	$y_{15} = 39$
	$y_{16} = 35$	$y_{17} = 39$	$y_{18} = 39$	$y_{19} = 40$	$y_{110} = 43$
Machine 2	$y_{21} = 39$	$y_{22} = 41$	$y_{23} = 41$	$y_{24} = 40$	$y_{25} = 43$
	$y_{26} = 41$	$y_{27} = 42$	$y_{28} = 38$	$y_{29} = 40$	
Machine 3	$y_{31} = 45$	$y_{32} = 42$	$y_{33} = 40$	$y_{34} = 41$	$y_{35} = 40$
	$y_{36} = 40$	$y_{37} = 41$	$y_{38} = 40$		
Machine 4	$y_{41} = 41$	$y_{42} = 44$	$y_{43} = 42$	$y_{44} = 41$	$y_{45} = 41$
	$y_{46} = 41$	$y_{47} = 43$	$y_{48} = 39$	$y_{49} = 41$	

four machines. In follows, consider two situations to understand the need of presenting fuzzy-decision-based approach:

1. Let us only change $y_{11} = 37$, in Table 1, to $y_{11} = 36$. Now we have $F1 = 2.907$ and $F_{0.95,3,32} = 2.901$. So we reject $H_0 : \mu_1 = \mu_2 = \mu_3 = \mu_4$, for $\gamma = 0.05$ because $F1 \geq F_{0.95,3,32}$ (i.e. there is a difference in the breaking strength of a monofilament fibre produced by four machines).
2. Reconsider the observations of the experiment in Table 1, if we change the crisp significance level $\gamma = 0.05$ to $\gamma = 0.06$ then the null hypothesis $H_0 : \mu_1 = \mu_2 = \mu_3 = \mu_4$ is rejected since $F1 = 2.789$ is greater than $F_{0.94,3,32} = 2.732$.

Therefore, we are not sure whether to accept or reject H_0, since the values of $F_{1-\gamma,k-1,n-k}$ and $F1$ are close to each other, based on data in Table 1 and $\gamma = 0.05$. We overcome the sensitivity of this test by using linear hypothesis testing based on a fuzzy test statistic and a fuzzy significance level. In this example we have $SS = SSE = 97.419$, $SS^* = SSTr = 25.469$, $r = (k - 1) = 3$, $F = F1 = 2.789$ and $n = 36$. Based on Theorem 4.1.1 a fuzzy estimate for σ^2 is a fuzzy number with α-cuts

$$
(\widetilde{\sigma^2})_\alpha = \begin{cases} \left[97.419/\chi^2_{1-\alpha+\alpha(0.533),32}, \, 97.419/\chi^2_{\alpha(0.533),32} \right] & 0.01 \leq \alpha \leq 1 \\ (\widetilde{\sigma^2})_{0.01} & 0 \leq \alpha < 0.01 \end{cases}
$$

where $p' = 0.533$ is obtained from the relation $\chi^2_{p',32} = 32$. By Lemma 4.1.1, the membership function of this fuzzy estimate can be given by

$$
\widetilde{\sigma^2}(u) = \begin{cases} \dfrac{1 - G(97.419/u)}{1 - 0.533} & \dfrac{97.419}{\chi^2_{0.99 + 0.01(0.533),32}} \leq u \leq \dfrac{97.419}{32} \\ \dfrac{G(97.419/u)}{0.533} & \dfrac{97.419}{32} \leq u \leq \dfrac{97.419}{\chi^2_{0.01(0.533),32}} \\ 0 & otherwise, \end{cases}
$$

where G is the cumulative distribution function of a χ^2 variable with 32 degrees of freedom, as depicted in Fig. 4. Also, under the null hypothesis $H_0 : \mu_1 = \mu_2 = \mu_3 = \mu_4$ an unbiased fuzzy estimate for σ^2 based on Theorem 4.1.2 is a fuzzy number with the following α - cuts:

$$
(\widetilde{\sigma^2_{H_0}})_\alpha = \begin{cases} \left[25.469/\chi^2_{1-\alpha+\alpha(0.608),3}, \, 25.469/\chi^2_{\alpha(0.608),3} \right] & 0.01 \leq \alpha \leq 1 \\ (\widetilde{\sigma^2_{H_0}})_{0.01} & 0 \leq \alpha < 0.01 \end{cases}
$$

and $p'' = 0.608$ is obtained from the relation $\chi^2_{p'',3} = 3$. Now by Theorem 4.1.3, the observed value of the fuzzy test statistic $\tilde{F}$ is a fuzzy number with the following α-cuts (see Fig. 5):

Fig. 4 The fuzzy estimator
for σ^2

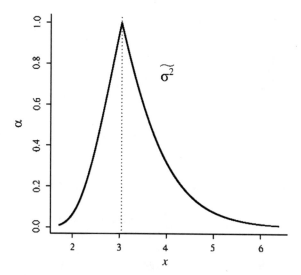

Fig. 5 The fuzzy numbers $\tilde{F}$
and $\widetilde{cv}$

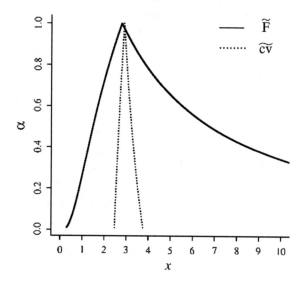

$$\tilde{F}_\alpha = \begin{cases} \left[\dfrac{\chi^2_{\alpha(0.533),32}}{\chi^2_{1-\alpha+\alpha(0.608),3}} 0.261, \dfrac{\chi^2_{1-\alpha+\alpha(0.533),32}}{\chi^2_{\alpha(0.608),3}} 0.261 \right] & 0.01 \leq \alpha \leq 1 \\ \tilde{F}_{0.01} & 0 \leq \alpha < 0.01. \end{cases}$$

By using Definition 4.2.1 we consider the fuzzy significance level as a fuzzy
number with the following α-cuts:

Table 2 Quantity of dissolved oxygen (%)

Location 1	$y_{11} = 7.8$	$y_{12} = 6.4$	$y_{13} = 8.2$	$y_{14} = 6.9$		
Location 2	$y_{21} = 6.7$	$y_{22} = 6.8$	$y_{23} = 7.1$	$y_{24} = 6.9$	$y_{25} = 7.3$	
Location 3	$y_{31} = 7.2$	$y_{32} = 7.4$	$y_{33} = 6.9$	$y_{34} = 6.4$	$y_{35} = 6.5$	
Location 4	$y_{41} = 6$	$y_{42} = 7.4$	$y_{43} = 6.5$	$y_{44} = 6.9$	$y_{45} = 7.2$	$y_{46} = 6.8$

$$\tilde{\gamma}_\alpha = \begin{cases} [0.02 + 0.03\alpha, 0.08 - 0.03\alpha] & 0.01 \le \alpha \le 1 \\ \tilde{\gamma}_{0.01} & 0 \le \alpha < 0.01, \end{cases}$$

and by Theorem 4.2.1 one can obtain α-cuts of the fuzzy critical value as follows:

$$(\widetilde{cv})_\alpha = \begin{cases} \left[F_{(0.92 + 0.03\alpha),3,32}, F_{(0.98 - 0.03\alpha),3,32} \right] & 0.01 \le \alpha \le 1 \\ (\widetilde{cv})_{0.01} & 0 \le \alpha < 0.01. \end{cases}$$

The graphs of the fuzzy numbers $\widetilde{cv}$ and $\tilde{F}$ are shown in Fig. 5. Finally, by Decision rule 4.2.1, since $A_R/(A_R + A_L) = 0.8971$ where $A_R = 9.7892$ and $A_L = 1.1230$, the null hypothesis $H_0 : \mu_1 = \mu_2 = \mu_3 = \mu_4$ is rejected for every credit level $\varphi \in (0, 0.8971]$. In fact it is possible for us to reject H_0 for a high level of credit, since high ratio of observed values of the test statistic lead to reject H_0.

Example 6.2 The quantity of oxygen dissolved in water is used as a measure of water pollution. Samples are taken at four locations in a lake and the quantity of dissolved oxygen is recorded in [12] as in Table 2 (lower reading corresponds to greater pollution). We would like to see that whether the data indicate a significant difference in the average amount of dissolved oxygen for the four location based on a fuzzy test statistic and the crisp significance level $\gamma = 0.05$. The one-way ANOVA model is $Y_{ij} = \mu_i + \epsilon_{ij}, j = 1, 2, \ldots, n_i; i = 1, 2, 3, 4$.

In this example we have $SS = SSE = 4.267$, $SS^* = SSTr = 0.718$, $r = (k - 1) = 3$, $F = F1 = 0.897$, $n = 20$ and $F_{0.95,3,16} = 3.239$. By Lemma 4.1.1, the membership function of the fuzzy estimate for σ^2 is given as follows:

$$\tilde{\sigma}^2(u) = \begin{cases} \dfrac{1 - G(4.267/u)}{1 - 0.547} & \dfrac{4.267}{\chi^2_{0.99 + 0.01(0.547),16}} \le u \le \dfrac{4.267}{16} \\ \dfrac{G(4.267/u)}{0.547} & \dfrac{4.267}{16} \le u \le \dfrac{4.267}{\chi^2_{0.01(0.547),16}} \\ 0 & otherwise, \end{cases}$$

where G is the cumulative distribution function of a χ^2 variable with 16 degrees of freedom, as depicted in Fig. 6.

By Theorem 4.1.3, the observed value of the fuzzy test statistic $\tilde{F}$ is a fuzzy number with the following α-cuts:

Fig. 6 The fuzzy estimator for σ^2

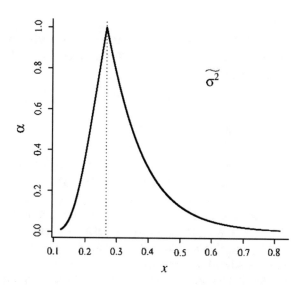

$$\tilde{F}_\alpha = \begin{cases} \left[\dfrac{\chi^2_{\alpha(0.547),16}}{\chi^2_{1-\alpha+\alpha(0.608),3}} 0.168, \quad \dfrac{\chi^2_{1-\alpha+\alpha(0.547),16}}{\chi^2_{\alpha(0.608),3}} 0.168 \right] & 0.01 \le \alpha \le 1 \\ \tilde{F}_{0.01} & 0 \le \alpha < 0.01. \end{cases}$$

The graph of the fuzzy test statistic $\tilde{F}$ is shown in Fig. 7. Finally, by Decision rule 4.1.1, since $A_R/(A_R + A_L) = 0.578$ where $A_R = 2.467$ and $A_L = 1.797$, the

Fig. 7 The fuzzy test statistic $\tilde{F}$

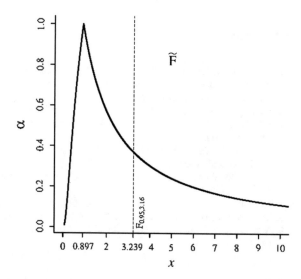

null hypothe-sis $H_0 : \mu_1 = \mu_2 = \mu_3 = \mu_4$ is accepted for every credit level $\varphi \in (0.578, 1]$. In other words, there is not any difference at significance level 0.05 in the average amount of dissolved oxygen for the four location for every credit level $\varphi \in (0.578, 1]$.

7 Conclusions

We have applied fuzzy techniques to linear hypothesis testing in this paper. Basically, in this method a set of $(1 - \alpha)$ 100 % confidence intervals, for all $0.01 \leq \alpha \leq 1$, are employed to produce the notion of the fuzzy test statistic. Also the concept of fuzzy critical value is derived based on α-cuts of a defined fuzzy significance level. Then, decision rules are provided based on these notions. Employing all the confidence intervals from the 99 % to the 0 % rather than only a single confidence interval results in using far more information in data for the statistical inference. Moreover, this method improves the statistical hypotheses testing when there is an uncertainty in accepting or rejecting the hypotheses. This issue is clarified by practical examples. As a simple case of the linear hypothesis testing, one-way analysis of variance based on fuzzy test statistic and fuzzy significance level is discussed. Nevertheless, as a matter of fact, the proposed method in this article is still applicable to other cases of linear hypothesis testing. An interesting topic for future research is the study of the proposed method on the linear hypothesis testing when the hypotheses are fuzzy rather than crisp. Also, one can consider this problem based on fuzzy data with crisp/fuzzy parameters.

References

1. Arefi, M., Taheri, S.M.: Testing fuzzy hypotheses using fuzzy data based on fuzzy test statistic. J. Uncertain Syst. **5**, 45–61 (2011)
2. Buckley, J.J.: Fuzzy statistics: hypothesis testing. Soft. Comput. **9**, 512–518 (2005)
3. Buckley, J.J.: Fuzzy Probability and Statistics. Springer, Berlin Heidelberg (2006)
4. Falsafain, A., Taheri, S.M.: On Buckley's approach to fuzzy estimation. Soft. Comput. **15**, 345–349 (2011)
5. Falsafain, A., Taheri, S.M., Mashinchi, M.: Fuzzy estimation of parameter in statistical model. Int. J. Comput. Math. Sci. **2**, 79–85 (2008)
6. Holena, M.: Fuzzy Hypothesis testing in a framework of fuzzy logic. Int. J. Intell. Syst. **24**, 529–539 (2009)
7. Kalpanapriya, D., Pandian, P.: Int. J. Mod. Eng. Res. **2**, 2951–2956 (2012)
8. Montgomery, D.C.: Design and Analysis of Experiments, 3rd edn. Wiley, New York (1991)
9. Nguyen, H.T., Walker, E.A.: A First Course in Fuzzy Logic, 3rd ed. Paris: Chapman Hall/CRC (2005)
10. Nourbakhsh, M.R., Parchami, A., Mashinchi, M.: Analysis of variance based on fuzzy observations. Int. J. Syst. Sci. **44**(4), 714–726 (2013)
11. Rizzo, M.L.: Statistical Computing with R. Chapman Hall/CRC, Paris (2008)

12. Rohatgi, V.K., Ehsanes Saleh, A.K.M.: An Introduction to Probability and Statistics, 2nd edn. Wiley, New York (2001)
13. Searl, S.R.: Linear Models. Wiley, New York (1971)
14. Taheri, S.M., Arefi, M.: Testing fuzzy hypotheses based on fuzzy test statistic. Soft. Comput. **13**, 617–625 (2009)
15. Taheri, S.M., Hesamian, G.: Goodman-Kruskal measure of assocition for fuzzy-categorized variables. Kybernetika **47**(110), 122 (2011)
16. Wu, H.C.: Analysis of variance for fuzzy data. Int. J. Syst. Sci. **38**, 235–246 (2007)
17. Zadeh, L.A.: Fuzzy sets. Inf. Control **8**, 338–359 (1965)

A Practical Application of Fuzzy Analysis of Variance in Agriculture

R. Ivani, S.H. Sanaei Nejad, B. Ghahraman, A.R. Astaraei and H. Feizi

Abstract For comparing several populations, the fuzzy analysis of variance has been summarized and reviewed where the collected data considered fuzzy rather than crisp numbers. As a practical work based on the real-word data, a case study was carried out to investigate effects of three concentrations (0, 50 and 100 ppm) $nanoSiO_2$ on seedling growth and dry matter weight of fenugreek (*Trigonella foenum-graceum* L.). All presented data in this study are fuzzy and therefore we need an extended version of analysis of variance to investigate on these fuzzy observations. Although, the presented analysis of variance approach based on vague data can causes to a fuzzy decision, but as an advantage of the proposed approach the vagueness of this fuzzy decision measured.

Keywords Fuzzy analysis of variance · $NanoSiO_2$ · Fenugreek seedling · Fuzzy numbers

R. Ivani (✉)
Faculty of Agriculture, Water Engineering Department,
International Branch of Ferdowsi University of Mashhad, Mashhad, Iran
e-mail: reyhane.ivani@gmail.com

S.H. Sanaei Nejad · B. Ghahraman
Faculty of Agriculture, Water Engineering Department,
Ferdowsi University of Mashhad, Mashhad, Iran
e-mail: sanaein@gmail.com

B. Ghahraman
e-mail: bijangh@ferdowsi.um.ac.ir

A.R. Astaraei
Faculty of Agriculture, Soil Science Department,
Ferdowsi University of Mashhad, Mashhad, Iran
e-mail: alirezaastaraei@yahoo.com

H. Feizi
Faculty of Agriculture and Natural Resources, Plant Production Department,
University of Torbat Heydarieh, Torbat Heydarieh, Iran
e-mail: hasanfeizi@yahoo.com

© Springer International Publishing Switzerland 2016
C. Kahraman and Ö. Kabak (eds.), *Fuzzy Statistical Decision-Making*,
Studies in Fuzziness and Soft Computing 343,
DOI 10.1007/978-3-319-39014-7_17

315

1 Background and Introduction

Nanotechnology is a science that has widely application in industrial and com-
mercial products in recent years [17]. Nowadays, the rapidly increasing applications
of engineered nanomaterials, with sizes smaller than 100 nm in the economy various
areas, such as electronics, cosmetics, textiles, pharmaceutics, and environmental
remediation, have received a lot of attention and concern [4]. Nanotechnology
allows wide advances in agriculture science, such as agricultural transfer, repro-
ductive science and technology, wastes of food to energy, disease prevention, and
treatment in plants using various nanocides [4]. Nanoparticles, because of their tiny
size show unique characteristics. They can change physic–chemical properties
compared to their bulk particles. Because of these larger surface areas, their solu-
bility and surface reactivity was higher than their bulk particles [5]. However, the
interaction mechanisms understanding at the molecular level between biological
systems and nanomaterials is largely unknown [3, 9].

Silicon is the second most abundant element in the earth's crust, yet its role in
plants biology has been poorly understood [20]. Although silicon has not been
listed among the essential elements for plants, but it can reach levels in plants
similar to those of macro elements [2]. Ahmed et al. [1] reported that silicon caused
to reduce the impacts of different stresses such as metal toxicity, various pests, high
temperature, nutrients imbalance, salt and drought stresses.

Other studies were reported that nanoSiO$_2$ can promote growth and dry matter
weights in borage seedling (*Borago officinalis* L.) [24], tomato (*Lycopersicum
esculentum* Mill. cv Super Strain B) [23] and tomato [12]. Lu et al. [16] indicated
that a combination of nanosized TiO$_2$ and SiO$_2$ could increase the nitrate reductase
enzyme in soybean (*Glycine max*) and its abilities of utilizing water and fertilizer
which in fact end up to accelerate its germination and growth.

The observations including vagueness can be treated using the concept of fuzzy
sets. In other words, the vagueness which is included in data can be expressed
exactly using fuzzy set membership function, and hence, the vague data can be
expressed by the membership functions. Such vague data is processed directly
using the membership function in the statistical analysis. The calculation process
becomes more complicated with respected to the traditional statistical analysis [15],
because it is necessary to perform the calculation precisely using the membership
functions. In many environmental and applied sciences such as social sciences,
agriculture and geology, there are several real-life populations where non-precise
values can be assigned to their experimental outcomes. In this way, fuzzy numbers
are suitable models to handle and formulize these populations in real cases, which is
the reason of our need to the fuzzy set theory in analysis of variance (ANOVA). In
recent years, some papers which have concentrated on different areas of ANOVA
using fuzzy set theory have been published. These areas are presented in follow:
investigating on the behavior of one-way fuzzy analysis of variance (FANOVA)
and comparing it with regression model [7], bootstrap method for approximation of
asymptotic one-sample tests by fuzzy random variables [18], developing a one-way

ANOVA approach for the functional data on a given Hilbert space [6], exact one-way ANOVA testing under normal fuzzy random variables [19], processing analysis of variance method using the moment correction for vague data [15], bootstrap asymptotic multi-sample testing of means assuming simple fuzzy random variables [10, 11], considering the cuts of fuzzy random variables for one-way analysis of variance on the basis of fuzzy data based on optimization approach [25], extending one-way ANOVA for fuzzy observations based on extension principle approach [21], generalizing one-way ANOVA for fuzzy random variables and using least squares method for estimating fuzzy parameters [14].

Mathematics of FANOVA is not as well developed as conventional ANOVA, and this may be one of the reasons why the use of FANOVA has not been reported in the environmental literature. This research can be considered as the first step to cover the gap between environmental practice and theoretical FANOVA. In this paper, the authors are going to test the mean absorption of Cadmium and Lead in aerial and bellow corn parts to conclude whether it is dependent on the added levels of organic fertilizers.

Organizing this paper is as follows. Several arithmetic operations on triangular fuzzy numbers reviewed in Sect. 2. As an alternative ANOVA method for fuzzy environments, a new approach to analysis of variance reviewed in Sect. 3 from [21], when the observed data are fuzzy rather than precise. As an application of this alternative ANOVA method, an agricultural case study presented in Sect. 4 about the effect of nanoparticle concentrations on fenugreek seedling based on the generated real-world data in a Lab. of Ferdowsi University of Mashhad. Conclusions given in the final section.

2 Arithmetic Operations on Fuzzy Numbers

The notion of a fuzzy set extends the concept of set membership to situations in which there are many continuum grades of membership [26]. Unlike a classical set which has a clearly defined boundary, in the sense that a real number is either a member of the set or it is not, a fuzzy set is a set without a precise and crisp boundary, and it can contain elements with only a partial degree of membership. In other words, a given element can simultaneously be a member of more than one set. For example, suppose one define the optimum range of Zn absorption in a plant as interval [15,80] mg kg^{-1} DM. Using traditional set theory, it is possible to define the equilibrium absorption amount as a single value set containing the element 25 mg kg^{-1} DM, or define it as a broader set containing the elements between 30 and 40 mg kg^{-1} DM. In both cases, a crisp set considered and any given absorption amount is either in or not in the equilibrium range. Fuzzy sets allow for partial membership and an absorption amount 25 mg kg^{-1} DM might regarded as having partial membership of the (fuzzy) equilibrium set and partial membership of a below-equilibrium set. In practice, this allows the resulting model a high degree of flexibility in dealing with uncertainty and imprecision which is in the nature of many real world problems.

An especial case of fuzzy numbers called triangular fuzzy number (TFN), where its membership function defined by

$$T_{a,b,c}(x) = \begin{cases} (x-a+b)/b & \text{if } a-b \leq x < a \\ (a+c-x)/c & \text{if } a \leq x < a+c \\ 0 & \text{elsewhere} \end{cases} \tag{1}$$

and it symbolically denoted by $T(a,b,c)$. The real number a called the core value and the positive real numbers b and c called left and right spreads of TFN, respectively. $F_T(R)$ and $F_T(R^+) = \{T_{a,b,c} \mid a,b,c \in R^+\}$, respectively, denote the set of all TFNs and the set of all positive TFNs, where R^+ is the set of all positive real numbers. Also, symmetric triangular fuzzy number (STFN) symbolically denoted by $T(a,b)$ in this paper where left and right spreads considered equal to b.

The following equations have been proved in [8] for any $T(a,b,c)$, $T(a',b',c') \in F_T(R)$ by Zadeh's extension principle:

$$T(a,b,c) \oplus T(a',b',c') = T(a+a',b+b',c+c'), \tag{2}$$

$$T(a,b,c) \ominus T(a',b',c') = T(a-a',b+c',c+b'). \tag{3}$$

Also, for $T(a,b,c)$, $T(a',b',c') \in F_T(R^+)$, operations $\otimes$ and ø are given by approximation as:

$$T(a,b,c) \otimes T(a',b',c') \cong T(aa', ab'+a'b, ac'+a'c), \tag{4}$$

and

$$T(a,b,c) \, \text{ø} \, T(a',b',c') \cong T(\frac{a}{a'}, \frac{ac'+a'b}{a'^2}, \frac{ab'+a'c}{a'^2}), \quad a' \neq 0. \tag{5}$$

Meanwhile, the scalar multiplication of $T(a,b,c) \in F_T(R)$ and $k \in \{0\} \cup R^+$ is defined by:

$$T(a,b,c) \odot k = k \odot T(a,b,c) = T(ka,kb,kc) \tag{6}$$

3   FANOVA Based on Triangular Fuzzy Numbers

In this section, we briefly review a new extended version of fuzzy ANOVA which we call it FANOVA from [21] which is needed for the real-word data agricultural case study in this chapter. Let r denotes the number of levels of the factor under study, any one of these levels is denoted by the index i, $i = 1, \ldots, r$. The number of cases for the ith factor level is denoted by n_i, and the total number of cases in the study is denoted by n_t, i.e. $n_t = \sum_{i=1}^{r} n_i$. The index j will be used to identify the given case or trial for a particular factor level. Therefore, y_{ij} denotes the jth observation on the response variable for the ith factor level and Y_{ij} denotes its corresponding random variable. For instance, Y_{ij} can be the sales volume of the jth store featuring of the ith type of shelf display, or the productivity of the jth employee in the ith plant. The number of cases or trials for the ith factor level is denoted by n_i, and so $j = 1, \ldots, n_i$. Similar to the classical ANOVA model, FANOVA model can state by:

$$Y_{ij} = \mu_i + \varepsilon_{ij}, \quad \text{for } i = 1, \ldots, r \text{ and } j = 1, \ldots, n_i, \qquad (7)$$

in which, Y_{ij}'s are the response variables in the jth trial for the ith factor level, μ_i's are the factor level means, and ε_{ij}'s are independent random variables having the normal distribution $N(0, \sigma^2)$. Therefore, one can expect that Y_{ij}'s are independent random variables having the normal distribution $N(\mu_i, \sigma^2)$ for any $i = 1, \ldots, r$. Let the total sum of squares (SST), the treatment sum of squares (SSTR), and the error sum of squares (SSE) be respectively defined by

$$SST = \sum_{i=1}^{r} \sum_{j=1}^{n_i} (Y_{ij} - \overline{Y}_{..})^2 = \sum_{i=1}^{r} \sum_{j=1}^{n_i} Y_{ij}^2 - \frac{Y_{..}^2}{n_t},$$

$$SSTR = \sum_{i=1}^{r} n_i (\overline{Y}_{i.} - \overline{Y}_{..})^2 = \sum_{i=1}^{r} \frac{Y_{i.}^2}{n_i} - \frac{Y_{..}^2}{n_t}$$

and

$$SSE = \sum_{i=1}^{r} \sum_{j=1}^{n_i} (Y_{ij} - \overline{Y}_{i.})^2 = \sum_{i=1}^{r} \sum_{j=1}^{n_i} Y_{ij}^2 - \sum_{i=1}^{r} \frac{Y_{i.}^2}{n_i},$$

where

$\overline{Y}_{..} = \frac{Y_{..}}{n_t} = \frac{1}{n_t} \sum_{i=1}^{r} \sum_{j=1}^{n_i} Y_{ij}$ and $\overline{Y}_{i.} = \frac{Y_{i.}}{n_i} = \frac{1}{n_i} \sum_{j=1}^{n_i} Y_{ij}$, for $i = 1, \ldots, r$. The mean of squares are obtained by $MSTR = \frac{SSTR}{r-1}$ and $MSE = \frac{SSE}{n_t - r}$ where stands for treatment mean square and error mean square, respectively. And finally, the FANOVA test statistic is $F = \frac{MSTR}{MSE}$.

It must be noted that the introduced FANOVA test statistic is a usual random variable and therefore similar to test statistic in classical ANOVA, has Fisher distribution with $r - 1$ and $n_t - r$ degrees of freedom. But, the only difference between classical ANOVA and FANOVA is in the kind of observed data. In other words, as a developed version of conventional ANOVA, FANOVA is able to handle with fuzzy-valued data on the basis of Zadeh's extension principle. Considering the above discussion, it is assumed that we are concerned with a conventional ANOVA, where the entire theoretical elements of the model such as parameters, random variables and statistical hypothesis are crisp, but only the observed values of the classical random variables can be considered as fuzzy numbers. In such cases, the recorded data and observations can be considered as triangular fuzzy numbers $\tilde{y}_{ij} = T(y_{ij}, a_{ij}, b_{ij})$, where $\tilde{y}_{ij}$ is interpreted as "approximately y_{ij}", for $i = 1, 2, \ldots, r$ and $j = 1, 2, \ldots, n_i$. Therefore, considering extension principle, the observed statistics in FANOVA can be presented by the following fuzzy sets [21]:

$$
\begin{aligned}
\widetilde{sst} &= \left[\overset{r}{\underset{i=1}{\oplus}} \overset{n_i}{\underset{j=1}{\oplus}} \tilde{y}_{ij}^2 \right] \ominus \left[\frac{1}{n_t} \odot \tilde{y}_{..}^2 \right] \\
&= \left[\overset{r}{\underset{i=1}{\oplus}} \overset{n_i}{\underset{j=1}{\oplus}} \left(\tilde{y}_{ij} \otimes \tilde{y}_{ij} \right) \right] \ominus \left[\frac{1}{n_t} \odot \left(\tilde{y}_{..} \otimes \tilde{y}_{..} \right) \right],
\end{aligned}
\tag{8}
$$

$$
\begin{aligned}
\widetilde{ssr} &= \left\{ \overset{r}{\underset{i=1}{\oplus}} \left[\frac{1}{n_i} \odot \tilde{y}_{i.}^2 \right] \right\} \ominus \left[\frac{1}{n_t} \odot \tilde{y}_{..}^2 \right] \\
&= \left\{ \overset{r}{\underset{i=1}{\oplus}} \left[\frac{1}{n_i} \odot \left(\tilde{y}_{i.} \otimes \tilde{y}_{i.} \right) \right] \right\} \ominus \left[\frac{1}{n_t} \odot \left(\tilde{y}_{..} \otimes \tilde{y}_{..} \right) \right]
\end{aligned}
\tag{9}
$$

and

$$
\begin{aligned}
\widetilde{sse} &= \left[\overset{r}{\underset{i=1}{\oplus}} \overset{n_i}{\underset{j=1}{\oplus}} \tilde{y}_{ij}^2 \right] \ominus \left[\overset{r}{\underset{i=1}{\oplus}} \left(\frac{1}{n_i} \odot \tilde{y}_{i.}^2 \right) \right] \\
&= \left[\overset{r}{\underset{i=1}{\oplus}} \overset{n_i}{\underset{j=1}{\oplus}} \left(\tilde{y}_{ij} \otimes \tilde{y}_{ij} \right) \right] \ominus \left[\overset{r}{\underset{i=1}{\oplus}} \left(\frac{1}{n_i} \odot \left(\tilde{y}_{i.} \otimes \tilde{y}_{i.} \right) \right) \right],
\end{aligned}
\tag{10}
$$

in which

$$
\begin{aligned}
\tilde{y}_{i.} &= \overset{n_i}{\underset{j=1}{\oplus}} \tilde{y}_{ij} = \tilde{y}_{i1} \oplus \tilde{y}_{i2} \oplus \cdots \oplus \tilde{y}_{in_i} = \overset{n_i}{\underset{j=1}{\oplus}} T\left(y_{ij}, a_{ij}, b_{ij}\right) \\
&= T\left(\sum_{j=1}^{n_i} y_{ij}, \sum_{j=1}^{n_i} a_{ij}, \sum_{j=1}^{n_i} b_{ij} \right) = T(y_{i.}, a_{i.}, b_{i.}),
\end{aligned}
$$

and

$$
\begin{aligned}
\tilde{y}_{..} &= \overset{r}{\underset{i=1}{\oplus}} \overset{n_i}{\underset{j=1}{\oplus}} \tilde{y}_{ij} = \tilde{y}_{11} \oplus \tilde{y}_{12} \oplus \cdots \oplus \tilde{y}_{rn_r} = \overset{r}{\underset{i=1}{\oplus}} \overset{n_i}{\underset{j=1}{\oplus}} T\left(y_{ij}, a_{ij}, b_{ij}\right) \\
&= T(y_{11}, a_{11}, b_{11}) \oplus T(y_{12}, a_{12}, b_{12}) \oplus \cdots \oplus T(y_{rn_r}, a_{rn_r}, b_{rn_r}) \\
&= T\left(\sum_{i=1}^{r} \sum_{j=1}^{n_i} y_{ij}, \sum_{i=1}^{r} \sum_{j=1}^{n_i} a_{ij}, \sum_{i=1}^{r} \sum_{j=1}^{n_i} b_{ij} \right) = T(y_{..}, a_{..}, b_{..}).
\end{aligned}
$$

Also, it is similarly obvious that

$$\widetilde{msr} = \frac{1}{r-1} \odot \widetilde{ssr}, \quad \widetilde{mse} = \frac{1}{n_t - r} \odot \widetilde{sse}, \tag{11}$$

and

$$\tilde{f} = \widetilde{msr} \ \text{ø} \ \widetilde{mse} = \frac{n_t - r}{r - 1} \odot (\widetilde{ssr} \ \text{ø} \ \widetilde{sse}). \tag{12}$$

Based on positive TFNs $\tilde{y}_{ij} = T(y_{ij}, a_{ij}, b_{ij})$, $i = 1, 2, \ldots, r$, $j = 1, 2, \ldots, n_i$, [21] calculate the observed value of fisher statistics in FANOVA model by

$$\begin{aligned}
\tilde{f} &= \widetilde{msr} \ \text{ø} \ \widetilde{mse} \\
&= \frac{n_t - r}{r - 1} \odot (\widetilde{ssr} \ \text{ø} \ \widetilde{sse}) \\
&\cong T(f_C, f_L, f_U),
\end{aligned} \tag{13}$$

in which

$$f_C = \frac{n_t - r}{r - 1} \left[\frac{\sum_{i=1}^r \frac{y_{i.}^2}{n_i} - \frac{y_{..}^2}{n_t}}{\sum_{i=1}^r \sum_{j=1}^{n_i} y_{ij}^2 - \sum_{i=1}^r \frac{y_{i.}^2}{n_i}} \right],$$

$$f_L = \frac{2(n_t - r)}{r - 1} \left[\frac{\left[\left(\sum_{i=1}^r \frac{y_{i.}^2}{n_i} - \frac{y_{..}^2}{n_t}\right) \left(\sum_{i=1}^r \sum_{j=1}^{n_i} y_{ij}b_{ij} + \sum_{i=1}^r \frac{y_{i.}a_{i.}}{n_i}\right) \right]}{\left(\sum_{i=1}^r \sum_{j=1}^{n_i} y_{ij}^2 - \sum_{i=1}^r \frac{y_{i.}^2}{n_i}\right)^2} \right.$$
$$\left. + \frac{\left[\left(\sum_{i=1}^r \sum_{j=1}^{n_i} y_{ij}^2 - \sum_{i=1}^r \frac{y_{i.}^2}{n_i}\right) \left(\sum_{i=1}^r \frac{y_{i.}a_{i.}}{n_i} + \frac{y_{..}b_{..}}{n_t}\right) \right]}{\left(\sum_{i=1}^r \sum_{j=1}^{n_i} y_{ij}^2 - \sum_{i=1}^r \frac{y_{i.}^2}{n_i}\right)^2} \right]$$

and

$$f_U = \frac{2(n_t - r)}{r - 1} \left[\frac{\left[\left(\sum_{i=1}^r \frac{y_{i.}^2}{n_i} - \frac{y_{..}^2}{n_t}\right) \left(\sum_{i=1}^r \sum_{j=1}^{n_i} y_{ij}a_{ij} + \sum_{i=1}^r \frac{y_{i.}b_{i.}}{n_i}\right) \right]}{\left(\sum_{i=1}^r \sum_{j=1}^{n_i} y_{ij}^2 - \sum_{i=1}^r \frac{y_{i.}^2}{n_i}\right)^2} \right.$$
$$\left. + \frac{\left[\left(\sum_{i=1}^r \sum_{j=1}^{n_i} y_{ij}^2 - \sum_{i=1}^r \frac{y_{i.}^2}{n_i}\right) \left(\sum_{i=1}^r \frac{y_{i.}b_{i.}}{n_i} + \frac{y_{..}a_{..}}{n_t}\right) \right]}{\left(\sum_{i=1}^r \sum_{j=1}^{n_i} y_{ij}^2 - \sum_{i=1}^r \frac{y_{i.}^2}{n_i}\right)^2} \right].$$

Fig. 1 The membership function of the observed FANOVA test statistic and the indicator function of the αth quantile of the Fisher distribution with $r-1$ and $n_t - r$ degrees of freedom

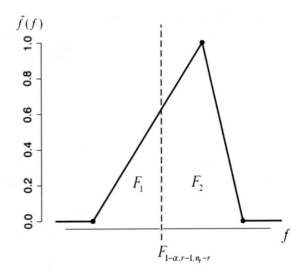

To test whether factor level means μ_i's are equal or not, FANOVA has been considered to decide whether to reject (or accept) the null hypothesis "$H_0 : \mu_1 = \mu_2 = \cdots = \mu_r$", against the alternative hypothesis "H_1 : not all μ_i's are equal", based on fuzzy data. FANOVA decision rule constructed on the basis of comparing the real numbers $F_1 = \int_0^{F_{1-\alpha,r-1,n_t-r}} \tilde{f}(f)\, df$ and $F_2 = \int_{F_{1-\alpha,r-1,n_t-r}}^{\infty} \tilde{f}(f)\, df$, in which $F_{1-\alpha;r-1,n_t-r}$ is the αth quantile of the fisher distribution with $r-1$ and $n_t - r$ degrees of freedom (see Fig. 1). Therefore, at the given significance level α, we accept the null hypothesis H_0 with degree of acceptance $D_{H_0} = F_1/(F_1 + F_2)$ if $F_1 > F_2$; otherwise we reject H_0 with degree of rejection $D_{H_1} = F_2/(F_1 + F_2)$ [21].

4   Agricultural Case Study

NanoSiO$_2$ powder was as nano silicon oxide that was supplied by Tecnan (Navarrean Nanoproducts Technology) Company. Specific surface area of nanoSiO$_2$, Average primary particle size, Purity and True density were 180–270 m^{-2} g^{-1}, 10–15 nm, >99.9 % and 2.2 g cc^{-1}, respectively. Also, Pore volume and Average pore size were 0.549 cm^3 g^{-1} and 110.13 Å, respectively. The effect of nanoSiO$_2$ in three different concentrations (0, 50 and 100 ppm) was investigated on fenugreek seedling growth and dry matter weight parameters. Seeds were surface sterilized with 0.1 % mercuric chloride for 30 s and washed several times with distilled water. They were soaked in solution of different concentrations of nanoSiO$_2$ for 24 h and dried by sterile paper. Then, 25 seeds were transferred into the each sterile Petri dish of approximately 9 cm diameter, and also 5 mM distilled water were added with different concentrations. The dishes were covered with Slophan paper and were placed in incubator under 20 ± 1 °C temperature and 70 % humidity. Seeds were considered as germinated when their radicle was at least 2 mm length [13]. After 14 days, final germination percentage and seedling growth and dry matter weight parameters were determined. The precise data for investigation on the effect of nanoSiO$_2$ on above and below parts of fenugreek seedling are presented in Tables 1 and 2 which are gathered in horticulture science laboratory at agriculture department of Mashhsd University.

Regarding to Tables 1 and 2, the added nanoSiO$_2$ to seeds are considered at three levels zero, 50 and 100 ppm in this study. For example, the first three lines of Table 1 are the effect of nanoSiO$_2$ concentrations on Shoot dry matter weight of fenugreek plant which can be considered as the factor levels zero, 50 and 100 ppm in an ANOVA test. Now, we are going to test whether the factor level means μ_i's are equal or not for each six measured parameters at the significance level 0.05, where $i = 1, 2, 3$. In other words, for each measured parameters we want to decide

Table 1 The effect of nanoSiO$_2$ concentrations on dry matter parameters of fenugreek

Dry matter parameters	NanoSiO$_2$ concentrations (ppm)	Weight (mg)			
Shoot dry matter	0	11.00	7.83	9.56	11.33
	50	10.00	10.91	11.10	10.60
	100	11.00	15.28	13.11	16.00
Root dry matter	0	2.50	2.50	2.50	2.50
	50	2.81	2.15	1.80	2.15
	100	2.55	1.44	3.42	1.92
Seedling dry matter	0	13.8	10.33	12.06	12.00
	50	12.86	12.91	12.90	12.80
	100	13.55	16.72	16.58	16.25

Table 2 The effect of nanoSiO$_2$ concentrations on growth parameters of fenugreek

Growth parameters	NanoSiO$_2$ concentrations (ppm)	Length (mm)			
Shoot length	0	57.50	58.36	57.91	58.30
	50	63.16	60.00	66.32	66.00
	100	56.92	52.81	60.87	52.00
Root length	0	42.00	43.33	42.66	42.00
	50	58.00	52.87	56.87	56.00
	100	55.00	53.06	54.42	54.00
Seedling length	0	109.5	91.66	100.57	109.00
	50	128.00	105.07	119.14	105.00
	100	111.00	112.06	110.00	111.00

to accept only one of the following hypotheses based on the observed data in
Tables 1 and 2:

$$H_0 : \ \mu_1 = \mu_2 = \mu_3,$$

$$H_1 : \text{ not all } \mu_i\text{'s are equal}, i = 1, 2, 3.$$

Considering Model [7], for example in testing the effect of different nanoSiO$_2$
concentrations on shoot dry matter weight of fenugreek, variable Y_{ij} denotes the
shoot dry matter weight for jth Petri dish in ith level of nanoSiO$_2$ where $i = 1, 2, 3$
and $j = 1, 2, 3, 4$. Although the validity of normality tests is very low for the small
observed sample size, but the normal distribution assumption for random variable
Y_{ij} comes from the essence of random variable which is rooted from nature.
Moreover, one can easily cheak the accuracy of the normality assumption for the
observations. For example considering the effect of nanoSiO$_2$ concentrations on
shoot dry matter of fenugreek, the Shapiro-Wilk test confirme normality assumption
with p-value > 0.5.

There are some unavoidable cases in experiment which could be cause the
vagueness in the recorded data. These are:

(1) the possibility of incubation errors and electric oven errors which can be
     causes non-precision for the observed data in Tables 1 and 2.
(2) the possibility of the laboratory errors in making the precise levels for
     nanoSiO$_2$ solution.
(3) the possibility of human errors in measuring, reading and keeping records.
(4) the limit of the digital laboratory scales of precision.

Therefore, one can conclude that a preferred way to record the observations is to
use vague and fuzzy numbers. In this study, we decide to rewrite the data by
symmetric triangular fuzzy numbers from now, to cover the unavoidable cases
described earlier. The core values of STFNs set equal to the precise recorded data in
Tables 1 and 2, and the vagueness of these STFNs can be considered as a coefficient of the precise recorded data in Tables 1 and 2. In other words, to cover the
above mentioned unavoidable elements, we convert the precise observations y_{ij}'s in
Tables 1 and 2 to the symmetric triangular fuzzy numbers $\tilde{y}_{ij} = T(y_{ij}, \frac{y_{ij}}{100})$, where
$i = 1, 2, 3$ and $j = 1, 2, 3, 4$. It is obvious that the traditional ANOVA cannot
analyze these fuzzy numbers, and we need to the extended version of ANOVA,
which presented in Sect. 3, to investigate on these fuzzy observations. Table 3 is

Table 3 The results of FANOVA on nanoSiO$_2$ concentrations for dry matter parameters of
fenugreek

Dry matter weigh parameters	Observed FANOVA test statistic ($\tilde{f}$)	F_1	F_2	Test result	Degree of acceptance
Shoot dry matter	$T(6.61, 30.3)$	12.9	17.4	accept H_1	$D_{H_1} = 0.575$
Root dry matter	$T(0.25, 4.64)$	4.59	0.042	accept H_0	$D_{H_0} = 0.991$
Seedling dry matter	$T(10.8, 107)$	47.5	60.2	accept H_1	$D_{H_1} = 0.559$

includes the results of three FANOVA tests based on the fuzzified data of Table 1. For example, the membership function of the observed test statistic in FANOVA model for investigation on the effect of three levels (0, 50 and 100 ppm) nanoSiO$_2$ on shoot dry matter weight of fenugreek has been calculated as $\tilde{f} = T(6.61, 30.3)$ by Eq. [12] which has been drown in Fig. 2. Therefore, after calculating $F_1 = 12.9$ and $F_2 = 17.4$ one can decide to reject the hypothesis "equality of the mean weights of shoot dry matter for three different levels of nanoSiO$_2$" with certainty $D_{H_1} = 0.575$ at significance level of 0.05 (see the first line of Table 3). In other words, one can claim that "the shoot dry matter weight" has depended on "the added levels of nanoSiO$_2$" with degree of certainty 0.575. Similarly, Table 4 are contain the results of several FANOVA tests based on the fuzzified data of Table 2 for three growth matter parameters of fenugreek.

Note that, although we faced a vague decision in FANOVA based on imprecise data, but this level of uncertainty can be measured in the proposed approach. It must be mentioned that all calculations of Table 3 done by a computer program in R software [22] which is available upon request on the basis of FANOVA presented approach in Sect. 3.

Fig. 2 The membership function of the observed test statistic and the indicator function of the αth quantile of Fisher distribution in FANOVA model for investigation on the effect of three nanoSiO$_2$ concentrations on the shoot dry matter weight

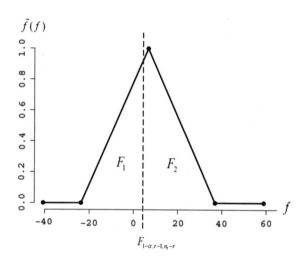

Table 4 The results of FANOVA on nanoSiO$_2$ concentration levels for growth matter parameters of fenugreek

Growth parameters	Observed FANOVA test statistic $(\tilde{f})$	F_1	F_2	Test result	Degree of acceptance
Shoot length	$T(8.4, 283)$	138	146	accept H_1	$D_{H_1} = 0.515$
Root length	$T(108, 7958)$	3876	4082	accept H_1	$D_{H_1} = 0.513$
Seedling length	$T(2.16, 64.13)$	34.1	30	accept H_0	$D_{H_0} = 0.532$

5 Conclusions and Future Works

The need for analysis of variance based on vague data emerged from the attempt of providing a rigorous mathematical framework for precisely dealing with uncertain phenomena expressed by non-precise numbers. As a practical problem, one may face fuzzy observation rather than crisp data. In analyzing such vague numbers, we use an extended version of one-way analysis of variance which can measure the fuzziness of decision and this can be important form the applied view. The fuzziness of decision, or the degree of acceptance/rejection of the null hypothesis belongs to [0, 1], but in the ordinary testing hypotheses it belongs to {0, 1}. This is one of the benefits of using fuzzy approaches instead of the traditional methods in routine environmental practices containing uncertainties from nature. The results showed that the proposed FANOVA test is a rational substitution for classical analysis of variance when the observed data are fuzzy. Theoretical and applied study on analysis of variance based on fuzzy hypotheses, fuzzy parameters and fuzzy observations are some potential subjects for further researches.

References

1. Ahmed, A.H.H., Harb, E., Higazy, M.A., Morgan, S.: Effect of silicon and boron foliar applications on wheat plants grown under saline soil conditions. Int. J. Agr. Res. **3**, 1–26 (2008)
2. Amirossadat, Z., Mohammadi Ghehsareh, A., Mojiri, A.: Impact of silicon on decreasing of salinity stress in greenhouse cucumber (Cucumis sativusL.) in soilless culture. J. Environ. Sci. **17**, 171–174 (2012)
3. Barrena, R., Casals, E., Colon, J., Font, X., Sanchez, A., Puntes, V.: Evaluation of the ecotoxicity of model nanoparticles. Chemosphere **75**, 850–857 (2009)
4. Carmen, I.U., Chithra, P., Huang, Q., Takhistov, P., Liu, S., Kokini, J.L.: Nanotechnology: a new frontier in food science. Food Technol **57**, 24–29 (2003)
5. Castiglione Monica, R., Cremonini, R.: Nanoparticles and higher plants. Caryologia **62**, 161–165 (2009)
6. Cuevas, A., Febrero, M., Fraiman, R.: An ANOVA test for functional data. Comput. Stat. Data Anal. **47**, 111–122 (2004)
7. De Garibay, V.G.: Behaviour of FANOVA. Kybernetes **16**(2), 107–112 (1987)
8. Dubois, D., Prade, H.: Fuzzy Sets and Systems: Theory and Application. Academic, New York (1980)
9. Feizi, H., Rezvani Moghaddam, P., Shahtahmassebi, N., Fotovat, A.: Impact of bulk and nanosized titanium dioxide (TiO2) on wheat seed germination and seedling growth. Biol. Trace Elem. Res. (2011)
10. Gil, M.A., Montenegro, M., González-Rodríguez, G., Colubi, A., Casals, M.R.: Bootstrap approach to the multi-sample test of means with imprecise data. Comput. Stat. Data Anal. **51**(1), 148–162 (2006)
11. González-Rodríguez, G., Colubi, A., Gil, M.A.: Fuzzy data treated as functional data: a one-way ANOVA test approach. Comput. Stat. Data Anal. **56**(4), 943–955 (2012)
12. Haghighi, M., Afifipour, Z., Mozafarian, M.: The effect of N–Si on tomato seed germination under salinity levels. J. Biol. Environ. Sci. **16**, 87–90 (2012)
13. ISTA: ISTA Rules. International Seed Testing Association, Zurich (2009)

14. Jiryaei, A., Parchami, A., Mashinchi, M.: One-way ANOVA and least squares method based on fuzzy random variables. Turk. J. Fuzzy Syst. **4**(1), 18–33 (2013)
15. Konishi, M., Okuda, T., Asai, K.: Analysis of variance based on fuzzy interval data using moment correction method. Int. J. Innovative Comput. Inf. Control **2**(1), 83–99 (2006)
16. Lu, C.M., Zhang, C.Y., Tao, M.X.: Research of the effect of nanometer on germination and growth enhancement of Glycine max and its mechanism. Soybean Sci. **21**, 168–172 (2002)
17. Manchikanti, P., Bandopadhyay, T.K.: Nanomaterials and effects on biological systems: development of effective regulatory norm. Nanoethics **4**, 77–83 (2010)
18. Montenegro, M., Colubi, A., Casals, M.R., Gil, M.A.: Asymptotic and bootstrap techniques for testing the expected value of a fuzzy random variable. Metrika **59**, 31–49 (2004)
19. Montenegro, M., Gonzalez-Rodriguez, G., Gil, M.A., Colubi, A., Casals, M.R.: Introduction to ANOVA with fuzzy random variables. In: López-Díaz, M.C., Angeles Gil, M., Grzegorzewski, P., Hryniewicz, O., Lawry, J. (eds.), Soft Methodology and Random Information Systems, pp. 487–494. Springer, Berlin (2004)
20. Nasseri, M., Arouiee, H., Kafi, M., Neamati, H.: Effect of silicon on growth and physiological parameters in fenugreek (Trigonella foenum graceum L.) under salt stress. Int. J. Agric. Crop Sci. **21**, 1554–1558 (2012)
21. Nourbakhsh, M.R., Parchami, A., Mashinchi, M.: Analysis of variance based on fuzzy observations. Int. J. Syst. Sci. (2011) doi:10.1080/00207721.2011.618640
22. Rizzo, M.L.: Statistical computing with R. Chapman Hall/CRC, Paris (2008)
23. Siddiqui, H.M., Al-Whaibi, H.M.: Role of nano-SiO2 in germination of tomato (Lycopersicum esculentum seeds Mill.). Saudi J. Biol. Sci. **21**, 13–17 (2014)
24. Torabi, F., Majd, A., Enteshari, S.: Effect of exogenous silicon on germination and seedling establishment in Borago officinalis L. J. Med. Plants Res. **6**, 1896–1901 (2012)
25. Wu, H.C.: Analysis of variance for fuzzy data. Int. J. Syst. Sci. **38**, 235–246 (2007)
26. Zadeh, L.A.: Fuzzy sets. Inf. Control **8**, 338–359 (1965)

A Survey of Fuzzy Data Mining Techniques

Tzung-Pei Hong, Chun-Hao Chen and Jerry Chun-Wei Lin

Abstract Data mining is very popular recently due to lots of analysis applications of big data. A well-known algorithm for mining association rules from transactions is the Apriori algorithm. Because transactions may include quantitative values, fuzzy sets which can be used to handle quantitative values are thus utilized to mine fuzzy association rules. Hence in this chapter, some useful fuzzy data mining techniques are introduced. Firstly, with the predefined membership functions, the Apriori-based fuzzy data mining algorithms that provide an easily way to mine fuzzy association rules are described. Since they may be time-consuming when dataset size is large, several tree-based fuzzy data mining methods are then stated to improve the mining efficiency. Besides, how to define appropriate membership functions for fuzzy data mining is important and it can be transferred into an optimization problem. Four types of genetic-fuzzy mining approaches are thus given to find both membership functions and fuzzy association rules. At last, some extended issues are discussed to provide future research directions.

Keywords Association rule · Fuzzy data mining · Membership function · Genetic algorithm · Genetic-fuzzy data mining

T.-P. Hong (✉)
Department of Computer Science and Information Engineering, National University of Kaohsiung, Kaohsiung 811, Kaohsiung, Taiwan, ROC
e-mail: tphong@nuk.edu.tw

T.-P. Hong
Department of Computer Science and Engineering, National Sun Yat-sen University, Kaohsiung, Taiwan, ROC

C.-H. Chen
Department of Computer Science and Information Engineering, Tamkang University, New Taipei City 25137, Taiwan, ROC

J.C.-W. Lin
School of Computer Science and Technology, Innovative Information Industry Research Center (IIIRC), Harbin Institute of Technology Shenzhen Graduate School, HIT Campus Shenzhen University Town, Xili, Shenzhen, People's Republic of China

© Springer International Publishing Switzerland 2016
C. Kahraman and Ö. Kabak (eds.), *Fuzzy Statistical Decision-Making*, Studies in Fuzziness and Soft Computing 343, DOI 10.1007/978-3-319-39014-7_18

329

1 Introduction

The goal of data mining is to extract useful knowledge and patterns for solving specific issues. Currently, many data mining techniques are applied to various applications [2, 6, 64], and basket analysis is one among them. One common usage is to mine association rules from a given dataset. An association rule can be represented as "$X \rightarrow Y$", where X and Y are itemsets. When used in analyzing purchase behavior, the rule means that if itemset X is bought, then itemset Y is bought as well. A famous approach for mining association rules is the Apriori algorithm [2], which consists of three phases including (1) generating candidate itemsets, (2) finding large itemsets above a given minimum support, and (3) inducing association rules above a given minimum confidence. Since traditional association rules only consider relationship among items, they are also called binary association rules. In the past, many mining approaches were proposed based on the Apriori algorithm [1–3].

However, in real applications, transactions usually have quantitative values. Hence, some variant approaches have been proposed to deal with such situations. Since the fuzzy set theory has good ability to process quantitative values and to represent linguistic meaning, fuzzy data mining problems were then formed, and several fuzzy data mining approaches were proposed. Thus, based on the Apriori algorithm, the evolution from binary association rules to fuzzy association rules for mining algorithms is illustrated in Fig. 1.

Figure 1 shows that existing approaches based on the Apriori algorithm can be divided into two groups according to a single minimum support and multiple minimum supports. The representative approaches with a single minimum support and multiple minimum supports were proposed by Agrawal et al. [2] and Liu et al. [63], respectively. Then several researchers extended these approaches with taxonomy [28, 65, 75, 77] and quantitative databases [31, 50, 53, 55, 75]. Based on the two types of minimum support, the fuzzy data mining algorithms can be divided into the following two classes:

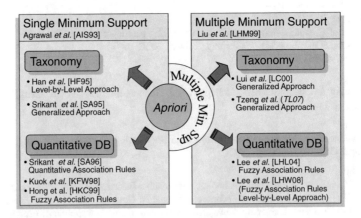

Fig. 1 The evolution of Apriori-based association rule mining

(1) Single-minimum-support fuzzy-mining approaches (*SMSFM*) [7, 23–26, 31, 33, 41, 50, 52, 66, 71, 72, 79, 82];

(2) Multiple-minimum-support fuzzy-mining approaches (*MMSFM*) [53–55].

For *SMSFM*, all items use only one minimum support to judge their importance and to mine fuzzy association rules. However, using a minimum support for mining fuzzy association rules may loss some high-valued items and rules. In order to solve this problem, some approaches were then proposed for *MMSFM*. Its main concept is that each item has its own minimum support to reflect its importance. Besides, since items may have taxonomy, some mining algorithms were also proposed for mining fuzzy generalized association rules and fuzzy multi-level association rules. Examples for *SMSFM* include [25, 26, 33, 52, 72] and for *MMSFM* contain [54, 55].

Besides the Apriori algorithm to mine fuzzy frequent itemsets in a level-wise way, Han et al. presented the frequent-pattern-tree (FP-tree) and FP-growth mining approach [29] to mine frequent itemsets from binary databases without candidate generation. It consists of two phases: first constructing an FP-tree structure by keeping only frequent 1-itemsets and then deriving fuzzy frequent itemsets from the constructed FP-tree structure. For fuzzy data mining with the tree concept, the processing is usually complicated due to the fuzzy operators, and thus extra information needs to be stored in the nodes of a tree. Papadimitriou et al. proposed fuzzy frequent pattern tree (FFPT) algorithm [73] to find fuzzy association rules based on the pattern-growth approach with an FP-tree-like mechanism. Lin et al. also proposed several different tree structures to mine fuzzy frequent item sets [57–59].

The fuzzy data mining algorithms mentioned above assume the membership functions are already known in advance. However, the membership functions may have a critical influence on the final mining results in fuzzy data mining. Designing effective and efficient approaches to get both the appropriate membership functions and fuzzy association rules are worth to be studied. Some scholars thus transform the problem of deriving appropriate membership functions as an optimization problem and use genetic algorithms (GAs) to solve it. The problem for finding membership functions and fuzzy association rules with GAs is called genetic-fuzzy mining (GFM) problem. The two types of fuzzy mining problems are *SMSFM* and *MMSFM*. The ways of processing items include processing all the items together (integrated approach) and processing them individually (divide-and-conquer approach). Therefore, according to the types of fuzzy data mining problems and the ways of processing items, the GFM problem can be divided into four kinds. Currently, many GFM approaches have been proposed for mining membership functions and fuzzy association rules, simultaneously. Some of them will be described in the following sections.

In addition to processing fuzzy association rules, some extended techniques for handling fuzzy web browsing patterns and fuzzy high-utility patterns are also interesting. Maintaining the mined fuzzy knowledge in an efficient way is important in dynamic environments. Speeding up the mining process needs to be concerned as well. These issues will be briefly introduced.

The remaining parts of this chapter are organized as follows. The Apriori-based fuzzy data mining is stated in Sect. 2. The tree-based fuzzy data mining is described in Sect. 3. The genetic-fuzzy data mining is explained in Sect. 4. Some extensions of fuzzy data mining are presented in Sect. 5. Conclusions and future works are given in Sect. 6.

2 Apriori-Based Fuzzy Data Mining

Knowledge discovery in databases (KDD) has emerging a critical issue in recent years since it can be used to discover potential relationships among items in databases. Depending on different types of the processed databases, the discovered information or knowledge can be generally classified as association rules, sequential patterns, classification, clustering, among others. Among them, association-rule mining (ARM) [1, 2, 10] is the most commonly seen in KDD. For ARM, most algorithms are performed to handle binary databases, in which the items are represented as binary attributes of 1 or 0. For fuzzy ARM, the data to be processed is generalized to linguistic ones. In this section, fuzzy mining techniques based on the Apriori algorithm will be introduced.

2.1 Apriori Algorithm

Agrawal et al. presented a generate-and-test methodology to find frequent (or called large) itemsets level by level for inducing association rules [1]. The pseudo code of the Apriori algorithm is given in Table 1.

In Algorithm 1, the database is first scanned to find the occurrence frequencies of items (Line 1) and the frequent items (1-itemsets) are derived by checking their counts over the minimum support count (Lines 2–6). The discovered frequent 1-itemsets are then used to generate the candidate 2-itemsets (Line 9). After that, the combined candidate 2-itemsets are then determined to find frequent 2-itemsets (Lines 10–15). This process is repeated until no candidate itemsets are generated (Lines 8–18). After that, the complete frequent itemsets are discovered for later generating process of association rules (Line 19).

2.2 Fuzzy Concept on Mining

In real-life situations, the same product may be purchased with multiple copies at the same time in transaction databases. It is a non-trivial task to mine association rules from this kind of quantitative databases. The fuzzy-set theory was introduced by Zadeh in 1965 [81] and very suitable to handle quantitative values and represent

Table 1 Apriori algorithm

Algorithm 1: Apriori
Input: D, a binary database; *minsup*, a pre-defined minimum support threshold.

Output: A set of frequent itemsets.
1.   **scan** D to find $f(i_j)$.
2.   **for** each item i_j in D **do**
3.   **if** $f(i_j) \geq minsup \times |D|$ **then**
4.   set $L_1 \leftarrow i_j$.
5.   **end if**
6.   **end for**
7.   set $k := 2$
8.   **while** $L_{k-1} \neq null$ **do**
9.   set $C_k \leftarrow \{a \cup b \mid a, b \in L_{k-1}, a \notin b\}$.
10.   **for** each transaction t in D **do**
11.   set $C_t \leftarrow \{z \mid z \in C_k \wedge z \subseteq t\}$.
12.   **for** each $z \in C_t$ **do**
13. $f(z) := f(z) + 1$.
14.   **end for**
15.   **end for**
16.   set $L_k \leftarrow \{z \mid z \in C_k \wedge f(z) \geq minsup\}$;
17. $k := k + 1$.
18. **end while**
19. **return** $\bigcup_k L_k$

linguistic meaning. Linguistic representation is popular and may help knowledge more understandable to human beings. In a fuzzy set, each element of a universal set can be assigned a membership grade to represent its degree of belonging to the set. Assume that x_1 to x_n are the elements in fuzzy set A and μ_1 to μ_n are their grades of membership in A. A can be represented as:

$$A = \frac{\mu_1}{x_1} + \frac{\mu_2}{x_2} + \cdots + \frac{\mu_n}{x_n}. \tag{1}$$

Three operations namely *complementation*, *union* and *intersection* are commonly used in the fuzzy-set theory, which are described below.

1. Complementation: the complementation operation of a fuzzy set A is denoted as $\ulcorner A$. The membership function of $\ulcorner A$ can be defined as:

$$\mu_A(x) = 1 - \mu_A(x), \quad \forall x \in X. \tag{2}$$

2. Union: the union operation of two fuzzy sets A and B is denoted as $A \cup B$. The membership function of $A \cup B$ for standard operation can be defined as:

$$\mu_{A \cup B}(x) = max\{\mu_A, \mu_B\}, \quad \forall x \in X. \tag{3}$$

3. Intersection: the intersection operation of two fuzzy sets A and B is denoted as $A \cap B$. The membership function of $A \cap B$ for standard operation can be defined as:

$$\mu_{A \cap B}(x) = min\{\mu_A, \mu_B\}, \quad \forall x \in X. \tag{4}$$

These three operations will be used in fuzzy data mining for deriving fuzzy association rules.

2.3   Apriori-Based Approaches

Chan and Au first presented an F-APACS algorithm [8] to mine fuzzy association rules. The values of the quantitative attributes are first converted into the representation of linguistic terms with their membership values according to the pre-defined membership functions. The adjusted difference analysis is adopted to reveal interesting association rules among attributes. In the F-APACS algorithm, the user-specified thresholds are unnecessarily required based on the designed statistical analysis. Besides, both positive and negative fuzzy association rules can be discovered through the F-APACS algorithm.

Kuok et al. then proposed an approach to handle quantitative attributes for discovering fuzzy association rules [50]. A significance factor is designed to mine all frequent (large) itemsets from the quantitative databases. It reflects not only occurrence frequencies of items in the databases but also the degree supports of itemsets. A certainty factor is also designed to generate possible rules from the frequent itemsets.

At the same time, Hong et al. adopted the fuzzy-set theory and presented a FDTA algorithm [31] to handle the quantitative databases. It is based on the Apriori algorithm to level-wisely mine fuzzy frequent itemsets for inducing fuzzy association rules. The proposed FDTA algorithm first transforms the quantitative values of items into linguistic-term representation based on the pre-defined membership functions. The cardinalities of the transformed linguistic terms are then calculated. Only a linguistic term with the maximum cardinality of each attribute is used for later mining process. This process can keep the number of items the same as that of the original attributes, thus reducing computational cost of combinational explosion. After that, the remaining fuzzy frequent itemsets can be used to induce the fuzzy association rules. The FDTA algorithm is described in Table 2.

Table 2 FDTA algorithm

Algorithm 2: FDTA

Input: D, a binary database; *minsup*, a pre-defined minimum support threshold; *minconf*, a pre-defined minimum confidence threshold, *MFs*, a pre-defined membership functions.

Output: A set of fuzzy association rules.

1. **for** each transaction t_i in D **do**
2.   **for** each item (attribute) A_j **do**
3.   **convert** quantity q_{ij} by *MFs* as $(f_{ij1}/A_j.R_1 + f_{ij2}/A_j.R_2 + ... + f_{ijn}/A_j.R_n)$.
4.   **end for**
5. **end for**
6. **calculate** $count(A_j.R_k) := \textbf{\textit{sum}} \ \{f_{ijk}\}$.
7. **set** $MAXCount(A_j.R_k) := \textbf{\textit{max}}\{count(A_j.R_k)\}$.
8. **set** $L_1 \leftarrow \{A_j.R_k \mid MAXCount(A_j.R_k) \geq minsup \times |D|\}$.
9. **set** $r := 2$.
10. **while** $L_{r-1} \neq null$ **do**
11. **set** $C_r \leftarrow \{a \cup b \mid a, b \in L_{r-1}, a \notin b\}$.
12. **for** each transaction t_i in D **do**
13.   **set** $C_{ti} \leftarrow \{z \mid z \in C_r \wedge z \subseteq t_i\}$.
14.   **for** each $z \in C_{ti}$ **do**
15.   **calculate** $count(t_i.z) = \{\textbf{\textit{min}}(f_{ijx}, f_{ijy}) \mid x, y \in z, x \notin y\}$.
16.   **end for**
17. **end for**
18. **calculate** $count(z) := \textbf{\textit{sum}} \ \{count(t_i.z)\}$.
19. **set** $L_r \leftarrow \{z \mid count(z) \geq minsup \times |D|\}$.
20. $r := r + 1$.
21. **end while**
22. **set** $C_{FARs} \leftarrow \{L_1 \wedge L_2 \wedge ... \wedge L_r \rightarrow L_q \mid q = 1 \ to \ r\}$.
23. **for** each $w \in C_{FARs}$ **do**
24. **calculate** $conf(w)$.
25. **set** $FARs \leftarrow \{w \mid conf(w) \geq minconf \times |D|\}$
26. **end for**
27. **return** *FARs*.

In the FDTA algorithm, the quantity value of each item (attribute) of each transaction in the databases is first converted based on the given membership functions (Lines 1–5). The membership values of a linguistic term of each item $(A_j.R_k)$ are summed together in the database (Line 6). After that, the linguistic term of each item with the maximum cardinality is found to represent this item for later mining process (Line 7). If the represented cardinality of the represented linguistic term is not less than the given minimum support count, it is then put into the set of fuzzy frequent (large) itemsets (Line 8). After that, the Apriori-like mechanism is

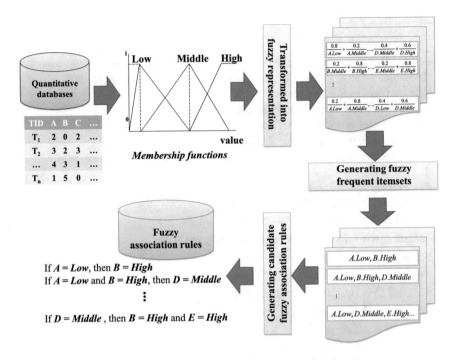

Fig. 2 Flowchart of the FDTA algorithm

performed to find the fuzzy frequent itemsets (Lines 10–21). This process is repeated until no candidate itemsets are generated. The discovered fuzzy frequent itemsets are then formed to calculate their confidence for inducing fuzzy association rules (Lines 22–27). Since each item uses only a linguistic term with the maximum cardinality, the number of items in the original database is the same as the revealed linguistic terms. Thus, the combinational explosion problem for mining fuzzy association rules can be avoided. The flowchart of the FDTA algorithm is shown in Fig. 2.

2.4 Some Variants

Several variant algorithms have been presented to mine fuzzy association rules. Gyenesei proposed an additional fuzzy normalization process to mine fuzzy association rules from quantitative databases [27]. In his approach, besides fuzzy support and confidence, a new fuzzy correlation factor was defined as a new measure to mine interesting fuzzy association rules. Hong et al. then enhanced the FDTA algorithm to design a new AprioriTid approach [36] for efficiently discovering fuzzy association rules. Yue et al. extended the FDTA approach for mining fuzzy association rules with weight constraint [44]. In their approach, each item was assigned a weight value in a range of [0, 1] to show its importance. The Kohonen self-organized mapping approach was also adopted to derive fuzzy sets for numerical attributes. Chen and Wei developed a generalized framework to mine fuzzy association rules based on fuzzy taxonomic structure [11]. The minimum support and the minimum confidence thresholds are generalized by means of sigma-counts. Hong et al. then designed a mining process for extracting interesting fuzzy association rules based on linguistic minimum support and minimum confidence thresholds [32]. Hong et al. also developed a fuzzy multiple-level mining algorithm to mine fuzzy association rules by integrating fuzzy-set concepts and multiple-level taxonomy [33].

3   Tree-Based Fuzzy Data Mining

Instead of finding fuzzy frequent itemsets level by level, another option is based on the FP-tree structure. Early studies addressed tree structures mainly for classification tasks like fuzzy decision tree. For example, Yuan and Shaw presented a fuzzy decision tree induction method to represent classification knowledge by considering cognitive uncertainties [80]. Janikow presented a decision tree with additional flexibility offered by fuzzy representation [43]. The symbolic tree structure was kept to maintain knowledge comprehensibility. This framework could be used to combine processing capabilities available in symbolic and fuzzy systems. Olaru and Wehenkel developed a new soft decision trees (SDT) to mine fuzzy classification rules by integrating the fuzzy-set theory [70]. The proposed SDT could reveal better accuracy than traditional decision tree from the experiments conducted.

3.1   Tree-Based Framework for Fuzzy Mining

For fuzzy association rule mining, the FDTA algorithm mentioned above adopts an Apriori-like mechanism to level-wisely mine fuzzy frequent itemsets for inducing fuzzy association rules. It requires multiple database scans for mining fuzzy frequent itemsets with time-consuming computation. To solve this problem, Papadimitriou et al. proposed the fuzzy frequent pattern tree (FFPT) algorithm [73]. In the algorithm, a threshold was set to remove unpromising linguistic terms in each transaction. This process might only discover local fuzzy frequent 1-itemsets but not global ones since some linguistic terms were removed from each transaction if their membership values were less than the pre-defined threshold. Besides, no operations from the fuzzy-set theory were adopted in discovering fuzzy frequent itemsets.

Lin et al. then presented another fuzzy mining framework for completely discovering fuzzy frequent itemsets based on the tree structure. Since the processing is usually complicated due to fuzzy operations, some extra information is stored in the

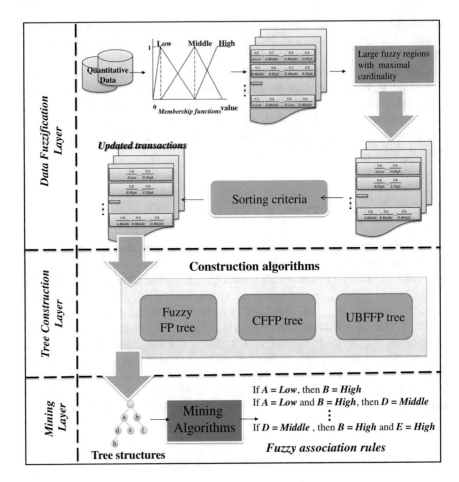

Fig. 3 The framework of tree-based fuzzy data mining

nodes of a tree for correctly performing the task. Three algorithms namely fuzzy frequent FP (FFP)-tree [58], compressed fuzzy frequent pattern (CFFP)-tree [59] and upper-bound fuzzy frequent pattern tree (UBFFP)-tree [57] were developed to mine fuzzy frequent itemsets from quantitative databases. They are different mainly in tree construction. In the framework, three layers are used to find fuzzy frequent itemsets as shown in Fig. 3.

In Fig. 3, the framework includes three layers to mine fuzzy frequent itemsets. The first layer is the data fuzzification layer. It is used to convert the quantitative values of items into the representation of linguistic terms and their membership values. This is the same process as that in the FDTA algorithm for transaction conversion except that different algorithms adopt different sorting strategies to transform transactions for later construction. The second layer is the tree construction layer, on which the fuzzified transactions are performed to construct tree structures. Different approaches will construct different tree structures. After that, the mining layer is designed to mine fuzzy frequent itemsets from the constructed tree structures based on FP-growth-like mechanisms.

3.2 Fuzzy FP-Tree Algorithm

The fuzzy FP (FFP)-tree algorithm is described in Table 3.

For the FFP-tree algorithm, it uses a similar approach to the FDTA algorithm to covert the quantities of items in original databases into the representation of linguistic terms and derive fuzzy frequent 1-itemsets (Line 1). Note that each item is

Table 3 Fuzzy FP (FFP)-tree algorithm

Algorithm 3: Fuzzy FP (FFP)-tree
Input: D, a binary database; *minsup*, a pre-defined minimum support threshold.
Output: A FFP-tree structure
1.　**find** L_1 from converted database.
2.　**build** *Htable* $\leftarrow L_1$.
3.　**create** *root* $\leftarrow$ *null*.
4.　**set** *root.next* $\leftarrow$ *ptr*.
5.　**for** each converted t_i in D **do**
6.　　**for** each $z \in L_1 \wedge z \subseteq t_i$ **do**
7.　　　**sort** z in descending order of $f_i(z)$.
8.　　**end for**
9.　　**for** each $z_k \subseteq t_i \wedge 1 \le k \le |t_i|$ **do**
10.　　　**if** $ptr := null \vee \forall \ ptr.name \neq z_k$.
11.　　　　**create** a node n.
12.　　　　**set** $(n.name, n.count) \leftarrow (z_k, f_i(z_k))$.
13.　　　　**set** $n.next \leftarrow ptr$.
14.　　　　**set** $n \leftarrow Htable.hyper$.
15.　　　　**set** $n.hyper \leftarrow lptr$.
16.　　　**else**
17.　　　　**set** $n.count := n.count + f_i(z_k)$.
18.　　　　**set** $n.next \leftarrow ptr$.
19.　　　　**find** $lptr$ from *Htable*.
20.　　　　**set** $n \leftarrow lptr$.
21.　　　　**set** $n.hyper \leftarrow lptr$.
22.　　　**end if**
23.　　**end for**
24.　**end for**

represented as its linguistic term with the maximum cardinality, thus reducing the processing time of possible combinational explosion. After that, the discovered fuzzy frequent 1-itemsets are used to build an index Hedaer_Table, which has the same function as the Header_Table of the FP-tree structure (Line 2). The converted transactions in the databases are then re-fined to keep only fuzzy frequent 1-itemsets. The local sorting strategy is adopted to sort the remaining fuzzy frequent 1-itemsets according to their transformed membership values in each transaction (Lines 6–8). The transaction is then processed one by one to buld the FFP-tree and each node in the tree keeps the membership value of the processed fuzzy frequent 1-itemsets in each transaction (Lines 9–23). An approach similar to the FP-growth is used to derive fuzzy frequent itemsets from the constructed FFP-tree structure.

Although the FFP-tree algorithm can be used to efficiently mine fuzzy frequent itemsets from the constructed FFP-tree, more nodes are required since the sorted order for constructing the FFP-tree is based on the membership values of items in each transaction. Thus, two transactions with the same linguistic terms may have different orders, thus generating different paths in the FFP-tree structure. This process may generate many extra nodes. For example, assume there are two transactions {$A.Low$:0.5, $B.High$:0.3} and {$A.Low$:0.2, $B.High$:0.4}. Two paths, $A.Low$:0.5 -> $B.High$:0.3 and $B.High$:0.4 -> $A.Low$:0.2, will be built in the FFP-tree. Although FFP-tree has to generate more tree nodes, the fuzzy frequent itemsets can be easily extracted by directly adopting FP-growth-like (proposed FFP-growth) approach to discover fuzzy frequent itemsets since the count of any super node is no less or equal to the sum of its child nodes. The proposed FFP-growth approach is shown in Table 4.

3.3   CFFP-Tree and UBFFP-Tree

To solve the above problem of the FFP-tree algorithm, Lin et al. then presented the compressed FFP (CFFP)-tree algorithm to reduce the number of tree nodes. The sorting strategy of the CFFP-tree algorithm is different from the FFP-tree algorithm. It adopts the global sorting strategy to sort the remaining fuzzy frequent 1-itemsets in descending order of their occurrence frequencies in all transactions. An extra array is attached to each node and updated in the CFFP-tree structure for keeping the membership values of the currently processed node with any of its prefix nodes by intersection operation. Based on the attached array of each node, a CFFP-growth algorithm is performed to mine the complete fuzzy frequent itemsets through a simplified intersection operation.

Although the CFFP-tree algorithm adopts an attached array in each node to reduce the number of tree nodes, it has the overhead of keeping the array. Since the attached array in each node keeps the membership values of the currently processed node with any of its prefix items in the path, the space complexity of each node is high if the size of the processed transactions is large. Lin et al. then proposed the upper-bound fuzzy frequent pattern (UBFFP)-tree algorithm to overestimate the

Table 4 FFP-growth algorithm

Algorithm 4: FFT-growth
Input: *FFP-tree*, a fuzzy FP-tree, *minsup*, a pre-defined minimum support threshold.

Output: complete fuzzy frequent itemsets.

Method: call **FFP-growth** (*FFP-tree, null*) as follows.

Procedure **FFP-growth** (*FFP-tree, α*).

1. **If** p is a single path, $p \in FFP$-*tree* **then**
2.   **for** each combination q of the nodes in p **do**
3.   **generate** $z \in (q \cup \alpha)$.
4.   **set** $z.count \leftarrow min\{n.count \mid n \in q\}$.
5.   **end for**
6. **else**
7.   **for** each i_j in Hedae_Table **do**
8.   **generate** $q \leftarrow (i_j \cup \alpha)$.
9.   **set** $q.count \leftarrow i_j.count$.
10.   **construct** q's conditional FFP-tree as $Tree_q$.
11.   **if** $Tree_q \neq null$ **then**
12.   **call FFP-growth**($Tree_q$, q).
13.   **end if**
14.   **end for**
15. **end if**

upper-bound membership values of fuzzy frequent itemsets for solving the overhead problem of the CFFP-tree structure. The UBFFP-tree construction algorithm uses the same global sorting strategy as the CFFP-tree algorithm for tree construction. Each item in the transactions is then fuzzified by keeping only the linguistic term with the maximum cardinality in later processes, which is the same procedure as the FFP-tree and CMFFP-tree. The transferred transactions are then processed tuple by tuple from the first transaction to the last one to construct the UBFFP tree. Each node in the tree keeps a fuzzy frequent 1-itemset with its accumulated fuzzy count, which is the same as the FFP-tree structure but different from the CFFP-tree structure. After the UBFFP-tree is constructed, an UBFFP-growth algorithm is used to recursively find the fuzzy frequent itemsets from the UBFFP-tree structure. After the construction process, the UBFFP-growth algorithm is then executed to mine fuzzy frequent itemsets. In the fuzzy data mining, the intersection operation is used to get the fuzzy counts of the derived fuzzy frequent itemsets. When n transactions ($n \geq 2$) are used to derive the fuzzy k-itemsets ($k \geq 2$), the actual count of non-redundant fuzzy k-itemset is less or equal to the minimum of the sum of the same fuzzy k-itemsets from n transactions. For example, supposing two transactions $\{f_1:s_1, f_2:s_2\}$ and $\{f_1:s_3, f_2:s_4\}$, and f_i represents as the transformed fuzzy region and s_i represents as the transformed fuzzy count.

Thus, $min(s_1, s_2) + min(s_3, s_4) \leq min(s_1 + s_3, s_2 + s_4)$. Hence, the merged branches of the same fuzzy regions in the UBFFP-tree can obtain the upper-bound property to mine fuzzy frequent itemsets. This property of the UBFFP-tree structure ensures that all possible fuzzy frequent itemsets (called upper-bound fuzzy frequent itemsets) can be discovered and unpromising candidate fuzzy itemsets can be pruned early. After the upper-bound fuzzy frequent itemsets are generated, an additional database scan is performed to find the actual fuzzy frequent itemsets from the kept upper-bound fuzzy frequent itemsets. The UBFFP-tree structure is a condensed tree structure when compared to the FFP-tree structure, and has the same number of tree nodes as the CMFFP-tree structure. Besides, it can solve the overhead problem of the CMFFP-tree algorithm.

3.4 Some Variants

Other variant algorithms were also presented to mine fuzzy frequent itemsets from quantitative databases. Hong et al. and Lin et al. considered all linguistic terms of an item, instead of only the one with the maximal scalar cardinality, in fuzzy mining [40, 62]. They extended the FFP-tree, CFFP-tree and UBFFP-tree structures to derive multi-term fuzzy frequent itemsets.

4 Genetic-Fuzzy Data Mining

In the previous sections, several fuzzy data mining approaches have been described, all of which assume the membership functions are predefined. However, in real applications, different items may have their own membership functions to reflect the importance of items. In this section, some genetic-fuzzy mining (GFM) approaches are introduced, which utilize genetic algorithms to mine appropriate membership functions for items and discover fuzzy association rules.

4.1 Four Types of Genetic-Fuzzy Mining Approaches

Since the membership functions have critical influence on final mining results, how to derive a set of membership functions for mining fuzzy association rules is an important issue. Utilizing the advantages of genetic algorithms, many researchers have proposed various GFM techniques for mining fuzzy association rules, fuzzy generalized association rules and fuzzy temporal association rules, among others. According to the encoding strategies and types of fuzzy mining problems, existing GFM approaches can be divided into four types. The encoding strategies are classified as integrated and divide-and-conquer strategies. The types of fuzzy

Table 5 The four types of genetic-fuzzy mining problems

	Integrated strategy	Divide-and-conquer strategy
Single minimum support	*IGFM-SMS*	*DGFM-SMS*
Multiple minimum supports	*IGFM-MMS*	*DGFM-MMS*

mining problems include SMSFM and MMSFM as mention in Sect. 1. The four categories of GFM approaches are shown in Table 5.

There are four categories in Table 5, including *IGFM-SMS*, *IGFM-MMS*, *DGFM-SMS* and *DGFM-MMS* problems. They are briefly described as follows.

(1) The Integrated Genetic-Fuzzy Mining Problem for Items with a Single Minimum Support (*IGFM-SMS*): The *IGFM-SMS* problem assumes there is only one minimum support for all items and the membership functions for all items are encoded into an integrated chromosome and then derived by genetic algorithms. At last, the derived membership functions are utilized to mine fuzzy association rules. Many approaches have been published for solving the *IGFM-SMS* problem [5, 12, 14, 18, 20–22, 34, 38, 45–49, 68, 76]. For example, Hong et al. proposed a genetic-fuzzy data-mining algorithm for extracting both association rules and membership functions from quantitative transactions [38]. Kaya et al. proposed a GA-based approach for mining membership functions and fuzzy rules, which was tried to derive membership functions that could reach a maximum profit within an interval of user-specified minimum support values [45]. Matthews et al. then took the temporal concept into consideration and proposed a temporal fuzzy association rule mining with the 2-tuples linguistic representation [68].

(2) The Integrated Genetic-Fuzzy Mining Problem for Items with Multiple Minimum Supports (*IGFM-MMS*): The previous subsection indicates there are lots of researches focusing on the *IGFM-SMS* problem. However, different items may have different properties. Thus, different criteria are needed to reflect their importance. For instance, assume there are some expensive items in a dataset. They are thus seldom bought because of their high prices, and thus their support values are low. However, a manager may still be interested in these products due to their high profits. Chen et al. thus proposed another genetic-fuzzy mining approach [15] for it, which was an extension of the approach proposed in [38]. The approach combines the clustering, fuzzy and genetic concepts to derive appropriate minimum support values and membership functions for items. Other extensions are proposed in [16, 19].

(3) The Divide-and-Conquer Genetic-Fuzzy Mining Problem for Items with a Single Minimum Support (*DGFM-SMS*): The advantages of the *IGFM* is that they are easy to use and with few constraints on fitness functions of GAs. However, if the number of items is large, the *IGFM* approaches may need lots of time to find a near-optimal solution because the length of a chromosome is long. The divide-and-conquer strategy can be used when only an approximate fitness function is adopted in GFM. Many approaches were also proposed in [13, 35, 37, 39]. For example, when the number of large 1-itemsets is used in fitness evaluation, the divide-and-conquer strategy becomes a good choice to deal with it since each item can be individually processed. Hong et al. thus proposed a GA-based framework with the divide-and-conquer strategy to search for membership functions with good converging effects [39].

(4) The Divide-and-Conquer Genetic-Fuzzy Mining Problem for Items with Multiple Minimum Supports (*DGFM-MMS*): The problem may be thought of as the combination of the *IGFM-MMS* and the *DGFM-SMS* problems. The framework for the *DGFM-MMS* problem can thus be easily designed from the previous frameworks for *IGFM-MMS* and *DGFM-SMS*. Since the *DGFM-MMS* is complex, only few literatures have been proposed for finding minimum supports and membership functions for items to mine fuzzy association rules [17].

Since the number of association rules is large in mining problems and cannot easily be coded in a chromosome, most approaches proposed learned the membership functions for items first and then derive fuzzy association rules according to the membership functions obtained.

According to the types of rules, the approaches can be divided into four types, including fuzzy association rule, fuzzy generalized association rule, fuzzy weighted association rule and fuzzy temporal association rule. Earlier approaches more focus on mining fuzzy association rules [6, 11–16, 18, 19, 32, 38–41, 44–47, 65] and fuzzy weighted association rules [47]. In recent years, genetic-fuzzy mining approaches are extended to mine fuzzy generalized association rules and fuzzy temporal association rules [21, 22, 49, 67, 68].

4.2 Genetic-Fuzzy Mining Framework

Many approaches were proposed for solving the *IGFSMS* problem. They are different in chromosome representation, genetic operators, and fitness functions [38]. A basic framework of genetic-fuzzy mining based on [38] is shown in Fig. 4.

The genetic-fuzzy mining framework for the IGFSMS consists of two phases that are membership function mining and fuzzy association rule mining. In the first phase, membership functions for items are encoded and generated according to different chromosome encoding methods. The genetic evolution process is then utilized to find appropriate membership functions. Then, fuzzy association rules are mined using the derived membership functions.

In 2003, Hong et al. used the following chromosome coding and fitness functions for GFM [40]. Each membership function is encoded as a pair (c, w), where c and w are center and half span (width) of the membership function as Parodi and Bonelli did [74].

The set of membership functions MF_j for the item I_j is then represented as a substring of $c_{j1}w_{j1}$, $c_{j2}w_{j2}$, ..., $c_{jl}w_{jl}$, where $|I_j|$ is the number of terms of I_j. The entire set of membership functions is then encoded by concatenating substrings of MF_1, MF_2,..., MF_l. Hong et al. evaluated chromosomes by the number of large 1-itemsets and the suitability of membership functions. The fitness value of a chromosome C_q was then defined as:

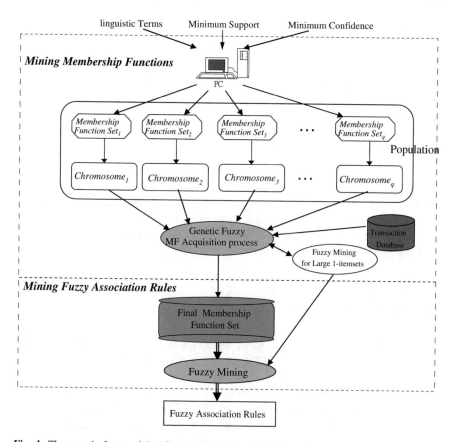

Fig. 4 The genetic-fuzzy mining framework for the IGFSMS problem

$$f(C_q) = \frac{|L_1|}{unsuitability(C_q)},$$

where $|L_1|$ is the number of large 1-itemsets obtained by using the set of membership functions in C_q. It is used to represent the knowledge amount under the set of membership functions decoded from a chromosome. Using the number of large 1-itemsets is a trade-off between execution time and rule interestingness because a larger number of 1-itemsets will result in a larger number of all itemsets with a higher probability, and then imply more interesting association rules. The evaluation by only 1-itemsets is much faster than that by all itemsets or interesting association rules. The other factor, unsuitability, in the fitness function, is designed to reduce the occurrence of bad types of membership functions. The two bad types of membership functions are shown in Fig. 5, where the first one is too redundant, and the second one is too separate.

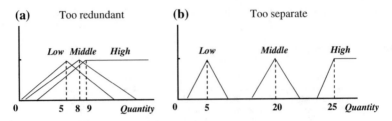

Fig. 5 The two bad types of membership functions

A formula is designed to calculate the value of unsuitability according to the overlap and the coverage degrees of a chromosome. After fitness evaluation, the genetic operations choose appropriate chromosomes for mating to create good offspring membership function sets. The offspring then undergo recursive evolution until a good set of membership functions is obtained.

4.3 Other Approaches

Kaya and Alhaji also proposed a GFM algorithm to derive membership functions and fuzzy association rules [46, 48] at nearly the same time. Their approach tried to derive membership functions that could reach a maximum profit within an interval of user-specified minimum support values. It then used the derived membership functions to mine fuzzy association rules. They used two criteria in the fitness function. The first criterion was to maximize the number of large itemsets, and the second one was to minimize the execution time. Since the two criteria conflicted, the genetic process evolved to achieve a trade-off solution. Kaya and Alhaji then extended the approach to mine fuzzy weighted association rules [47, 49].

Based on Hong's approach [38], Alcala-Fdez et al. also proposed a GFM algorithm for mining membership functions and fuzzy association rules [4]. The main difference between the two approaches is that the latter used the 2-tuple fuzzy linguistic representation model to encode membership functions into a chromosome. The coding is shown in Fig. 6.

Figure 6a shows that there are three base membership functions for attribute A, including *Low*, *Middle* and *High*. Assume there is a chromosome C_q as shown at the top of Fig. 6b. Then the actual membership functions will be derived from the modification of the base membership functions according to the chromosome. For example, the value -0.4 of *Middle* in C_q means the membership function of *Middle* will be moved 0.4 toward the left of the base membership function of Middle, and the resulting membership functions are shown at the bottom of Fig. 6b. The advantage of the encoding scheme is that each membership function only needs one parameter instead of three parameters (in triangular MF). They used the fitness function defined in [38] to evaluate the chromosomes.

(a)   **(b)**

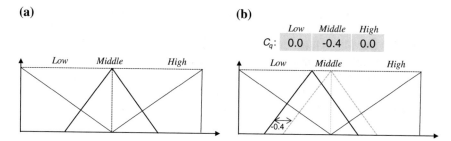

Fig. 6 The encoding method by 2-tuple fuzzy linguistic representation model. **a** Base membership functions of attribute *A*, **b** Chromosome for membership functions of attribute *A*

Matthews et al. then took the temporal concept into consideration and proposed a temporal fuzzy association rule mining with the 2-tuples linguistic representation [68]. Since the purpose of the proposed approach was to mine fuzzy temporal association rule, the Michigan method was adopted to encode a rule into a chromosome. They took the temporal concept into consideration and used fuzzy support and fuzzy confidence to evaluate a chromosome. Furthermore, they also extended their proposed approach to web usage mining [69].

4.4 Genetic-Fuzzy Mining with Taxonomy

In the previous section, the GFM algorithms introduced for mining membership functions and fuzzy association rules are executed under one single-concept level. In general, items may have taxonomy. Assume a predefined taxonomy is given in Fig. 7. There are two item classes, "Food" and "Drinks". Terminal nodes on the trees represent actual items appearing in transactions, while internal nodes represent classes or concepts formed from lower-level nodes. For example, "Apples" and "Oranges" are two kinds of "Fruit". "Fruit" is thus a higher level of concept than "Apples" or "Oranges".

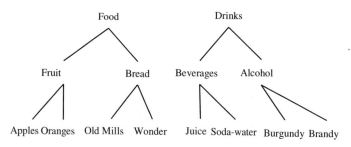

Fig. 7 The given taxonomy

Chen et al. thus proposed a multiple-level GFM algorithm for mining membership functions and fuzzy association rules on multiple-concept levels [21]. It used a slightly different fitness function from the previous one to evaluate the chromosomes and get the membership functions, and then adopted a fuzzy multi-level association rule mining approach [55] to mine fuzzy rules. Chen et al. then modified the approach by using multi-objective GAs to find Pareto solutions, each of which is a set of membership functions for mining fuzzy association rules [22].

5 Some Extended Issues

In this section, will discuss some extended issues to fuzzy data mining. They include fuzzy web mining, fuzzy utility mining, incremental fuzzy mining, and parallel fuzzy mining. These issues are briefly introduced below.

5.1 Fuzzy Web Mining

WWW has been a very popular media used in all over the world. Web mining thus plays an important role in discovering knowledge embedded in WWW. It can be roughly divided into three categories, including Web usage mining, Web content mining, and Web structure mining. The fuzzy-set theory can be similarly adopted in these techniques to find linguistic and meaningful patterns for human beings to easily understand and use [56]. Fuzzy Web usage mining focuses on analyzing the Web browsing behavior of users from Web logs and other data. Some techniques were developed for finding browsing patterns, personalization, and recommendation, among others. Fuzzy Web content mining mainly derives useful information or knowledge from Web page contents. It may include directly mining contents from Web documents or pages and improving the efficiency of search engines. Syntax and Semantic analysis and ontology are also used to increase the accuracy. Fuzzy Web structure mining analyzes the relationships among linked Web pages. The information retrieved from Fuzzy Web structure mining can be used to improve the organization of hyperlinks.

5.2 Fuzzy Utility Mining

A company is always measured by its earning. Thus, when analyzing the transaction data from a company, not only the sale frequencies of items are considered, but also their quantities and profits. Chan et al. thus proposed another new issue, namely utility mining, to consider these factors [9]. They multiplied the profit and the quantity of an item (or itemset) in a transaction and summed up the values in all the transactions as the utility of the item (itemset). If the utility of an item (itemset) was larger than or equal to a predefined threshold, then it was a desired itemset.

The fuzzy concept can also be put in utility mining. For example, Wang et al. defined a fuzzy utility function to evaluate utility of an item in databases [78]. For a fuzzy term of an item, its utility in a transaction was evaluated by the product of the three factors: the centroid value and the membership value of the fuzzy term, and the profit of the item. Its total utility was then the sum of the values from all transactions. The utility of an itemset in a transaction was the summation of the individual utility values of the fuzzy terms in the itemset. The approach did not adopt the combined membership degree of an itemset. Lan et al. then presented a new fuzzy utility function, which considered not only quantities and profits of items but also the combined membership value, to evaluate the utility of an itemset [51]. Since the downward-closure property did not exist, an effective fuzzy utility upper bound was adopted in the approach.

5.3 Incremental Fuzzy Mining

Up to now, the approaches introduced for fuzzy mining are executed in a batch way. When new transactions are inserted into a database, and old transactions are removed or modified from a database, these approaches must re-process the entire updated database to form new fuzzy knowledge. Much computational time is needed in this way. For solving this problem, Hong et al. proposed a maintenance approach on the tree structure [40]. When new transactions were inserted into a database, their approach first judged the fuzzy regions generated from the inserted transactions into four cases according to whether they were frequent or not in the original database and in the new transactions. It then processed the cases in different ways to update the Header_Table and the tree whenever necessary. Lin et al. also proposed a maintenance algorithm for fuzzy frequent itemsets based on the Apriori-based processing way [61]. Lin et al. also proposed an integrated algorithm for efficiently integrating multiple fuzzy pattern trees derived from distributed databases [60].

5.4 Parallel Fuzzy Mining

Due to dramatic increases in available computing power and concomitant decreases in computing costs over last decades, learning or mining by applying parallel processing techniques has become a feasible way of overcoming the problem of slow learning [30]. Some researches use parallel or cloud computing to speed up the execution of data mining tasks. It is, however, seldom seen for fuzzy data mining. Hong et al. adopted the master-slave architecture to speed up genetic-fuzzy data mining [42]. The fitness evaluation process was the most time-consuming part in the entire genetic-fuzzy mining process, and was thus processed in parallel by the slave processors. The other part was processed by the master processor. The master processor generated a single population, in which each individual represented a possible set of membership functions as mentioned before. It then distributed the tasks of fitness evaluation to slave processors. Each slave processor then evaluated the allocated chromosomes and then sent the results back to the master processor. The master processor collected the results and did appropriate genetic operations to generate the next generation. The same steps were repeated until the convergence condition was achieved.

6 Conclusions

In this chapter, different types of fuzzy data mining approaches have been described, including Apriori-based fuzzy data mining, tree-based fuzzy data mining, genetic-fuzzy data mining approaches, and some extended issues. The main concept in these approaches is to utilize the fuzzy-set theory to handle quantitative databases. In Apriori-based fuzzy data mining approaches, quantitative values are transformed into fuzzy sets according to the predefined membership functions. Then, fuzzy frequent itemsets and fuzzy association rules can be generated based on the Apriori execution process. Since the execution time of Apriori-based approaches is time-consuming, tree-based fuzzy data mining approaches are then described to speed up the mining process. Basically, these approaches are modified from the FP-tree for processing fuzzy itemsets. In general, the membership functions may be given by experts. However, the experts may not always be available. Genetic-fuzzy mining approaches are then proposed to automatically mine appropriate membership functions. Four types of genetic-fuzzy mining approaches are introduced, including *IGFM-SMS*, *IGFM-MMS*, *DGFM-SMS* and *DGFM-MMS*. Most of the genetic-fuzzy approaches process with two phases. In first phase, the genetic process is used to derive membership functions for items. In the second phase, the fuzzy association rules are mined by the fuzzy mining approach based on the derived membership functions. Some extended issues including fuzzy web mining, fuzzy utility mining, incremental fuzzy mining, and parallel fuzzy mining, are also discussed.

In recent years, along with the increase of data size, big data mining becomes an emerging research field. Big data means that the data amount is very large, the data types are heterogeneous, and the data variation is extremely dynamic. Extending the efficient fuzzy mining algorithms for the big data era is thus an important issue. As a result, we can know that fuzzy data mining will continuously grow but with a variety of forms. Another interesting work is to develop visual tools for showing the derived fuzzy association rules. It can help a decision maker clearly understand or get useful information quickly. Besides, applying cloud computing or other parallel and distributed architectures to speed up the fuzzy mining process is worth studying as well.

References

1. Agrawal, R., Srikant, R.: Fast algorithms for mining association rules in large databases. In: The International Conference on Very Large Data Bases, pp. 487–499 (1994)
2. Agrawal, R., Imielinski, T., Swami, A.: Database mining: A performance perspective. IEEE Trans. Knowl. Data Eng. **5**, 914–925 (1993)
3. Agrawal, R., Srikant, R., Vu, Q.: Mining association rules with item constraints. In: The Third International Conference on Knowledge Discovery in Databases and Data Mining (1997)
4. Alcala-Fdez, J., Alcala, R., Gacto, M.J., Herrera, F.: Learning the membership function contexts for mining fuzzy association rules by using genetic algorithms. Fuzzy Sets Syst. **160** (7), 905–921 (2009)
5. Alhajj, R., Kaya, M.: Multi-objective genetic algorithms based automated clustering for fuzzy association rules mining. J. Intell. Inf. Syst. **31**(3), 243–264 (2008)
6. Au, W.H., Chan, K.C.C.: Mining fuzzy association rules in a bank-account database. IEEE Trans. Fuzzy Syst. **11**(2), 238–248 (2003)
7. Ayouni, S., Yahia, S.B.: Extracting compact and information lossless set of fuzzy association rules. In: IEEE International Fuzzy Systems Conference, pp. 1–6 (2007)
8. Chan, K.C.C., Au, W.H.: Mining fuzzy association rules. In: International Conference on Information and Knowledge Management, pp. 209–215 (1997)
9. Chan, R., Yang, Q., Shen, Y.D.: Mining high utility itemsets. In: The Third IEEE International Conference on Data Mining, pp. 19–26 (2003)
10. Chen, M.S., Han, J., Yu, P.S.: Data mining: An overview from a database perspective. IEEE Trans. Knowl. Data Eng. **8**, 866–883 (1996)
11. Chen, G., Wei, Q.: Fuzzy association rules and the extended mining algorithms. Inf. Sci. **147**, 201–228 (2002)
12. Chen, C.H., Hong, T.P., Tseng, V.S.: A cluster-based fuzzy-genetic mining approach for association rules and membership functions. In: The IEEE International Conference on Fuzzy Systems, pp. 1411–1416 (2006)
13. Chen, C.H., Hong, T.P., Tseng, V.S.: A modified approach to speed up genetic-fuzzy data mining with divide-and-conquer strategy. In: The IEEE Congress on Evolutionary Computation, pp. 1–6 (2007)
14. Chen, C.H., Hong, T.P., Tseng, V.S., Chen, L.C.: A multi-objective genetic-fuzzy mining algorithm. In: The IEEE International Conference on Granular Computing (2008)
15. Chen, C.H., Hong, T.P., Tseng, V.S.: A cluster-based genetic-fuzzy mining approach for items with multiple minimum supports. Lect Notes Comput. Sci. **5012**, 864–869 (2008)
16. Chen, C.H., Hong, T.P., Tseng, V.S., Lee, C.S.: A genetic-fuzzy mining approach for items with multiple minimum supports. Soft Comput. **13**(5), 521–533 (2009)
17. Chen, C.H., Hong, T.P., Tseng, V.S.: An improved approach to find membership functions and multiple minimum supports in fuzzy data mining. Expert Syst. Appl. **36**(6), 10016–10024 (2009)
18. Chen, C.H., Hong, T.P., Tseng, V.S.: A SPEA2-based genetic-fuzzy mining algorithm. In: The IEEE International Conference on Fuzzy Systems (2010)
19. Chen, C.H., Hong, T.P., Tseng, V.S.: Genetic-fuzzy mining with multiple minimum supports based on fuzzy clustering. Soft Comput. **15**(12), 2319–2333 (2011)
20. Chen, C.H., Hong, T.P., Tseng, V.S.: Finding Pareto-front membership functions in fuzzy data mining. Int. J. Comput. Intell. Syst. **5**(2), 343–354 (2012)
21. Chen, C.H., Hong, T.P., Lee, Y.C.: Genetic-fuzzy mining with taxonomy. Int. J. Uncertainty Fuzziness Knowl. Based Syst. **20**(2), 187–205 (2012)
22. Chen, C.H., He, J.S., Hong, T.P.: MOGA-based fuzzy data mining with taxonomy. Knowl. Based Syst. **54**, 53–65 (2013)
23. Delgado, M., Marin, N., Sanchez, D., Vila, M.A.: Fuzzy association rules: general model and applications. IEEE Trans. Fuzzy Syst. **11**(2), 214–225 (2003)

24. Dubois, D., Prade, H., Sudkamp, T.: On the representation, measurement, and discovery of fuzzy associations. IEEE Trans. Fuzzy Syst. **13**(2), 250–262 (2005)
25. Farzanyar, Z., kangavari, M., Hashemi, S.: A new algorithm for mining fuzzy association rules in the large databases based on ontology. In: IEEE International Conference on Data Mining Workshops, pp. 65–69 (2006)
26. Gautam, P., Khare, N., Pardasani, K.R.: A model for mining multilevel fuzzy association rule in database. *CoRR* abs/1001.3488 (20100
27. Gyenesei, A.: A fuzzy approach for mining quantitative association rules. Acta Cybern. **15**, 305–320 (2001)
28. Han, J., Fu, Y.: Discovery of multiple-level association rules from large database. In: The Twenty-first International Conference on Very Large Data Bases, pp. 420–431 (1995)
29. Han, J., Pei, J., Yin, Y., Mao, R.: Mining frequent patterns without candidate generation: a frequent-pattern tree approach. Data Min. Knowl. Disc. **8**, 53–87 (2004)
30. Hong, T.P.: A study of parallel processing and noise management on machine learning. Ph.D. dissertation, National Chiao-Tung University, Taiwan, R.O.C. (1992)
31. Hong, T.P., Kuo, C.S., Chi, S.C.: Mining association rules from quantitative data. Intell. Data Anal. **3**, 363–376 (1999)
32. Hong, T.P., Chiang, M.J., Wang, S.L.: Mining from quantitative data with linguistic minimum supports and confidences. In: IEEE International Conference on Fuzzy Systems, pp. 494–499 (2002)
33. Hong, T.P., Lin, K.Y., Chien, B.C.: Mining fuzzy multiple-level association rules from quantitative data. Appl. Intell. **18**, 79–90 (2003)
34. Hong, T.P., Chen, C.H., Wu, Y.L., Lee, Y.C.: Mining membership functions and fuzzy association rules. In: The Joint Conference on AI, Fuzzy System, and Grey System (2003)
35. Hong, T.P., Chen, C.H., Wu, Y.L., Lee, Y.C.: Using divide-and-conquer GA strategy in fuzzy data mining. In: The Ninth IEEE Symposium on Computers and Communications (2004)
36. Hong, T.P., Kuo, C.S., Wang, S.L.: A fuzzy aprioritid mining algorithm with reduced computational time. Appl. Soft Comput. **5**, 1–10 (2004)
37. Hong, T.P., Chen, C.H., Wu, Y.L., Tseng, S.M.: Finding active membership functions in fuzzy data mining. In: The Workshop on Foundations of Data Mining in The Fourth IEEE International Conference on Data Mining (2004)
38. Hong, T.P., Chen, C.H., Wu, Y.L., Lee, Y.C.: A GA-based fuzzy mining approach to achieve a trade-off between number of rules and suitability of membership functions. Soft Comput. **10** (11), 1091–1101 (2006)
39. Hong, T.P., Chen, C.H., Lee, Y.C., Wu, Y.L.: Genetic-fuzzy data mining with divide-and-conquer strategy. IEEE Trans. Evol. Comput. **12**(2), 252–265 (2008)
40. Hong, T.P., Lin, C.W., Lin, T.C., Wang, S.L.: Incremental multiple fuzzy frequent pattern tree. In: The IEEE International Conference on Fuzzy Systems, pp. 2051–2055 (2012)
41. Hong, T.P., Lin, C.W., Lin, T.C.: The MFFP-tree fuzzy mining algorithm to discover complete linguistic frequent itemsets. Comput. Intell. **30**, 145–166 (2012)
42. Hong, T.P., Lee, Y.C., Wu, M.T.: An effective parallel approach for genetic-fuzzy data mining. Expert Syst. Appl. **41**(2), 655–662 (2014)
43. Janikow, C.Z.: Fuzzy decision trees: issues and methods. IEEE Trans. Syst. Man Cybern. B Cybern. **28**, 1–14 (1998)
44. Joyce, S.Y., Tseng, E., Yeung, D., Shi, D.: Mining fuzzy association rules with weighted items. In: IEEE International Conference on Systems, Man, and Cybernetics, pp. 1906–1911 (2000)
45. Kaya, M.: Multi-objective genetic algorithm based approaches for mining optimized fuzzy association rules. Soft Comput. **10**(7), 578–586 (2006)
46. Kaya, M., Alhajj, R.: Facilitating fuzzy association rules mining by using multi-objective genetic algorithms for automated clustering. In: IEEE International Conference on Data Mining, pp. 561–564 (2003)

47. Kaya, M., Alhajj, R.: Genetic algorithms based optimization of membership functions for fuzzy weighted association rules mining. In: IEEE Symposium on Computers and Communications, pp. 110–115 (2004)
48. Kaya, M., Alhajj, R.: Genetic algorithm based framework for mining fuzzy association rules. Fuzzy Sets Syst. **152**(3), 587–601 (2005)
49. Kaya, M., Alhajj, R.: Effective mining of fuzzy multi-cross-level weighted association rules. Lect. Notes Comput. Sci. **4203**, 399–408 (2006)
50. Kuok, C.M., Fu, A.W.-C., Wong, M.H.: Mining fuzzy association rules in databases. ACM SIGMOD Rec. **27**(1), 41–46 (1998)
51. Lan, G.C., Hong, T.P., Lin, Y.H., Wang, S.L.: Fuzzy utility mining with upper-bound measure. Appl. Soft Comput. **30**, 767–777 (2015)
52. Lee, K.M.: Mining generalized fuzzy quantitative association rules with fuzzy generalization hierarchies. In: IFSA World Congress and 20th NAFIPS International Conference, vol. 5, pp. 2977–2982 (2001)
53. Lee, Y.C., Hong, T.P., Lin, W.Y.: Mining fuzzy association rules with multiple minimum supports using maximum constraints. Lect. Notes Comput. Sci. **3214**, 1283–1290 (2004)
54. Lee, Y.C., Hong, T.P., Wang, T.C.: Mining multiple-level association rules under the maximum constraint of multiple minimum supports. In: IEA/AIE, pp. 1329–1338 (2006)
55. Lee, Y.C., Hong, T.P., Wang, T.C.: Multi-level fuzzy mining with multiple minimum supports. Expert Syst. Appl. **34**(1), 459–468 (2008)
56. Lin, C.W., Hong, T.P.: A survey of fuzzy web mining. Wiley Interdiscip. Rev. Data Min. Knowl. Discov. **3**, 190–199 (2013)
57. Lin, C.W., Hong, T.P.: Mining fuzzy frequent itemsets based on UBFFP trees. J. Intell. Fuzzy Syst. **27**, 535–548 (2014)
58. Lin, C.W., Hong, T.P., Lu, W.H.: Linguistic data mining with fuzzy FP-trees. Expert Syst. Appl. **37**, 4560–4567 (2010)
59. Lin, C.W., Hong, T.P., Lu, W.H.: An efficient tree-based fuzzy data mining approach. Int. J. Fuzzy Syst. **12**, 150–157 (2010)
60. Lin, C.W., Hong, T.P., Lin, T.C., Chen, Y.F., Pan, S.T.: An integrated MFFP-tree algorithm for mining global fuzzy rules from distributed databases. J. Univ. Comput. Sci. **19**(4), 521–538 (2013)
61. Lin, C.W., Wu, T.Y., Lin, G., Hong, T.P.: Maintenance algorithm for updating the discovered multiple fuzzy frequent itemsets for transaction deletion. In: The International Conference on Machine Learning and Cybernetics, pp. 475–480 (2014) (China)
62. Lin, J.C.W., Hong, T.P., Lin, T.C.: A cmffp-tree algorithm to mine complete multiple fuzzy frequent itemsets. Appl. Soft Comput. **28**, 431–439 (2015)
63. Liu, B., Hsu, W., Ma, Y.: Mining association rules with multiple minimum supports. In: ACM SIGKDD International Conference on Knowledge Discovery and Data Mining, pp. 337–341 (1999)
64. Lopez, F.J., Blanco, A., Garcia, F., Marin, A.: Extracting biological knowledge by fuzzy association rule Mining. In: IEEE International Fuzzy Systems Conference, pp. 1–6 (2007)
65. Lui, C.L., Chung, F.L.: Discovery of generalized association rules with multiple minimum supports. Lect. Notes Comput. Sci. **1910**, 510–515 (2000)
66. Mangalampalli, A., Pudi, V.: Fuzzy association rule mining algorithm for fast and efficient performance on very large datasets. In: IEEE International Conference on Fuzzy Systems, pp. 1163–1168 (2009)
67. Matthews, S.G., Gongora, M.A., Hopgood, A.A.: Evolving temporal fuzzy itemsets from quantitative data with a multi-objective evolutionary algorithm. In: The IEEE International Workshop on Genetic and Evolutionary Fuzzy Systems (2011)
68. Matthews, S.G., Gongora, M.A., Hopgood, A.A., Ahmadi, S.: Temporal fuzzy association rule mining with 2-tuple linguistic representation. In: The IEEE International Conference on Fuzzy Systems, pp. 1–8 (2012)

69. Matthews, S.G., Gongora, M.A., Hopgood, A.A., Ahmadi, S.: Web usage mining with evolutionary extraction of temporal fuzzy association rules. Knowl. Based Syst. **54**, 66–72 (2013)
70. Olaru, C., Wehenkel, L.: A complete fuzzy decision tree technique. Fuzzy Sets Syst. **138**, 221–254 (2003)
71. Ouyang, W., Huang, Q.: Mining direct and indirect weighted fuzzy association rules in large transaction databases. Int. Conf. Fuzzy Syst. and Knowl. Discov. **3**, 128–132 (2009)
72. Ouyang, W., Huang, Q., Luo, S.: Mining direct and indirect fuzzy multiple level sequential patterns in large transaction databases. In: ICIC, pp. 906–913 (2008)
73. Papadimitriou, S., Mavroudi, S.: The frequent fuzzy pattern tree. In: The WSEAS International Conference on Computers (2005)
74. Parodi, A., Bonelli, P.: A new approach of fuzzy classifier systems. In: Proceedings of Fifth International Conference on Genetic Algorithms, Morgan Kaufmann, Los Altos, CA, pp. 223–230 (1993)
75. Srikant, R., Agrawal, R.: Mining generalized association rules. In: The Twenty-first International Conference on Very Large Data Bases, pp. 407–419 (1995)
76. Thilagam, P.S., Ananthanarayana, V.S.: Extraction and optimization of fuzzy association rules using multi-objective genetic algorithm. Pattern Anal. Appl. **12**(2), 159–168 (2008)
77. Tseng, M.C., Li, W.Y.: Efficient mining of generalized association rules with non-uniform minimum support. Data Knowl. Eng. **62**(1), 41–64 (2007)
78. Wang, C.M., Chen, S.H., Huang, Y.F.: A fuzzy approach for mining high utility quantitative itemsets. In: The IEEE International Conference on Fuzzy Systems, pp. 20–24 (2009)
79. Watanabe, T.: Mining fuzzy association rules of specified output field. IEEE Int. Conf. Syst. Man Cybern. **6**, 5754–5759 (2004)
80. Yuan, Y., Shaw, M.J.: Induction of fuzzy decision trees. Fuzzy Sets Syst. **69**, 125–139 (1995)
81. Zadeh, L.A.: Fuzzy sets. Inf. Control **8**, 338–353 (1965)
82. Zaheeruddin, Anwer, J.: A simple technique for generation and minimization of fuzzy rules fuzzy systems. In: IEEE International Conference on Fuzzy Systems, pp. 489–494 (2005)

Index

© Springer International Publishing Switzerland 2016
C. Kahraman and Ö. Kabak (eds.), *Fuzzy Statistical Decision-Making*,
Studies in Fuzziness and Soft Computing 343,
DOI 10.1007/978-3-319-39014-7

Printed in the United States
By Bookmasters